Computational Lithography

WILEY SERIES IN PURE AND APPLIED OPTICS

Founded by Stanley S. Ballard, University of Florida

EDITOR: Glenn Boreman, University of Central Florida, CREOL & FPCE

BARRETT AND MYERS · *Foundations of Image Science*
BEISER · *Holographic Scanning*
BERGER-SCHUNN · *Practical Color Measurement*
BOYD · *Radiometry and The Detection of Optical Radiation*
BUCK · *Fundamentals of Optical Fibers*, Second Edition
CATHEY · *Optical Information Processing and Holography*
CHUANG · *Physics of Optoelectronic Devices*
DELONE AND KRAINOV · *Fundamentals of Nonlinear Optics of Atomic Gases*
DERENIAK AND BOREMAN · *Infrared Detectors and Systems*
DERENIAK AND CROWE · *Optical Radiation Detectors*
DE VANY · *Master Optical Techniques*
ERSOY · *Diffraction, Fourier Optics and Imaging*
GASKILL · *Linear Systems, Fourier Transform, and Optics*
GOODMAN · *Statistical Optics*
HOBBS · *Building Electro-Optical Systems: Making It All Work*, Second Edition
HUDSON · *Infrared System Engineering*
IIZUKA · *Elements of Photonics, Volume I: In Free Space and Special Media*
IIZUKA · *Elements of Photonics, Volume II: For Fiber and Integrated Optics*
JUDD AND WYSZECKI · *Color in Business, Science, and Industry*, Third Edition
KAFRI AND GLATT · *The Physics of Moire Metrology*
KAROW · *Fabrication Methods for Precision Optics*
KLEIN AND FURTAK · *Optics*, Second Edition
MA AND ARCE · *Computational Lithography*
MALACARA · *Optical Shop Testing*, Third Edition
MILONNI AND EBERLY · *Lasers*
NASSAU · *The Physics and Chemistry of Color: The Fifteen Causes of Color*, Second Edition
NIETO-VESPERINAS · *Scattering and Diffraction in Physical Optics*
OSCHE · *Optical Detection Theory for Laser Applications*
O'SHEA · *Elements of Modern Optical Design*
OZAKTAS · *The Fractional Fourier Transform*
PRATHER, SHI, SHARKAWY, MURAKOWSKI, AND SCHNEIDER · *Photonic Crystals: Theory, Applications, and Fabrication*
SALEH AND TEICH · *Fundamentals of Photonics*, Second Edition
SCHUBERT AND WILHELMI · *Nonlinear Optics and Quantum Electronics*
SHEN · *The Principles of Nonlinear Optics*
UDD · *Fiber Optic Sensors: An Introduction for Engineers and Scientists*
UDD · *Fiber Optic Smart Structures*
VANDERLUGT · *Optical Signal Processing*
VEST · *Holographic Interferometry*
VINCENT · *Fundamentals of Infrared Detector Operation and Testing*
WEINER · *Ultrafast Optics*
WILLIAMS AND BECKLUND · *Introduction to the Optical Transfer Function*
WYSZECKI AND STILES · *Color Science: Concepts and Methods, Quantitative Data and Formulae*, Second Edition
XU AND STROUD · *Acousto-Optic Devices*
YAMAMOTO · *Coherence, Amplification, and Quantum Effects in Semiconductor Lasers*
YARIV AND YEH · *Optical Waves in Crystals*
YEH · *Optical Waves in Layered Media*
YEH · *Introduction to Photorefractive Nonlinear Optics*
YEH AND GU · *Optics of Liquid Crystal Displays*, Second Edition

Computational Lithography

XU MA AND GONZALO R. ARCE

WILEY
A JOHN WILEY & SONS, INC., PUBLICATION

Published by John Wiley & Sons, Inc., Hoboken, New Jersey
Published simultaneously in Canada

Library of Congress Cataloging-in-Publication Data:

Ma, Xu, 1983-
Computational lithography / Xu Ma and Gonzalo R. Arce.
p. cm. – (Wiley series in pure and applied optics)
Includes bibliographical references and index.
ISBN 978-0-470-59697-5 (cloth)
1. Microlithography–Mathematics. 2. Integrated circuits–Design and construction–Mathematics. 3. Photolithography–Mathematics. 4. Semiconductors–Etching–Mathematics. 5. Resolution (Optics) I. Arce, Gonzalo R. II. Title.
TK7872.M3C66 2010
621.3815'31–dc22

2009049250

To Our Families

Contents

Preface

Moore's law and the integrated circuit industry have led the electronics industry to make technological advances that have transformed the society in many ways. Wireless communications, the Internet, and the astonishing new modalities in medical imaging have all been realized by the availability of the computational power inside IC processors. At this pace, if Moore's law continues to hold for the next couple of decades, the computational power of integrated circuits will play a key role in unveiling the secrets of the working mechanisms behind the living brain, it will also be the enabler in the advances of health informatics and of the solutions to other grand challenges singled out by the National Academy of Engineering. Maintaining this pace, however, requires a constant search by the semiconductor industry for new approaches to reduce the size of transistors. At the heart of Moore's law is optical lithography by which ICs are patterned, one layer at a time. By steadily reducing the wavelength of light in optical lithography, the IC industry has kept pace with the Moore's law. In the past two decades, the wavelength used in optical lithography has shrunk down to today's standard of 193 nm. This strategy, however, has become less certain as wavelengths shorter than 193 nm cannot be used without a major overhaul of the lithographic process, since shorter wavelengths are absorbed by the optical elements in lithography. While new lithography methods are under development, such as extreme ultraviolet (EUV) at the wavelength of 13 nm, the semiconductor industry is relying more on resolution enhancement techniques (RETs) that aim at coaxing light into resolving IC features that are smaller than its wavelength. RETs are becoming increasingly important since their implementation does not require significant changes in fabrication infrastructure.

The laws of optical wave propagation determine that the smallest resolvable features in optical lithography are proportional to the wavelength used and inversely proportional to the numerical aperture of the underlying optical system. Reducing the optical wavelength in optical lithography and exploring new methods to increase the numerical aperture are the two ways in which the semiconductor industry has made advances to keep up with the Moore's law. A third approach is that of reducing the proportionality constant k through resolution enhancement techniques. RETs manipulate the amplitude, phase, and direction of light propagation impinging on the lithographic mask to reduce the proportionality constant. In particular, optical proximity correction (OPC) modifies the wavefront amplitude, off-axis illumination (OAI) modifies the light wave direction of propagation, and phase-shifting masks manipulate the phase. OPC methods add assisting subresolution features on the mask pattern to correct the distortion of the optical projection systems. PSM methods modify both the amplitude and phase of the mask patterns. OAI methods exploit various illumination configurations to enhance the resolution. Used individually or in combination, RETs have proven effective in subwavelength lithography.

The literature on RET methods has been growing rapidly in journals and conference articles. Most of the methods used in RET exploit the rule-based principles developed and refined by practicing lithographers. Several excellent books on optical lithography have appeared in print recently. Wong provides a tutorial reference focusing on RET technology in optical lithography systems [92]. Wong subsequently extended this previous work and provided an integrated mathematical view of the physics and numerical modeling of optical projection lithography [93]. Levinson addressed and discussed an overall view of lithography, from the specific technical details to economical costs [36]. Mack captured the fundamental principles of the incredibly fast-changing field of semiconductor microlithography from the underlying scientific principles of optical lithography [49]. While the rule-based RET methods will continue to provide a valuable tool set for mask design in optical lithography, the new frontier for RETs will be on the development of tools and methods that capitalize from the ever rapid increase of computational power available for the RET design.

This book first aims at providing an adequate summary of the rule-based RET methodology as well as a basic understanding of optical lithography. It can thus serve as a tutorial for those who are new to the field. Different from the above-mentioned textbooks, this book is also the first to address the computational optimization approaches to RETs in optical lithography. Having vast computational resources at hand, computational lithography exploits the rich mathematical theory and practice of inverse problems, mathematical optimization, and computational imaging to develop optimization-based resolution enhancement techniques for optical lithography. The unique contribution of the book is thus a unified summary of the models and the optimization methods used in computational lithography. In particular, this book provides an in-depth and elaborate discussion on OPC, PSM, and OAI RET tools that use model-based mathematical optimization in their design. The book starts with an introduction of optical lithography systems, electric magnetic field principles, and fundamentals of optimization. Based on this preliminary knowledge, this book describes different types of optimization algorithms to implement RETs in detail. Most of the optimization algorithms developed in this book are based on the application of the OPC, PSM, and OAI approaches and their combinations. In addition, mathematical derivations of all the optimization frameworks are presented as appendices at the end of the book.

The Matlab's m-files for all the RET methods described in the book are provided at ftp://ftp.wiley.com/public/sci_tech_med/computational_lithography. All the optimization tools are made available at ftp://ftp.wiley.com/public/sci_tech_med/computational_lithography as Matlab's m-files. Readers may run and investigate the codes to understand the algorithms. Furthermore, these codes may be used by readers for their research and development activities in their academic or industrial organizations. The contents of this book are tailored for both entry-level and experienced readers.

XU MA AND GONZALO R. ARCE

Department of Electrical and Computer Engineering, University of Delaware

Acknowledgments

We are thankful to many colleagues for their advice and contributions. It has been our good fortune to have had the opportunity to interact and have received the guidance of some of the world's leaders in optical lithography from the Intel Corporation. In particular, we are indebted to Dr. Christof Krautschik, Dr. Yan Borodovsky, Dr. Vivek Singh, and Dr. Jorge Garcia, all from the Intel Corporation, for their guidance and support. Our contributions to this field and the elaboration of this book would not have been possible without their support. We thank Dr. Dennis Prather from the University of Delaware for insightful discussions on optics, polarization, and optical wavefront propagation. The discussions on optimization and inverse problems as applied to RET design with Dr. Yinbo Li, Dr. David Luke, Dr. Javier Garcia-Frias, and Dr. Ken Barner, all from the University of Delaware, are greatly appreciated. We also thank Dr. Avideh Zakhor from the University of California, Berkeley, and Dr. Stephen Hsu from AMSL Corporation for insightful discussions on RETs. The material in this textbook has benefited greatly from our interactions with many bright students at the University of Delaware, with special appreciation to Dr. Zhongmin Wang, Peng Ye, Yuehao Wu, Dr. Lu Zhang, Dr. Bo Gui, and Xiantao Sun. We are particularly grateful to Prof. Glenn Boreman from CREOL at the University of Central Florida for his support in including this book in the Wiley Series in Pure and Applied Optics. We would like to thank our editor George Telecki and the staff at Wiley for supporting this project from the beginning stage through that at the printing press.

XU MA AND GONZALO R. ARCE

Department of Electrical and Computer Engineering, University of Delaware

Acronyms

ACAA	Average Coherence Approximation Algorithm
BL	Boundary Layer
CD	Critical Dimension
CMTF	Critical Modulation Transfer Function
DCT	Discrete Cosine Transform
DEL	Double Exposure Lithography
DPL	Double Patterning Lithography
DUVL	Deep Ultraviolet Lithography
EBL	E-Beam Lithography
EUVL	Extreme Ultraviolet Lithography
FDTD	Finite-Difference Time-Domain Method
FFT	Fast Fourier Transform
IC	Integrated Circuit
ILT	Inverse Lithography Technique
ITRS	International Technology Roadmap for Semiconductors
MOS	Metal Oxide Silicon
MoSi	Molybdenum Silicide
MTF	Modulation Transfer Function
NA	Numerical Aperture
OAI	Off-Axis Illumination
OPC	Optical Proximity Correction
PAC	Photoactive Compound
PCI	Partially Coherent Illumination
PSF	Point Spread Function
PSM	Phase-Shifting Mask
RET	Resolution Enhancement Technique
SMO	Simultaneous Source and Mask Optimization
SNR	Signal-to-Noise Ratio
SOCS	Sum of Coherent System
SR1	Symmetric Rank One
SVD	Singular Value Decomposition
WG	Waveguide Method

1

Introduction

1.1 OPTICAL LITHOGRAPHY

Complex circuitries of modern microelectronic devices are created by building and wiring millions of transistors together. At the heart of this technology is optical lithography. Optical lithography technology is similar in concept to printing, which was invented more than 3000 years ago [92]. In optical lithography systems, a mask is used as the template, on which the target circuit patterns are carved. A light-sensitive polymer (photoresist) coated on the semiconductor wafer is used as the recording medium, on which the circuit patterns are projected. Light is used as the writing material, which is transmitted through the mask, thus optically projecting the circuit patterns from the mask to the wafer. The lithography steps are typically repeated 20–30 times to make up a circuit, where each underprinting pattern must be aligned to the previously formed patterns. After a lengthy lithography process, a complex integrated circuit (IC) structure is built from the interconnection of basic transistors. Moore's law, first addressed by Intel cofounder G. E. Moore in 1965, describes a long-term trend in the history of computing hardware. Moore's law predicted that the critical dimension (CD) of the IC would shrink by 30% every 2 years. This trend has continued for almost half a century and is not expected to stop for another decade at least. As the dimension of IC reduces following Moore's law, optical lithography has become a critical driving force behind microelectronics technology. During the past few decades, our contemporary society has been transformed by the dramatic increases in electronic functionality and lithography technology. Two main factors of optical lithography attract the attention of scientists and engineers. First, since lithography is the cardinal part of the IC fabrication process, around 30% of the cost of IC manufacturing is attributed to the lithography steps. Second, the advance and ultimate performance of lithography determine further advances of the critical size reduction in IC and thus transistor speed and silicon area. Both of the above aspects drive optical lithography into one of the most challenging places in current IC manufacturing technology. Current commercial optical lithography systems are able to image features smaller than 100 nm (about one-thousandth the thickness of human hair) of the IC pattern. As the dimension of features printed on the wafer continuously

Computational Lithography By Xu Ma and Gonzalo R. Arce

shrinks, the diffraction and interference effects of the light become very pronounced resulting in distortion and blurring of the circuit patterns projected on the wafer. The resolution limit of the optical lithography system is related to the wavelength of light and the structure of the entire imaging system. Due to the resolution limits of optical lithography systems, the electronics industry has relied on *resolution enhancement techniques* (RETs) to compensate and minimize mask distortions as they are projected onto semiconductor wafers. There are three RET techniques: optical proximity correction (OPC), phase-shifting masks (PSMs), and off-axis illumination (OAI). OPC methods add assisting subresolution features on the mask pattern to correct the distortion of the optical projection systems. PSM methods modify both the amplitude and phase of the mask patterns. OAI methods exploit various illumination configurations to enhance the resolution.

1.1.1 Optical Lithography and Integrated Circuits

Optical lithography is at the heart of integrated circuit manufacturing. Generally, three stages are involved in the IC creation process: design, fabrication, and testing [92]. The flow chart of the IC creation process is illustrated in Fig. 1.1.

First, the IC products are defined and designed. In this stage, the abstract functional units such as amplifiers, inverters, adders, flip-flops, and multiplexers are translated into physically connected elements such as metal-oxide-silicon (MOS) transistors. Subsequently, the design results of the physically connected elements are exploited in the second stage of fabrication, where the desired circuit patterns are carved on the masks, which are to be replicated onto the wafers through an optical lithography process. After a series of development processes applied to the exposed wafer such as etching, adding impurities, and so on, the ICs are packaged and tested for functional

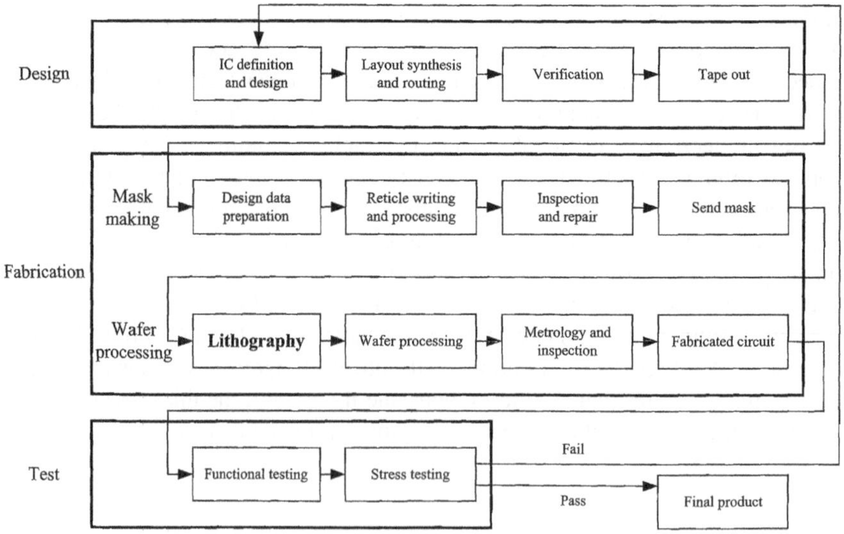

Figure 1.1 The flow chart of the IC creation process.

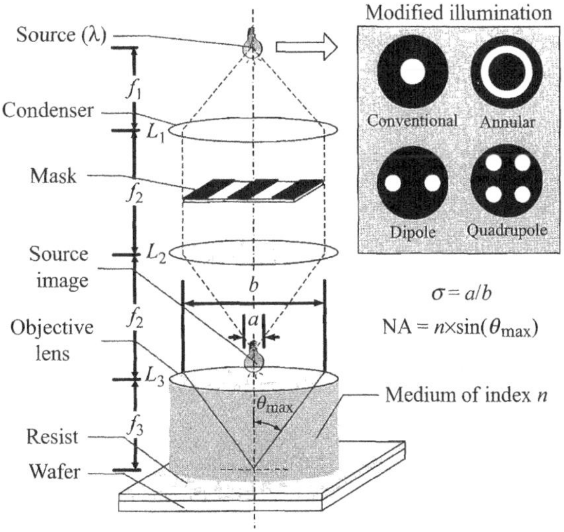

Figure 1.2 The scheme of a typical optical lithography system.

correctness and durability. During the entire IC creation process, optical lithography plays a significant role and is mainly responsible for the miniaturization of IC sizes.

Similar to printing, optical lithography uses light to print circuit patterns carried by the mask onto the wafer. The optical lithography system comprises four basic parts: an illumination system, a mask, an exposure system, and a wafer [92]. A typical optical lithography processing system is shown in Fig. 1.2. In Fig. 1.2, n is the diffraction index of the medium surrounding the lens. $\theta_{\max}$ is the maximum acceptable incident angle of the light exposed onto the wafer. The numerical aperture of the optical lithography system is defined as $\text{NA} = n \sin \theta_{\max}$. The partial coherence factor $\sigma = \frac{a}{b}$ is defined as the ratio between the size of the source image and that of the pupil. Partial coherence factor measures the physical extent of the illumination. Larger partial coherence factor represents larger illumination and lower degree of coherence of the light source [92].

In the optical lithography process, the output pattern sought on the wafer is carved on the mask. Light emitted from the illumination system is transmitted through the mask, where the electric field is modulated by the transparent clear quartz areas and opaque chrome areas on the mask. Subsequently, the modulated electric field propagates through the exposure system and is finally projected onto the light-sensitive photoresist layer coated on the wafer, which is then partially dissolved by the solvents. The details of the photoresist processes and characteristics are discussed in Section 1.3.

1.1.2 Brief History of Optical Lithography Systems

Early optical lithography systems used contact lithography methods, where the mask is pressed against the photoresist-coated wafer during the exposure [11]. Since neither

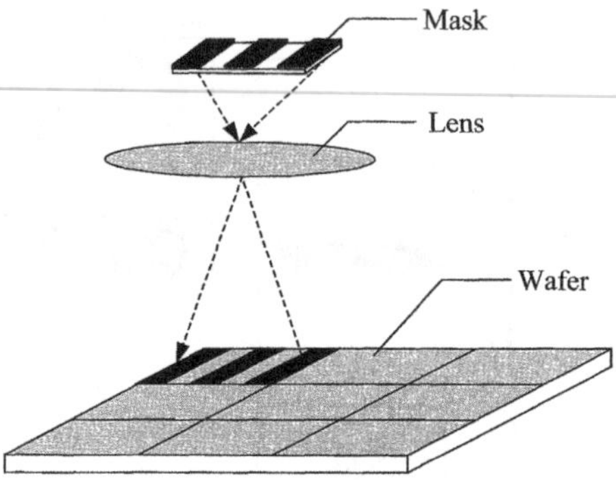

Figure 1.3 The configuration of wafer stepper.

the mask nor the wafer is perfectly flat, the hard contact method was used to push the mask into the wafer by applying a pressure ranging from 0.05 to 0.3 atm [11]. The advantage of contact lithography is that small features can be imaged using relatively cheap equipment. However, defects were generated on both wafer and mask due to the hard contact. In order to avoid defects, proximity lithography was introduced [11]. In proximity lithography, a gap, typically ranging from 10 to 50 μm, was maintained between mask and wafer. The primary disadvantage of proximity lithography is the resolution reduction due to the divergent light. Subsequently, projection lithography was developed so as to obtain high resolution without the defects associated with contact lithography. The projection optical lithography system is illustrated in Fig. 1.2. As shown in Fig. 1.3, in order to replicate a pattern onto a wafer of large scale, a wafer stepper is applied to repeat the lithography process for each small portion of the wafer. There are two configurations of wafer stepper: "step and repeat" and "step and scan" [4]. In the "step and repeat" configuration, the wafer is moved after each exposure until the total wafer has been exposed. Thus, the image size is limited by the largest size of lens field of sufficient imaging quality. In the "step and scan" configuration, the size of the lens field just covers a portion of the mask. The mask and wafer are scanned by the light source until the entire mask pattern is imaged on the wafer. In the past decade, optical lithography systems developed from deep ultraviolet lithography (DUVL) systems employing radiation with a wavelength of 300 nm to the ArF laser lithography system employing radiation with a wavelength of 193 nm.

As the critical dimension of the IC continuously reduces, lithography technology will be pushed into the next-generation lithography systems, including extreme ultraviolet lithography (EUVL), e-beam lithography (EBL), X-ray lithography, and ion-beam projection lithography [6]. Current research mainly concentrates on EUVL and EBL. In EUVL, radiation with a wavelength in the range from 10 to 14 nm is used to carry out projection imaging. In this wavelength range, the radiation is strongly absorbed in almost all materials. Thus, the EUVL systems must operate in near-vacuum environments. In addition, EUVL systems are entirely reflective, including masks [6]. On the other hand, EBL uses electron beam to directly write patterns on the wafer.

The primary advantage of the EBL is that it overcomes the diffraction limit of light. However, the major disadvantage is the low throughput.

1.2 RAYLEIGH'S RESOLUTION

The International Technology Roadmap for Semiconductors (ITRS) (2007 Edition), driven by Moore's law, shows the trend as depicted in Table 1.1 [1]. The critical dimension, which is the minimum feature size to be printed on the wafer, is limited by the Rayleigh's resolution [92].

According to the Fourier optics and the properties of lenses, the light energy passing through a mask forms a distribution in the pupil plane, which is proportional to the mask spectrum [25, 92]. The set of discrete spatial frequency of the mask pattern is referred to as the diffraction orders. Observing from the center of the mask, rays of low spatial frequency components travel with small angles, while those of high spatial frequency components travel with large angles. Therefore, lower frequency components pass closer to the center of the pupil. Higher frequency components are out of scope of the pupil and cannot be collected by the lens. Thus, the effect of the lens in the optical lithography system is equivalent to a low-pass filter, cutting off some high spatial frequency components of the mask pattern.

Consider the optical lithography system under coherent illumination. The spectrum of an isolated opening in the spatial domain is continuous such that some components of the signal spectrum always pass through the low-pass filter and image the pattern [4]. Figure 1.4 shows the imaging process of the isolated opening with a width of d.

Figure 1.5 shows the imaging process of a periodic pattern. The periodic patterns has discrete spatial frequency spectrum at intervals $\Delta k = \frac{2\pi}{p}$, where p is the period of the periodic pattern, referred to as the pitch. The periodic pattern shown in Fig. 1.5 depicts a pitch of p and a width of d. Let the diffraction index $n = 1$, the mth diffraction order diverges from the mask at an angle of θ_m, where

$$\sin\theta_m = m\frac{\lambda}{p}, \quad m = 0, \pm 1, \pm 2, \ldots, \tag{1.1}$$

Table 1.1 The International Technology Roadmap for Semiconductors

Technology Node	2009	2010	2011	2012	2013	2014
DRAM						
1/2 pitch (nm)	50	45	40	36	32	28
MPU						
1/2 pitch (nm)	52	45	40	36	32	28
Gate in resist (nm)	34	30	27	24	21	19
Physical gate length (nm)	20	18	16	14	13	11
Mask minimum features						
Nominal image size (nm)	135	120	107	95	85	76
Minimum primary feature size (nm)	94	84	75	67	59	53
Subresolution feature size (nm), opaque	67	60	54	48	42	38

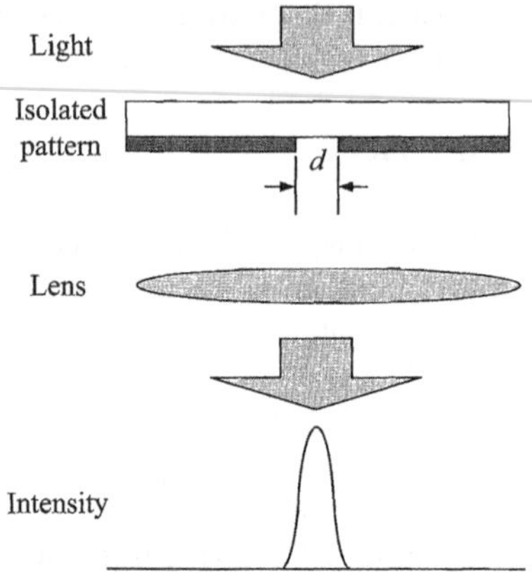

Figure 1.4 The imaging process of the isolated opening with a width of d.

where λ is the wavelength. It has been proven that at least two diffraction orders are needed to image the periodic pattern with distinguishable intensity variation [92].

According to Eq. (1.1), to allow the ± 1 diffraction orders to pass the low-pass filter,

$$\sin \theta_{\max} \geq \frac{\lambda}{p}. \tag{1.2}$$

Thus, the minimum distinguishable pitch is

$$p_{\min} = \frac{\lambda}{\sin \theta_{\max}} = \frac{\lambda}{\text{NA}}. \tag{1.3}$$

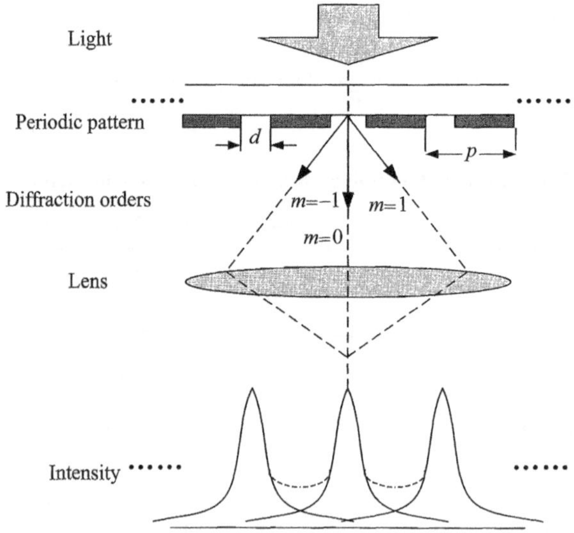

Figure 1.5 The imaging process of a periodic pattern with a pitch of p and a width of d.

In addition, the CD is defined as

$$\mathrm{CD} = \frac{p_{\min}}{2} = \frac{\lambda}{2\mathrm{NA}}. \tag{1.4}$$

In a partially coherent illumination system, the pitch and CD limits change as expected. Partially coherent illumination introduces the partial coherence factor $0 < \sigma < 1$ and thus enhances the resolution. Intuitively, the partially coherent illumination has nonzero line width. Thus, the partially coherent illumination system allows the ± 1 diffraction orders to pass through the low-pass filter; this is more difficult with a coherent illumination system. Therefore, the minimum distinguishable pitch in a partially coherent illumination system can be smaller than that in a coherent illumination system. In partially coherent illumination systems, the minimum distinguishable pitch is [92]

$$p_{\min} = \frac{1}{1+\sigma}\frac{\lambda}{\mathrm{NA}} \tag{1.5}$$

and

$$\mathrm{CD} = \frac{1}{1+\sigma}\frac{\lambda}{2\mathrm{NA}}. \tag{1.6}$$

When the dimension of the partially coherent illumination continuously extends, σ may become larger than 1. However, $\sigma > 1$ does not contribute to the enhancement of the resolution. Therefore, when $\sigma > 1$,

$$p_{\min} = \frac{\lambda}{2\mathrm{NA}} \tag{1.7}$$

and

$$\mathrm{CD} = \frac{\lambda}{4\mathrm{NA}}. \tag{1.8}$$

The above discussion of the Rayleigh's resolution takes only the diffraction effect into account. In order to incorporate the photoresist effect, resolution enhancement techniques, and so on, a process constant k is introduced to describe the comprehensive resolution limit:

$$R = \mathrm{CD} = k\frac{\lambda}{\mathrm{NA}}. \tag{1.9}$$

1.3 RESIST PROCESSES AND CHARACTERISTICS

Photoresist, or simply resist, is a photosensitive compound coated on the wafer, whose properties are changed by the impinging light radiation transmitted through the mask. The photoresist is first exposed in the optical lithography system. Subsequently, the wafer is immersed in a developer solution, then removed from the solution, rinsed off, and dried [11]. After the photoresist process and development, the mask pattern is replicated on the surface of the wafer. Photoresist used in IC fabrication normally consists of three components: a resin or base material, a photoactive compound (PAC), and a solvent, which is used to control the mechanical properties, such as the viscosity of the base or keeping it in a liquid state [11].

There are two types of photoresist processes. These differ in the polarity. The thickness of the remaining photoresist after development is nonlinearly related to the exposure dose (referred to as the aerial image) exceeding a given threshold intensity. In a positive photoresist process, the PAC acts as an inhibitor before the exposure to reduce the dissolving rate of the photoresist when it is developed. Under exposure, chemical reaction occurs in the photoresist and changes the inhibitor to a sensitizer. Thus, the dissolving rate of the photoresist is increased. Because of these properties, almost all the photoresist material remains in the low-exposure areas on the wafer and is removed in the high-exposure areas. Between these two extremes is the transition region. The negative photoresist responds in the opposite manner. Positive photoresist tends to have the best resolution and is therefore much more popular for IC fabrication [11]. On the other hand, compared to positive photoresist, negative photoresist tends to exhibit better adhesion to various substrates such as Si, GaAs, InP, and glass, as well as metals, including Au, Cu, and Al. In addition, the current generation of G-, H-, and I-line negative photoresists exhibit higher temperature resistance over positive photoresists. The steps involved in a typical lithography process are shown in Fig. 1.6.

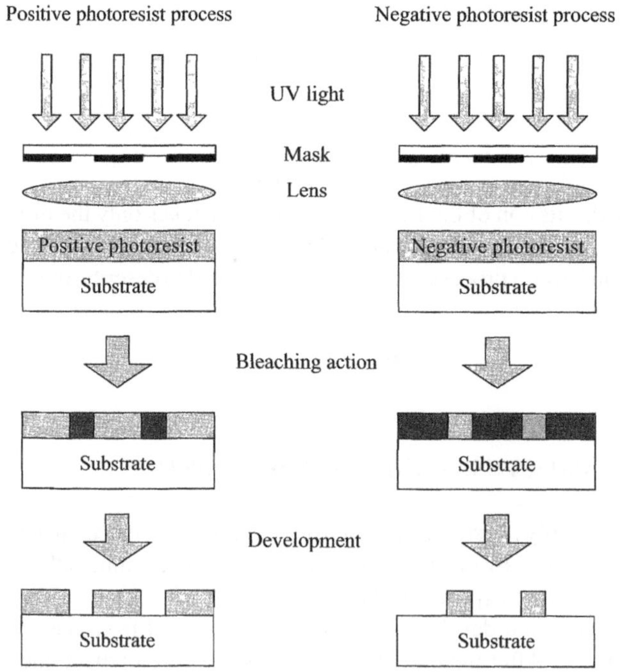

Figure 1.6 The steps involved in a typical optical lithography with positive photoresist process (left) and negative photoresist process (right).

Because of the photoresist absorption, the radiation intensity is decreased with increasing depth into the photoresist. The relationship between the radiation intensity I and the depth z is described by the logarithmic function as follows [11]:

$$I(z) = I_0 e^{-\alpha z}, \tag{1.10}$$

where α is the optical absorption coefficient of the photoresist with unit of inverse of length. I_0 is the intensity on the top of the photoresist. The absorbance A is defined as

$$A = \frac{\int_0^T I_0 - I(z) \mathrm{d}z}{I_0 T} = 1 - \frac{1 - e^{-\alpha T}}{\alpha T}, \tag{1.11}$$

where T is the thickness of the photoresist.

The positive photoresist is taken as an example to discuss the characteristics of photoresist in detail. For positive photoresist, the resist process can be characterized by the relationship between the thickness of the developed resist and the exposure dose for a fixed time. The exposure dose is defined as the light intensity multiplied by the exposure time. The plot of the normalized remaining thickness of resist versus the exposure dose is illustrated in Fig. 1.7. In Fig. 1.7, the x-axis represents exposure dose with unit mJ/cm^2 in logarithmic scale. The y-axis represents the normalized remaining thickness of resist in linear scale. The actual relationship is shown in solid line, which may be approximated by a piecewise linear curve shown in dashed line. There are three regions in the piecewise linear curve. Below the dose D_l is the low-exposure region. Beyond the dose D_h is the high-exposure region. Between the doses D_l and D_h is the transition region. Because of the logarithmic dependence on the exposure dose, the resist development is a nonlinear function of the dose. It is this nonlinearity that transforms sloped aerial images into relatively vertical photoresist profiles [92].

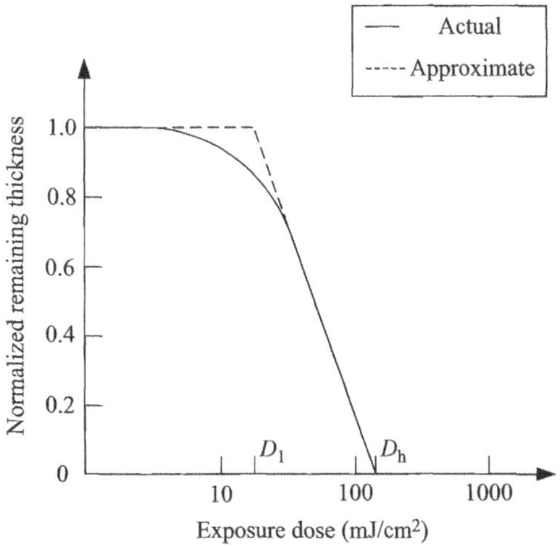

Figure 1.7 The plot of the normalized remaining thickness of resist versus the exposure dose.

The two extensively used metrics to measure the performance of the photoresist are the contrast and the critical modulation transfer function. The contrast is defined as

$$\gamma = \frac{1}{\log_{10}(D_h/D_l)}. \tag{1.12}$$

Thus, the contrast is just the slope of the piecewise curve shown in Fig. 1.7. The contrast depends on the resist material, the development process, the post-exposure bake processes, the wavelength of the light, the surface reflectivity of the wafer, and several other factors [11]. The lithographic image quality in general improves as the contrast is increased. In the ideal case, $D_l = D_h$; thus, $\gamma = \infty$. Therefore, the piecewise curve in Fig. 1.7 will reduce to a hard threshold function. Although this ideal limit cannot be reached in practice, the hard threshold approximation is used in the following chapters to simplify the photoresist model.

On the other hand, the modulation transfer function of an image is defined as

$$\text{MTF} = \frac{I_{max} - I_{min}}{I_{max} + I_{min}}, \tag{1.13}$$

where I_{max} and I_{min} are the maximum and minimum intensities of the aerial image, respectively. The critical modulation transfer function (CMTF) is defined as

$$\text{CMTF} = \frac{D_h - D_l}{D_h + D_l}, \tag{1.14}$$

where D_h and D_l are shown in Fig. 1.7. The CMTF can be explained as the approximate minimum MTF necessary to obtain a pattern [11]. In order to print an aerial image on the photoresist, the MTF of the aerial image must be larger than the CMTF of the resist.

1.4 TECHNIQUES IN COMPUTATIONAL LITHOGRAPHY

Due to the resolution limits of optical lithography systems, the electronics and photonics industry has relied on 2D and 3D resolution enhancement techniques to compensate and minimize mask distortions as they are projected onto semiconductor wafers [78, 92, 95]. According to Eq. (1.9), resolution in optical lithography obeys the Rayleigh's criterion: resolution$(R) = k\frac{\lambda}{\text{NA}}$, where λ is the wavelength, NA is the numerical aperture taking on values around 0.9 for most lithography systems used today, and k is the process constant. In order to extend the limits of the resolution, the wavelength of the employed illumination is continuously reduced to enhance the resolution. On the other hand, NA is also increased by immersion lithography, where a liquid medium is filled in the space between the front lens and the photoresist. Different from these methods, RETs are applied to minimize the process constant k [11, 37, 76, 77]. RET methods manipulate the local amplitude and phase features of the optical wavefront to precompensate for image distortions. There are mainly three traditional kinds of techniques included in RETs such as optical proximity correction, phase-shifting masks, and off-axis illumination. In addition, second-generation RETs have been proposed as the supplements of the traditional RETs. Second-generation RETs include a variety of techniques such as multiple mask exposure lithography, simultaneous source and mask optimization, photoresist tone reversing methods, and so on. The introduction of these approaches is described next.

1.4.1 Optical Proximity Correction

For low-k imaging, a sizable fraction of the transmitted light energy is concentrated in the high spatial frequency components of the mask spectrum. However, the low-pass filtering properties of the lens in the exposure systems cut off the high-frequency components and lead to image distortion. In general, there are four types of image distortion [92]. The first is the variation of the printed image under different environments with the same nominal critical dimension. The second type of distortion occurs when the changes of the nominal CD are not reflected linearly in the printed image. The third is the line shortening, and the final one is the corner rounding. In order to compensate for these image distortions, OPC methods modify the mask amplitude by the addition of subresolution features to the mask pattern such that the output patterns are as close to the desired pattern as possible. The typical scheme of optical proximity correction is shown in Fig. 1.8. The two types of OPC are rule-based approaches and model-based approaches. Rule-based approaches, which are simple to implement, can just compensate for the warping in local features. On the other hand, the model-based approaches used in this book rely on mathematical models to represent the image formation process of the optical lithography system and seek the global minimization of the cost function to improve the output pattern fidelity on the wafer. Model-based approaches include inverse and forward methods. In inverse methods, the optimization algorithms start from the desired output pattern and iteratively obtain the optimized layout. In the forward methods, the original layouts are continuously modified until the output patterns and the mask manufacturability properties are acceptable.

1.4.2 Phase-Shifting Masks

The application of the OPC methods encounters energy diffusing problems, where unwanted energy in the opaque (chrome) regions appears due to the close proximity of the neighboring transparent (quartz) features [69]. Figure 1.9 illustrates the imaging process of the OPC methods. In Fig. 1.9, the binary mask just includes clear area (quartz) and opaque area (chrome). Note that the energy diverges into the gaps between exposed areas on the wafer and reduces the resolution. Phase-shifting masks, commonly attributed to Levenson [35], induce phase shifts in the transmitted field that have a favorable constructive or destructive interference effect to remove the unwanted energy in the opaque regions. Three types of PSMs are extensively used in the IC fabrication industry: alternating phase-shifting masks, attenuated phase-shifting masks, and chromeless phase-shifting masks.

Alternating PSMs modulate the phases of the adjacent features on the mask by 180° out of phase with each other. The phase adjustment is implemented by the

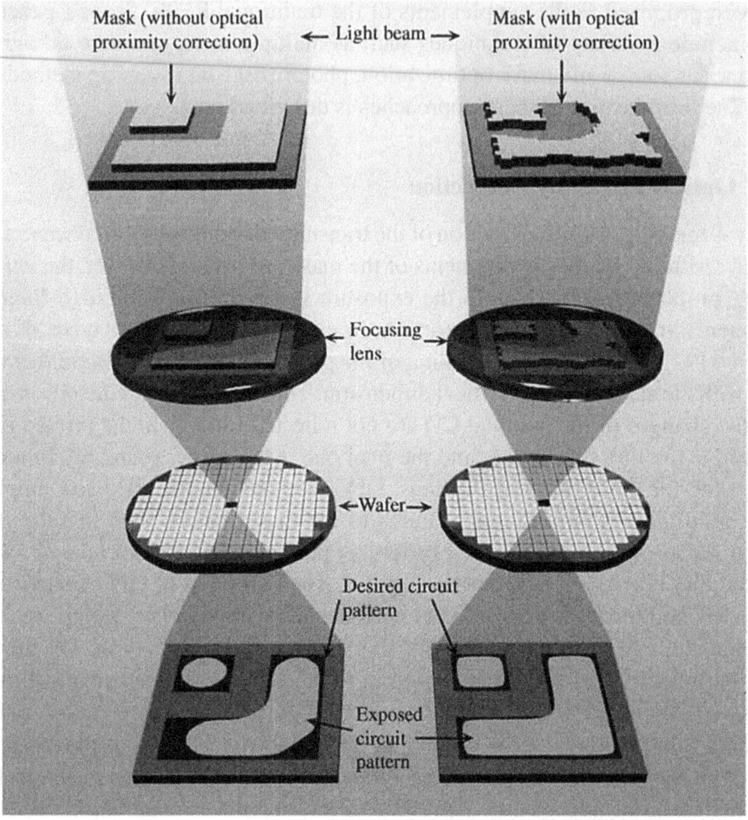

Figure 1.8 The typical scheme of optical proximity correction.

quartz etching. The features with 0° phase are referred to as clear areas, while shifting areas have 180° phase. Phase difference on the alternating PSM leads to destructive interference, removing the diffused energy in the opaque areas and resulting in better contrast and resolution of the printed image compared to binary masks. The imaging process of the alternating PSM is shown in Fig. 1.10.

The attenuated PSMs replace the chrome (opaque area) on the binary masks with molybdenum silicide (MoSi), through which the light partially transmits. The thickness of the MoSi layer introduces a phase shift of 180°. Unlike alternating PSMs where all transparent regions image onto the wafer, the background due to the partially transparent MoSi layer is not printed on the wafer [92]. Phase difference between the transparent regions and partially transparent regions leads to destructive interference. Compared to the alternating PSMs, attenuated PSMs achieve the process latitude improvement of sparse spaces such as an isolated contact without phase-shifting assisting features. However, the alternating PSMs outperform attenuated PSMs when printing narrow dark images [71, 92]. The imaging process of the attenuated PSM is shown in Fig. 1.11.

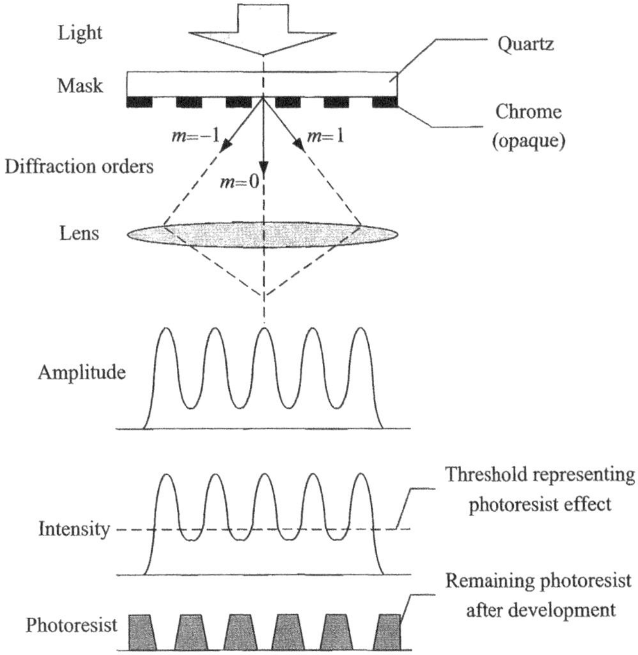

Figure 1.9 The imaging process of the OPC methods.

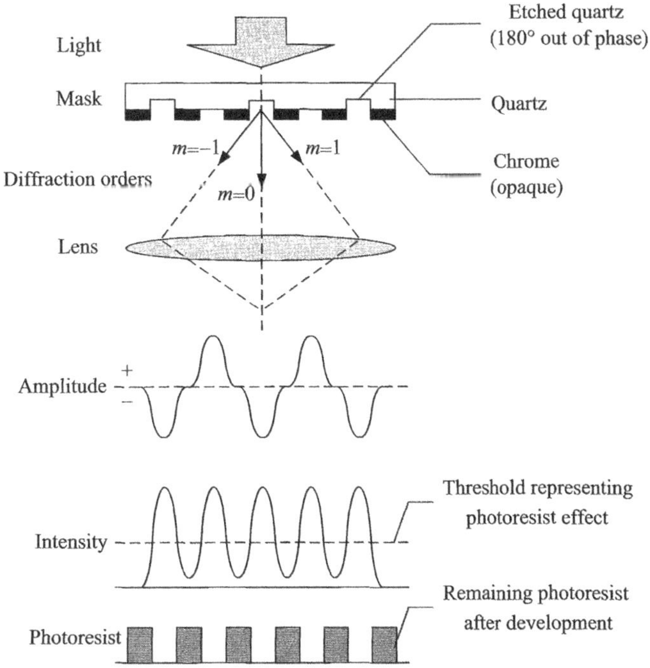

Figure 1.10 The imaging process of alternating PSM.

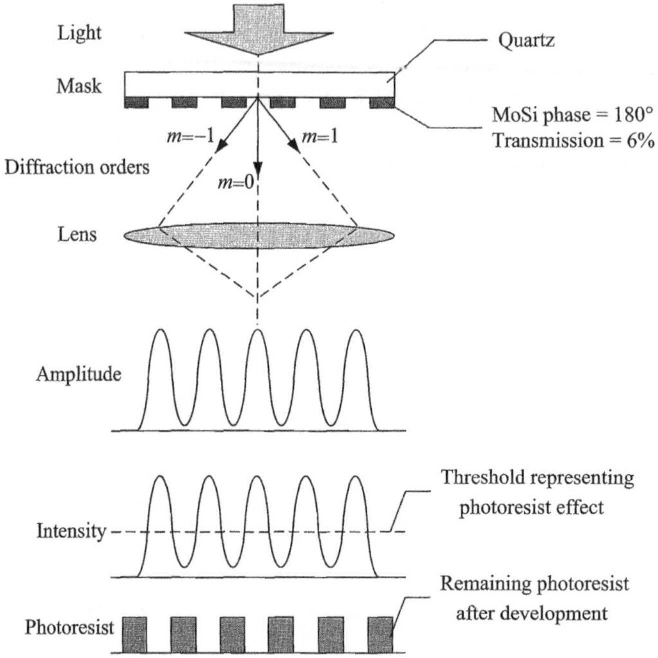

Figure 1.11 The imaging process of attenuated PSM.

Chromeless PSMs replace the MoSi layers of the attenuated PSMs with the etched transparent quartz layers, which preserve 180° phase. The interaction between quartz layers with 0° and 180° phases introduces destructive interference. The imaging process of the chromeless PSM is shown in Fig. 1.12.

1.4.3 Off-Axis Illumination

While OPC methods and PSMs modulate the amplitudes and phases of the features on the mask, off-axis illuminations modify the direction of the impinging light onto the wafer, thus influencing the diffraction orders captured by the lens. In the on-axis illumination system, all the 0 and ± 1 diffraction orders are collected by the lens. These collected light sources carry a lot of information of the background, rather than contributing to image formation. However, in OAI systems, just 0 and either one of ± 1 diffraction orders are collected. The diffraction between the two collected diffraction orders improves the imaging resolution. The OAI is implemented by designing the geometric pattern of the illumination. Common OAI configurations include dipole, quadrupole, and annular illuminations, among others [92]. The conventional illumination and a variety of OAIs are shown in Fig. 1.13. The conventional OAI methods derive the illumination pattern geometries by modifying the accepted diffraction orders by the lens to enhance the resolution limit and contrast. This book formulates the design of OAI pattern as an optimization problem, where both illumination and mask patterns are divided into pixels, each of which is optimized by gradient-based algorithms.

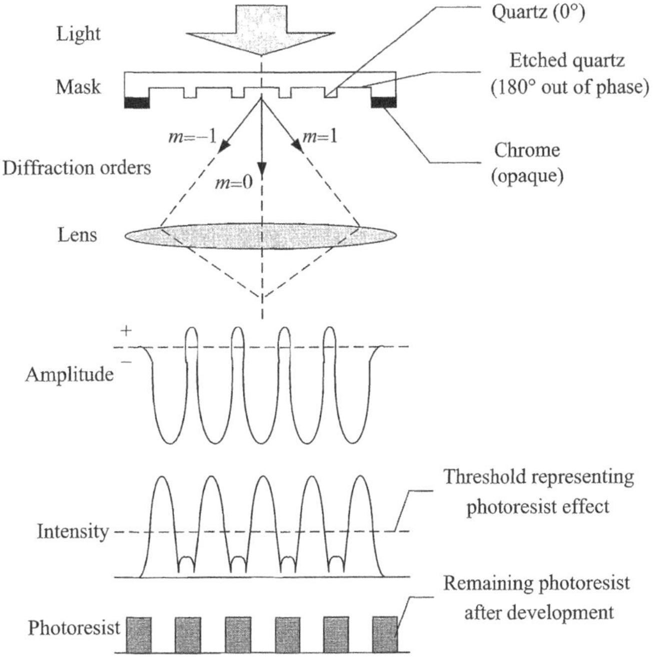

Figure 1.12 The imaging process of chromeless PSM.

1.4.4 Second-Generation RETs

During the past two decades, the continuous decrease in the critical dimension has motivated the development of second-generation RETs, which are not constrained in the range of the traditional RETs discussed above [8, 73, 84, 92]. Multiple mask exposure methods expose the coated photoresist layer several times with different mask

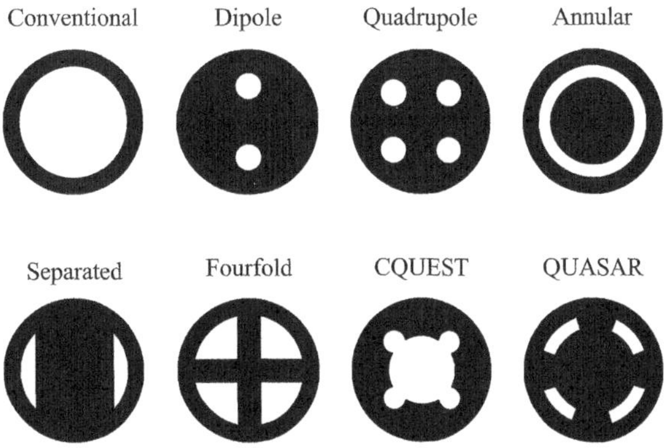

Figure 1.13 The conventional illumination and a variety of off-axis illuminations. *Top row* (from left to right): Conventional, dipole, quadrupole, and annular illuminations. *Bottom row* (from left to right): Separated, fourfold, CQUEST, and QUASAR illuminations.

patterns, capable of printing images even impossible by single exposure methods. In the multiple mask exposure methods, a dense circuit pattern is split into several relative sparse patterns. Masks are fabricated according to each of the sparse patterns and separately exposed on the wafer. There are two typical models of multiple mask exposure methods: double patterning lithography (DPL) and double exposure lithography (DEL). In DPL, two masks are exploited and each exposure is followed by its own etching process of the photoresist. On the other hand, DEL uses just one etching process after two exposures. Although multiple mask exposure methods are suitable for printing dense patterns, the drawback of these methods is the reduction of throughput.

Traditional RETs fix the illumination shape, thus limiting the degrees of freedom during the optimization of the mask pattern. In order to overcome this restriction, the simultaneous source and mask optimization methods have been developed recently where the illumination configuration and the mask pattern are designed simultaneously. The resulting source and mask patterns fall well outside the realm of known design forms. Usually, the optimized illumination is far from the OAIs discussed in Section 1.4.3.

Another new RET method is that of photoresist tone reversing, which exploits both positive and negative photoresist materials on the wafer and improves the lithography performance of small spaces in the output pattern. In addition to these three approaches, there are numerous other potential second-generation RETs, which may be found in relative literature.

1.5 OUTLINE

The organization of the book is as follows.

In Chapter 2, Abbe's formulation of the partially coherent imaging system is first summarized. As the simplified version of Abbe's formulation, the Hopkins diffraction model is used to represent the optical lithography system. Subsequently, three kinds of decompositions of the Hopkins diffraction model are discussed. First is the Fourier series expansion model, where the partially coherent imaging system is represented as the sum of several coherent systems. The accuracy of the Fourier series expansion model is the same as the direct discretization of the Hopkins diffraction model. Subsequently, two kinds of approximation models, referred to as the average coherent approximation model and the singular value decomposition (SVD) model, are summarized and used to reduce the computational complexity of the Fourier series expansion model. As the two limits of the partially coherent imaging system, coherent and incoherent imaging systems are discussed at the end of Chapter 2.

In Chapter 3, the rule-based RETs are described. First, the RET approaches are classified into rule-based, model-based, and hybrid RET approaches. Subsequently, the chapter focuses on the rule-based RETs, where rule-based OPC, PSM, and OAI approaches are described in detail. The rule-based OPC includes catastrophic OPC, one-dimensional OPC, line-shortening reduction OPC, and two-dimensional OPC. The rule-based PSM includes dark-field application and light-field application. For all these rule-based RET methods, the rules to modify the masks and illuminations are summarized.

In Chapter 4, the fundamentals of optimization are discussed. First, the definition and the classification of different optimization problems are summarized. According to the classification, the inverse lithography optimization can be transformed to a continuous, unconstrained, nonlinear, and deterministic problem. Subsequently, the unconstrained optimization problems are discussed in detail. The methods to recognize minimizers are presented as several theorems. Two strategies, such as line search strategy and trust region strategy, are discussed to solve the unconstrained optimization problem. Particularly, the line search strategy includes the steepest descent method, the Newton method, the quasi-Newton method, and the conjugate gradient method. The trust region strategy includes the dogleg method, the two-dimensional subspace minimization method, and so on. In the following chapters, the steepest descent algorithm is applied to the gradient-based inverse lithography optimization.

In Chapter 5, the OPC and PSM optimizations for inverse lithography is developed under coherent imaging systems. The forward imaging process of the optical lithography systems is approximated as the Hopkins diffraction model followed by a sigmoid function. The Hopkins diffraction model represents the formation process of the aerial image. The sigmoid function represents the photoresist effect. The MSE between the desired pattern and the output pattern after the photoresist process is used as the cost function for the mask optimization. Based on this model, OPC and two-phase PSM optimization algorithms are developed. Next, generalized gradient-based PSM optimization methods are developed. These generalized algorithms provide highly effective four-phase PSMs capable of generating mask patterns with arbitrary Manhattan geometries.

In Chapter 6, a set of regularization frameworks for the OPC and PSM optimizations is discussed. The pole penalty is developed to reduce the pattern errors resulting from the discretization of the amplitude and phase of the optimized complex-valued mask. In order to influence the solution patterns to have more desirable manufacturability properties, a wavelet penalty is introduced. The wavelet penalty offers more localized flexibility than total variation penalty, which is traditionally employed in inverse problems. Furthermore, the comparison between wavelet penalty and total variation penalty is discussed.

In Chapter 7, OPC optimization approaches are developed for the partially coherent imaging systems based on the Fourier series expansion model. In order to reduce the computational complexity, the average coherent approximation model is applied to develop effective and more computationally efficient OPC optimization algorithms for inverse lithography. The advantages and the disadvantages of both algorithms are discussed and analyzed. Subsequently, the SVD model is used to develop computationally efficient PSM optimization algorithms under partially coherent illuminations for inverse lithography. These PSM optimization algorithms are most effective with small to medium partial coherence factors.

In Chapter 8, a variety of techniques to improve the performance of OPC and PSM optimizations are described. A double patterning optimization method for general inverse lithography is described where each exposure uses an optimized two-phase mask. Furthermore, a novel DCT post-processing is derived to reduce the mask complexity and the output pattern error by cutting off the high-frequency components of the optimized masks in DCT domain. Finally, a photoresist tone reversing technique is exploited to improve the resolution limit.

In Chapter 9, simultaneous source and mask optimization (SMO) algorithms are described for both OPC and PSM designs. In this chapter, the SOCS model was first applied to decompose the partially coherent imaging systems. Then, the simultaneous source and mask design was formulated as an optimization problem, where the cost function was the square of the l^2-norm of the difference between the desired output pattern and the aerial image. Cost sensitivity was calculated and applied to drive the cost function in the descent direction during the optimization process. In order to influence the solution patterns to have more desirable manufacturability properties, topological constraints are added to the optimization framework.

In Chapter 10, the thick-mask effects are taken into account, as the CD printed on the wafer shrinks into the subwavelength regime, and the mask topography is considered as a 3D object. The OPC and PSM optimization methods are developed based on the boundary layer (BL) model to compensate for the thick-mask effects. In these algorithms, the model-based lithography methods are exploited to obtain the desired binary and phase-shifting masks.

In Chapter 11, the contributions of this book are concluded and the new future directions of RETs are outlined. A software guide for the accompanying Matlab codes is included in Appendix H.

2

Optical Lithography Systems

2.1 PARTIALLY COHERENT IMAGING SYSTEMS

Most practical illumination sources in optical lithography systems have a nonzero line width and their radiation is more generally described as partially coherent [75]. Partially coherent illumination (PCI) is desired, since it can improve the theoretical resolution limit. PCI is thus introduced in practice through modified illumination sources having large coherent factors or through off-axis illumination. In partially coherent imaging, the mask is illuminated by light traveling in various directions. The source points giving rise to these incident rays are incoherent with one another, such that there is no interference that could lead to nonuniform light intensity impinging on the mask [92, 93]. Common partially coherent illumination modes include dipole, quadrupole, and annular illumination. Partially coherent imaging models are discussed in this section.

In partially coherent optical lithography systems, the Köhler illumination configuration is assumed, which is shown in Fig. 2.1. In the Köhler illumination configuration, the light source is considered to be located at the focal plane of the condenser and the object plane is located at the condenser exit pupil [4]. Each point source on the illumination emits a coherent, linearly polarized plane wave with a spatial frequency determined by the position of the point source related to the optical axis. Under the assumption of the Köhler illumination configuration, two kinds of partially coherent imaging models are discussed in the following. The first one is the Abbe's model, and the second one is the Hopkins diffraction model, where the Hopkins diffraction model is a simplified and approximate version of the Abbe's model.

2.1.1 Abbe's Model

The Abbe's model, also referred to as source integration method, decomposes the partially coherent imaging system into the superposition of a set of coherent imaging systems [4, 7]. Each of these coherent imaging systems is based on the contribution of each point within the numerical aperture of the condenser (NA_c). Figure 2.2 illustrates the scheme of optical projection system. The source accepted by the condenser

Computational Lithography By Xu Ma and Gonzalo R. Arce

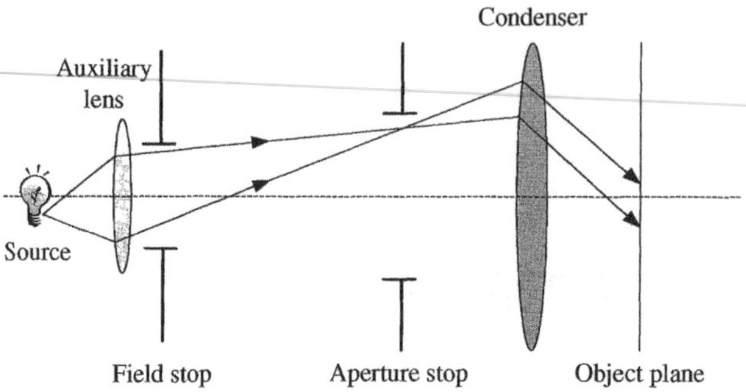

Figure 2.1 Köhler illumination configuration, where each source point generates a coherent, linearly polarized plane wave of spatial frequency determined by the position of the source point related to the optical axis [4].

numerical aperture is actually the image of the illumination shape in the lens pupil, referred to as the effective source. Each point on the effective source generates incident plane waves on the object plane with direction represented by the unit vector $\hat{\mathbf{p}} = (p_x, p_y, p_z)^T$. Therefore, (p_x, p_y) are sufficient to identify each point on the effective source. Different point sources are assumed to be incoherent with each other. The mask is located in the object plane. $\mathbf{E}_0(x', y'; p_x, p_y)$ and $\mathbf{H}_0(x', y'; p_x, p_y)$ are the electric and magnetic fields at the location of (x', y') on the exit surface of the mask, contributed by the effective source point of (p_x, p_y). The incident ray emitted from the mask to the entrance pupil is represented by the unit vector $\hat{\mathbf{r}}' = (r'_x, r'_y, r'_z)^T$. $\mathbf{E}_{\text{entrance}}$ denotes the electric field diffracted by the mask at the entrance pupil. θ is the included angle between $\hat{\mathbf{r}}'$ and the entrance pupil. The outgoing ray is represented by the unit vector $\hat{\mathbf{s}} = (s_x, s_y, x_z)^T$, pointing from the exit pupil to the origin. The radius of the spherical wavefront outgoing from the exit pupil is denoted by L. $\mathbf{E}_{\text{exit}}$ denotes the electric field at the exit pupil. θ' is the included angle between $\hat{\mathbf{s}}$ and the exit pupil. M denotes the demagnification of the lens, satisfying $\sin(\theta) = M\sin(\theta')$. The wafer

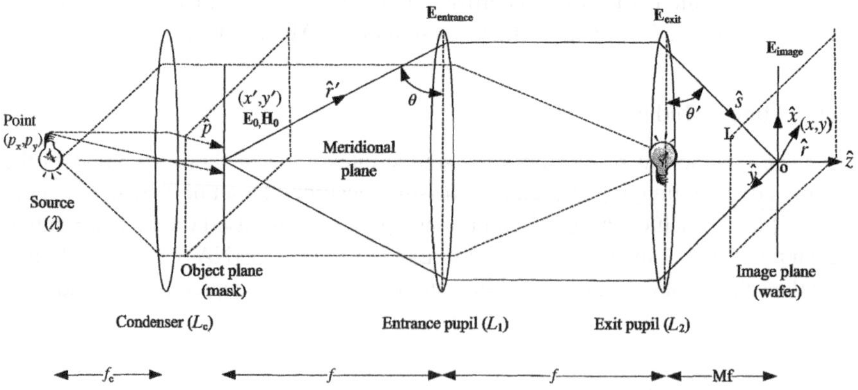

Figure 2.2 An optical projection system.

is located in the image plane. The location of (x, y) on the image plane corresponds to the vector $\mathbf{r}$, and $\hat{\mathbf{r}} = \mathbf{r}/|\mathbf{r}| = (r_x, r_y, r_z)^T$. $\mathbf{E}_{\text{image}}(x, y; p_x, p_y)$ denotes the electric field at the location of (x, y) on the wafer, contributed by the effective point source of (p_x, p_y). $\hat{\mathbf{x}}$, $\hat{\mathbf{y}}$, and $\hat{\mathbf{z}}$ represent the unit vectors along the x, y, and z axes, respectively. The x–z plane is referred to as the Meridional plane.

The electric field on the wafer contributed by the effective source point of (p_x, p_y) is formulated as [87]

$$\mathbf{E}_{\text{image}}(x, y; p_x, p_y) = \frac{j}{\lambda} \iint_{s_x^2+s_y^2 \le \text{NA}^2} \frac{\mathbf{a}(s_x, s_y; p_x, p_y)}{s_z} \mathrm{e}^{-jk[C+\Phi(s_x,s_y;p_x,p_y)+\hat{\mathbf{s}}\cdot\hat{\mathbf{r}}]} \mathrm{d}s_x \mathrm{d}s_y, \tag{2.1}$$

where the temporal term $\mathrm{e}^{j\omega t}$ has been dropped. $\mathbf{a}(s_x, s_y; p_x, p_y)$ is an amplitude function. The phase term e^{-jkC} is the constant representing the phase accumulated while propagating through the lens. The term $\Phi(s_x, s_y; p_x, p_y)$ denotes the aberration function with respect to the ideal spherical wavefront converging toward the focal point. Assuming that the polarization direction of the electric field vector maintains an approximately constant angle with respect to the Meridional plane, the amplitude function $\mathbf{a}(s_x, s_y; p_x, p_y)$ can be formulated as

$$\begin{aligned}\mathbf{a}(s_x, s_y; p_x, p_y) = \frac{1}{j2\lambda} M \sqrt{\frac{\cos\theta'}{\cos\theta}} \mathbf{T}\mathbf{F}\Big\{ & \left[\eta(\hat{\mathbf{z}} \times \mathbf{H}_0(x', y'; p_x, p_y)) \right.\\ & \left. - \eta(\hat{\mathbf{z}} \times \mathbf{H}_0(x', y'; p_x, p_y)) \cdot \hat{\mathbf{r}}\right] \hat{\mathbf{r}} \\ & - [(\hat{\mathbf{z}} \times \mathbf{E}_0(x', y'; p_x, p_y)) \times \hat{\mathbf{r}}]; \frac{Ms_x}{\lambda}, \frac{Ms_y}{\lambda}\Big\},\end{aligned} \tag{2.2}$$

where λ is the wavelength. $\mathbf{T}$ is the polarization tensor accounting for the polarization rotation between the incident electric field before the entrance pupil and the outgoing electric field after the exit pupil. Specifically,

$$\mathbf{T} = \begin{pmatrix} T_{xx} & T_{yx} & T_{zx} \\ T_{yx} & T_{yy} & T_{zy} \\ T_{zx} & T_{zy} & T_{zz} \end{pmatrix}, \tag{2.3}$$

where

$$T_{xx} = \frac{s_y^2 + s_x^2 \left[s_z r_z' - M\left(s_x^2 + s_y^2\right)\right]}{s_x^2 + s_y^2},$$

$$T_{yy} = \frac{s_x^2 + s_y^2 \left[s_z r_z' - M\left(s_x^2 + s_y^2\right)\right]}{s_x^2 + s_y^2},$$

$$T_{yx} = \frac{-s_x s_y \left[1 - s_z r_z' + M\left(s_x^2 + s_y^2\right)\right]}{s_x^2 + s_y^2},$$

$$T_{zz} = -\left(s_x^2 + s_y^2\right) M + r_z' s_z,$$

$$T_{zx} = -s_x \left(r_z' + M s_z\right),$$

$$T_{zy} = -s_y \left(r_z' + M s_z\right). \tag{2.4}$$

In Eq. (2.2), $\mathbf{F}$ denotes the Fourier transform evaluated at the spatial frequencies $(\frac{Ms_x}{\lambda}, \frac{Ms_y}{\lambda})$. η is the intrinsic impedance of the propagation medium. Further, assume that the mask is laid in the x'–y' plane. The main polarization directions of $\mathbf{E}_0(x', y'; p_x, p_y)$ and $\mathbf{H}_0(x', y'; p_x, p_y)$ are along x-axis and y-axis, respectively. E_{0x} and H_{0y} are the components of $\mathbf{E}_0(x', y'; p_x, p_y)$ and $\mathbf{H}_0(x', y'; p_x, p_y)$ along the main polarization directions. After mathematical simplification at length, Eq. (2.1) can be approximated as

$$\mathbf{E}_{\text{image}}(x, y; p_x, p_y) = -\frac{M}{\lambda^2} \iint_{s_x^2+s_y^2 \le \text{NA}^2} \sqrt{\frac{\cos\theta'}{\cos\theta}} \mathbf{T} \cdot \hat{\mathbf{x}} \mathbf{F} \left\{ E_{0x}; \frac{Ms_x}{\lambda}, \frac{Ms_y}{\lambda} \right\} \mathrm{e}^{-jk[C+\Phi(s_x,s_y;p_x,p_y)+\hat{\mathbf{s}}\cdot\mathbf{r}]} \mathrm{d}s_x \mathrm{d}s_y. \tag{2.5}$$

Based on Abbe's method, the intensity on the wafer is the superposition of all components contributed by every effective source point (p_x, p_y). Let the effective source point of (p_x, p_y) generate a time-average intensity of $I_{\text{source}}(p_x, p_y)$. Then, the aerial image at the location of (x, y) on the wafer is

$$I(x, y) = \iint_{p_x^2+p_y^2 \le \text{NA}_c^2} I_{\text{source}}(p_x, p_y) \mathbf{E}_{\text{image}}(x, y; p_x, p_y) \mathbf{E}^*_{\text{image}}(x, y; p_x, p_y) \mathrm{d}p_x \mathrm{d}p_y. \tag{2.6}$$

2.1.2 Hopkins Diffraction Model

Hopkins diffraction model is a simplified and approximate version of the Abbe's model, where the integration over the source is carried out before summing up the diffraction angles accepted by the lens [7, 28, 29]. To derive the Hopkins diffraction model, $\mathbf{E}_{\text{image}}(x, y; p_x, p_y)$ in Eq. (2.6) has to be decomposed into two terms. One term, depending on (p_x, p_y), denotes the effect due to different effective source points. The other term, depending on the coordinate on the object plane (x', y'), denotes the effect due to the mask. The condition of the decomposition is that the included angle between the incident ray and the normal direction is small enough. Typical optical lithography systems involve reduction factors of $4\times$ or $5\times$, and partial coherence factors of σ between 0.3 and 0.8. Thus, the incident angles are smaller than $10°$ with respect to the normal direction. It has been proven that the diffracted harmonics at the exit pupil of the mask remain approximately constant [63, 64, 86, 94]. In this case,

$$\mathbf{E}_0(x', y'; p_x, p_y) \approx \mathbf{E}_0(x', y'; p_x = 0, p_y = 0)\mathrm{e}^{-jk(p_x x' + p_y y')}, \tag{2.7}$$

$$\mathbf{H}_0(x', y'; p_x, p_y) \approx \mathbf{H}_0(x', y'; p_x = 0, p_y = 0)\mathrm{e}^{-jk(p_x x' + p_y y')}. \tag{2.8}$$

Substituting Eqs. (2.7) and (2.8) into Eq. (2.5), we have

$$\mathbf{E}_{\text{image}}(x, y; p_x, p_y) = -\frac{M}{\lambda^2}\iint_{-\infty}^{\infty} \mathbf{K}\left(\frac{s_x}{\lambda}, \frac{s_y}{\lambda}\right) \cdot \mathbf{F}\left\{\tilde{\mathbf{E}}_0; \frac{Ms_x - p_x}{\lambda}, \frac{Ms_y - p_y}{\lambda}\right\} e^{-jk(s_x x + s_y y)} ds_x ds_y, \quad (2.9)$$

where $\tilde{\mathbf{E}}_0$ is the Fourier transform integrand of Eq. (2.5), and

$$\mathbf{K}\left(\frac{s_x}{\lambda}, \frac{s_y}{\lambda}\right) = \sqrt{\frac{\cos\theta'}{\cos\theta}}\mathbf{T}e^{-jk[C+\Phi(s_x,s_y)-s_z\Delta_z]}\text{circ}\left(\frac{\sqrt{s_x^2 + s_y^2}}{\text{NA}}\right), \quad (2.10)$$

where Δ_z denotes the defocus distance in the direction along the optical axis. The filtering effects of the entrance pupil are represented by the circular step function circ.

Assume that the optical systems are isoplanatic or space invariant. Further, suppose that the critical dimension on the mask is much larger than wavelength λ, the mask can be treated as a 2D object, where the thick-mask effect is neglected. In this case, substitute Eqs. (2.9) and (2.10) into Eq. (2.6). After some parameter transformations and derivations, the light intensity distribution exposed on the wafer with partially coherent illumination is shown to be bilinear and described as [74, 92, 97]

$$I(\mathbf{r}) = \iint_{-\infty}^{+\infty} M(\mathbf{r_1})M^*(\mathbf{r_2})\gamma(\mathbf{r_1} - \mathbf{r_2})h(\mathbf{r} - \mathbf{r_1})h^*(\mathbf{r} - \mathbf{r_2})d\mathbf{r_1}d\mathbf{r_2}, \quad (2.11)$$

where $\mathbf{r} = (x, y)$, $\mathbf{r_1} = (x_1, y_1)$, and $\mathbf{r_2} = (x_2, y_2)$. $M(\mathbf{r})$ is the mask pattern, $\gamma(\mathbf{r_1} - \mathbf{r_2})$ is the complex degree of coherence, and $h(\mathbf{r})$ represents the amplitude impulse response of the optical system. The complex degree of coherence $\gamma(\mathbf{r_1} - \mathbf{r_2})$ is generally a complex number, whose magnitude represents the extent of optical interaction between two spatial locations $\mathbf{r_1} = (x_1, y_1)$ and $\mathbf{r_2} = (x_2, y_2)$ of the light source [92]. The complex degree of coherence in the spatial domain is the inverse 2D Fourier transform of the effective source shape. In the frequency domain, Eq. (2.11) is translated as

$$I(x, y) = \iiiint_{-\infty}^{+\infty} \text{TCC}(f_1, g_1; f_2, g_2)\tilde{M}(f_1, g_1)\tilde{M}^*(f_2, g_2) \times \exp\{-i2\pi[(f_1 - f_2)x + (g_1 - g_2)y]\}df_1 dg_1 df_2 dg_2, \quad (2.12)$$

where $\tilde{M}(f_1, g_1)$ and $\tilde{M}(f_2, g_2)$ are the Fourier transforms of $M(x_1, x_2)$ and $M(x_2, y_2)$ in Eq. (2.11), respectively. $\text{TCC}(f_1, g_1; f_2, g_2)$ is the transmission cross-coefficient, which indicates the interaction between $\tilde{M}(f_1, g_1)$ and $\tilde{M}(f_2, g_2)$. Specifically,

$$\text{TCC}(f_1, g_1; f_2, g_2) = \iint_{-\infty}^{+\infty} \tilde{\gamma}(f, g)\tilde{h}(f + f_1, g + g_1)\tilde{h}^*(f + f_2, g + g_2)df dg, \quad (2.13)$$

where $\tilde{\gamma}(f, g)$, referred to as the effective source, is the Fourier transform of $\gamma(x, y)$. $\tilde{h}(f, g)$ is the Fourier transform of $h(x, y)$. The transmission cross-coefficient is

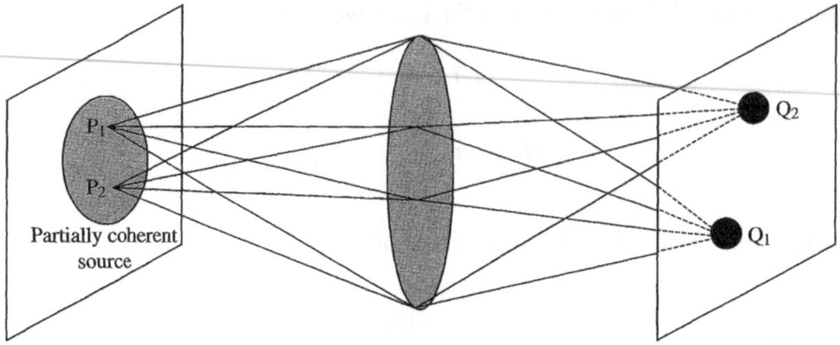

Figure 2.3 A partially coherent imaging system.

independent of the mask pattern. Therefore, given an optical system with fixed illumination, numerical aperture, defocus, and other aberrations, the transmission cross-coefficient needs to be calculated only once. Changing mask patterns does not change the value of $\mathrm{TCC}(f_1, g_1; f_2, g_2)$. Thus, the Hopkins diffraction model described in Eq. (2.12) is more computationally efficient than the Abbe's model shown in Eq. (2.6).

2.1.3 Coherent and Incoherent Imaging Systems

The partially coherent imaging system, illustrated in Fig. 2.3, reduces to simple forms in the two limits of complete coherence or complete incoherence. For the completely coherent case, the illumination source is at a single point; thus, $\gamma(\mathbf{r}) = 1$. The principal scheme of the completely coherent imaging system is shown in Fig. 2.4. In this case, the intensity distribution in Eq. (2.11) is separable on $\mathbf{r_1}$ and $\mathbf{r_2}$, and thus

$$I(\mathbf{r}) = |M(\mathbf{r}) \otimes h(\mathbf{r})|^2, \tag{2.14}$$

where $\otimes$ is the convolution operation. For the completely incoherent case, the illumination source is of infinite extent, and thus, $\gamma(\mathbf{r}) = \delta(\mathbf{r})$. The principal scheme of the

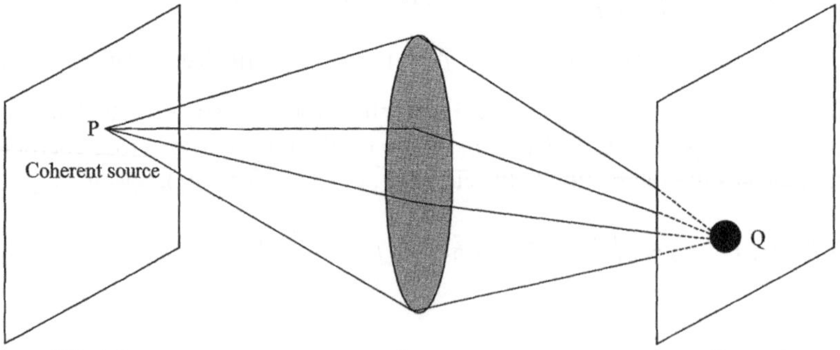

Figure 2.4 A coherent imaging system.

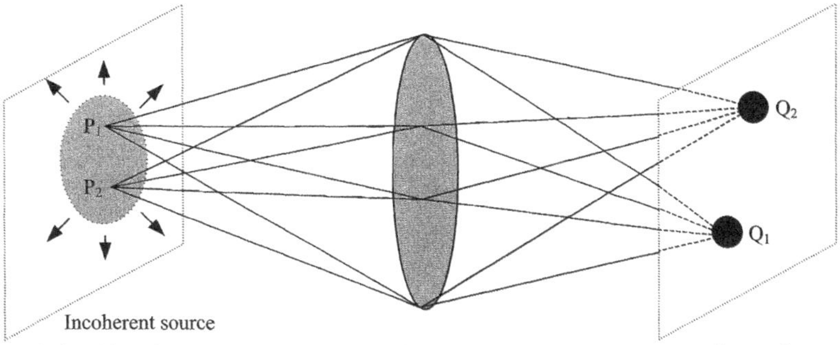

Figure 2.5 An incoherent imaging system.

completely incoherent imaging system is shown in Fig. 2.5. In this case, the intensity distribution reduces to

$$I(\mathbf{r}) = |M(\mathbf{r})|^2 \otimes |h(\mathbf{r})|^2. \tag{2.15}$$

The imaging synthesis and analysis of partially coherent systems are thus more complex than the coherent or incoherent imaging systems.

2.2 APPROXIMATION MODELS

Although the Hopkins diffraction model is more computationally efficient than the Abbe's model, the intensity distribution described in Eqs. (2.11) and (2.12) is tedious to evaluate. To reduce the computational complexity of the Hopkins diffraction model, a set of approximation models of partially coherent imaging systems are discussed in this section. The first model is the Fourier series expansion model, which approximates the partially coherent imaging system as a sum of coherent systems [42, 43, 74] based on 2D Fourier series expansion. The second model is the singular value decomposition (SVD) model, which decomposes the partially coherent imaging system into several coherent systems based on the eigenvalue decomposition [13, 44, 47]. When the partial coherence factor is small, the eigenvalues decay very fast, and the partially coherent systems can be approximated by several coherent components corresponding to the largest eigenvalues. The third model is the average coherent approximation model, which decomposes the partially coherent imaging system into the superposition of a coherent illumination component and another incoherent illumination component [75].

2.2.1 Fourier Series Expansion Model

Radiation of partially coherent light has been shown to be described as an expansion of coherent modes added incoherently in the image plane [58, 74]. Therefore, the bilinear Hopkins diffraction model of partially coherent imaging systems can be represented by a sum of coherent system (SOCS) model based on a Fourier series expansion. In typical imaging applications, the size of the mask is much larger than the width of the complex degree of coherence $\gamma(\mathbf{r})$. In this case, the Fourier series expansion model is applied to reduce the computation cost of partially coherent imaging.

Assume the mask is constrained in the square area A defined by $x, y \in \left[-\frac{D}{2}, \frac{D}{2}\right]$. Thus, for the computations involved in Eq. (2.11), the only values of $\gamma(\mathbf{r})$ needed are those inside the square area A_γ defined by $x, y \in [-D, D]$. Applying the 2D Fourier series expansion, $\gamma(\mathbf{r})$ can be rewritten as

$$\gamma(\mathbf{r}) = \sum_{\mathbf{m}} \Gamma_{\mathbf{m}} \exp(j\omega_0 \mathbf{m} \cdot \mathbf{r}) \tag{2.16}$$

and

$$\Gamma_{\mathbf{m}} = \frac{1}{D^2} \int_{A_\gamma} \gamma(\mathbf{r}) \exp(j\omega_0 \mathbf{m} \cdot \mathbf{r}) d\mathbf{r}, \tag{2.17}$$

where $\omega_0 = \pi/D$, $\mathbf{m} = (m_x, m_y)$, m_x and m_y are integers, and $\cdot$ represents the inner-product operation. Substituting Eq. (2.16) into Eq. (2.11), the light intensity on the wafer is given by

$$I(\mathbf{r}) = \sum_{\mathbf{m}} \Gamma_{\mathbf{m}} |M(\mathbf{r}) \otimes h^{\mathbf{m}}(\mathbf{r})|^2, \tag{2.18}$$

where

$$h^{\mathbf{m}}(\mathbf{r}) = h(\mathbf{r}) \exp(j\omega_0 \mathbf{m} \cdot \mathbf{r}). \tag{2.19}$$

It is observed from Eqs. (2.18) and (2.19) that the partially coherent imaging system is equal to the superposition of coherent systems. Since the Fourier series expansion model is based on direct discretization of the Hopkins diffraction model, they have the same accuracy. Taking the annular illumination as an example, the complex degree of coherence is

$$\gamma(\mathbf{r}) = \frac{J_1(2\pi r/2D_{\text{cu}})}{2\pi r/2D_{\text{cu}}} - \frac{D_{\text{cu}}^2}{D_{\text{cl}}^2} \frac{J_1(2\pi r/2D_{\text{cl}})}{2\pi r/2D_{\text{cl}}}, \tag{2.20}$$

where $r = \sqrt{x^2 + y^2}$. The corresponding Fourier series coefficients are

$$\Gamma_{\mathbf{m}} = \begin{cases} \frac{4D_{\text{cu}}^2 D_{\text{cl}}^2}{\pi D^2 (D_{\text{cl}}^2 - D_{\text{cu}}^2)}, & \text{for } D/2D_{\text{cl}} \le |\mathbf{m}| \le D/2D_{\text{cu}}, \\ 0, & \text{elsewhere}, \end{cases} \tag{2.21}$$

where D_{cl} and D_{cu} are the coherent lengths of the inner and outer circles, respectively. $\sigma_{\text{inner}} = \frac{\lambda}{2D_{\text{cl}}\text{NA}}$ and $\sigma_{\text{outer}} = \frac{\lambda}{2D_{\text{cu}}\text{NA}}$ are the corresponding inner and outer partial coherence factors. The convolution kernel $h(\mathbf{r})$ is defined as the Fourier transform of the circular lens aperture with cutoff frequency NA/λ [7, 85]; therefore,

$$h(\mathbf{r}) = \frac{J_1(2\pi r \text{NA}/\lambda)}{2\pi r \text{NA}/\lambda}. \tag{2.22}$$

The scheme of the SOCS decomposition by Fourier series expansion is depicted in Fig. 2.6, where each coherent system corresponds to one 2D Fourier series expansion

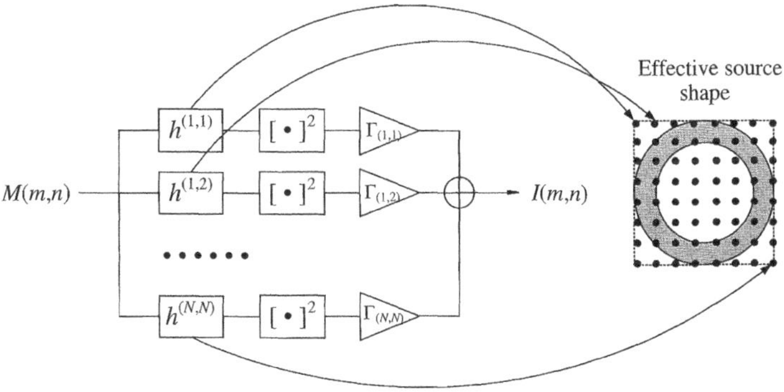

Figure 2.6 A partially coherent system represented by a Fourier series expansion model as a sum of coherent systems.

coefficient of $\gamma(\mathbf{r})$. In Fig. 2.6, $M(m, n)$ and $I(m, n)$ are the discretization of $M(x, y)$ and $I(x, y)$. The 2D Fourier series expansion $\Gamma_{\mathbf{m}}$ is the effective source shape.

In the following, annular illuminations are taken as examples to evaluate the Fourier series expansion model. Annular illuminations are classified by the sizes of their inner and outer partial coherence factors. The larger the partial coherence factor, the higher the resolvable spatial frequency. Thus, the large partial coherence factors lead to improvements on resolution and contrast. Small partial coherence factors, on the other hand, have the advantage to form sparse patterns, which can be exploited effectively by phase-shifting masks. Medium partial coherence factors are preferred for mask pattern containing both sparse and dense patterns [92]. Figure 2.7 illustrates annular illumination sources having large, medium, and small partial coherence factors. For the large partial coherence factor illumination in Fig. 2.7, $\sigma_{\text{inner}} = 0.8$ and $\sigma_{\text{outer}} = 0.975$. For the medium illumination, $\sigma_{\text{inner}} = 0.5$ and $\sigma_{\text{outer}} = 0.6$. For the small illumination, $\sigma_{\text{inner}} = 0.3$ and $\sigma_{\text{outer}} = 0.4$. The dashed lines represent the dimension of the pupil. The number of terms used in the Fourier series expansion in Eq. (2.18) plays a critical role in the computational complexity of the model. The number of terms in the expansion of Eq. (2.18) will be referred to as T. According to Eq. (2.21),

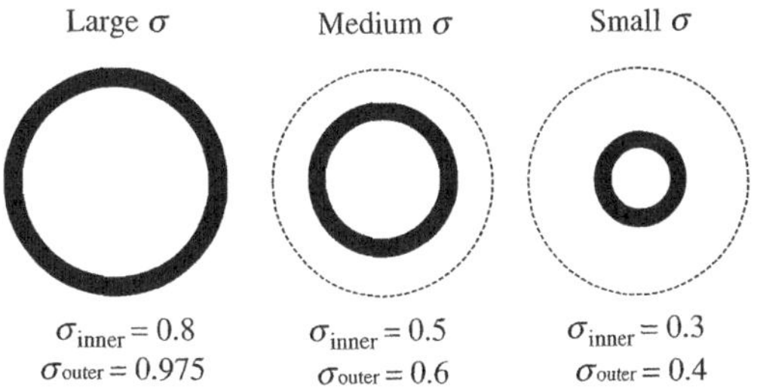

Figure 2.7 Annular illuminations with large, medium, and small partial coherence factors. The dashed lines represent the dimension of the pupil.

$D/2D_{\text{cl}} \leq |\mathbf{m}| \leq D/2D_{\text{cu}}$. In addition, $D_{\text{cl}} = \frac{\lambda}{2\sigma_{\text{inner}}}$, $D_{\text{cu}} = \frac{\lambda}{2\sigma_{\text{outer}}}$, and $D = N \times p$, where $p \times p$ is the pixel size. Thus,

$$T \sim \pi[(D/2D_{\text{cu}})^2 - (D/2D_{\text{cl}})^2] \sim CN^2, \tag{2.23}$$

where the constant C is $C = \frac{\pi p^2 \text{NA}^2(\sigma_{\text{outer}}^2 - \sigma_{\text{inner}}^2)}{\lambda^2}$. The parameter T is larger for sources with larger partial coherence factors. As an example, the values of T for the sources in Fig. 2.7 are 12, 12, and 52 as σ increases from smaller to larger partial coherence factors.

Figure 2.8 illustrates a mask of dimensions 1035 nm × 1035 nm and the corresponding aerial images formed by the annular illuminations having large, medium, and small partial coherence factors. The mask consists of 45 nm features. The pitch $p =$ 90 nm is indicated by dashed lines. The aerial images are synthesized by the Fourier series expansion model. In these simulations, NA = 1.25, $\lambda = 193$ nm, and $h(\mathbf{r})$ is assumed to vanish outside the area A_h defined by $x, y \in [-56.25\text{ nm}, 56.25\text{ nm}]$. The pixel size is 5.63 nm × 5.63 nm. Note that the aerial images increasingly become more blurred, as the partial coherence factor is decreased.

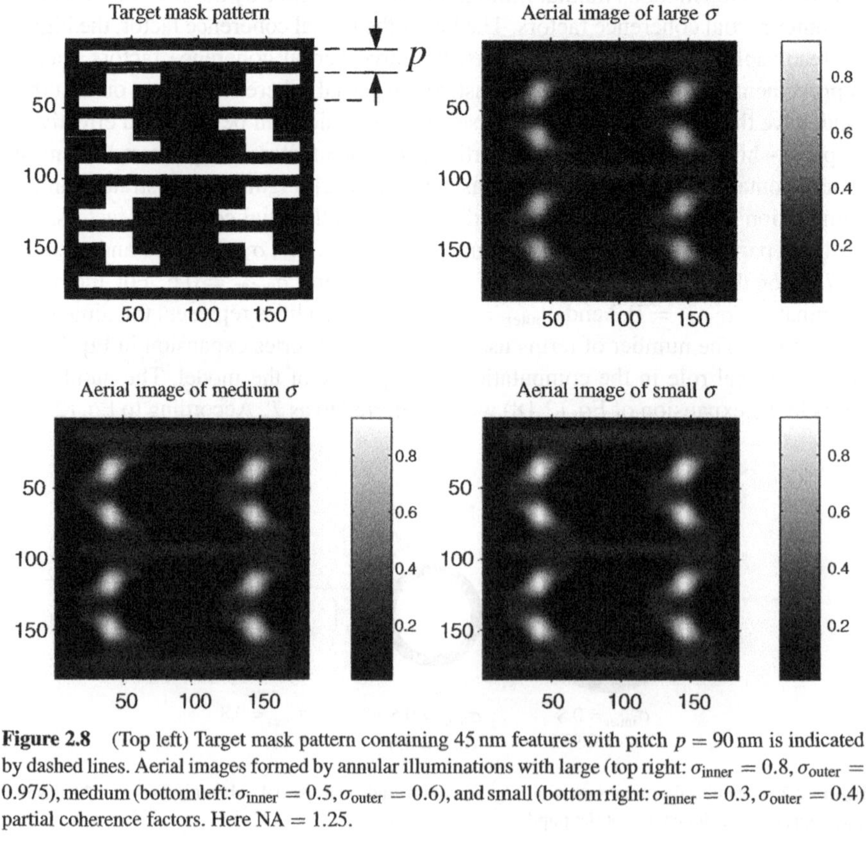

Figure 2.8 (Top left) Target mask pattern containing 45 nm features with pitch $p = 90$ nm is indicated by dashed lines. Aerial images formed by annular illuminations with large (top right: $\sigma_{\text{inner}} = 0.8, \sigma_{\text{outer}} = 0.975$), medium (bottom left: $\sigma_{\text{inner}} = 0.5, \sigma_{\text{outer}} = 0.6$), and small (bottom right: $\sigma_{\text{inner}} = 0.3, \sigma_{\text{outer}} = 0.4$) partial coherence factors. Here NA = 1.25.

2.2.2 Singular Value Decomposition Model

The SVD model, described in this section, decomposes the Hopkins diffraction model into a sum of coherent systems based on eigenvalue decomposition [13]. The result is a bank of linear systems whose outputs are squared, scaled, and summed. In this sense, the SVD model also belongs to the realm of the SOCS models. The SVD model is summarized as follows.

Given the discretization of the mask pattern $M(x, y)$, referred to as $M(m, n)$, $m, n = 1, 2, \ldots, N$, the intensity distribution on the wafer shown in Eq. (2.12) can be reformulated as a function of matrices

$$I(m, n) = \tilde{\mathbf{s}}^H A\tilde{\mathbf{s}}, \quad m, n = 1, 2, \ldots, N, \tag{2.24}$$

where H is the conjugate transposition operator, $\tilde{\mathbf{s}}$ is an $N^2 \times 1$ vector, and the ith entry of $\tilde{\mathbf{s}}$ is

$$\tilde{\mathbf{s}}_i = \tilde{M}(p, q)\exp[i2\pi(pm + qn)], \quad i = 1, 2, \ldots, N^2, \tag{2.25}$$

where $\tilde{M}(p, q) = \text{FFT}\{M(m, n)\}$, and $\text{FFT}\{\cdot\}$ is the FFT operator, $p = i \bmod N$, $q = \lceil \frac{i}{N} \rceil$, and $\lceil \cdot \rceil$ is the smallest integer larger than the argument. A is an $N^2 \times N^2$ matrix including the information of the transmission cross-coefficient TCC. Specifically, the ith row and jth column entry of A is $A_{ij} = \text{TCC}(p, q; r, u)$, where $p = i \bmod N$, $q = \lceil \frac{i}{N} \rceil$, $r = j \bmod N$, and $u = \lceil \frac{j}{N} \rceil$. To reformulate Eq. (2.24) into the sum of coherent systems, the variable pairs of (p, q) and (r, u) in the argument of TCC should be separated by the SVD. The result of the SVD of A is $A = \sum_{k=1}^{N^2} \alpha_k V_k V_k^*$, where α_k is the kth eigenvalue and $\alpha_1 > \alpha_2 > \cdots > \alpha_{N^2}$. The $N^2 \times 1$ vector V_k is the eigenfunction corresponding to α_k. Thus, Eq. (2.24) becomes

$$I(m, n) = \sum_{k=1}^{N} \alpha_k |\tilde{\mathbf{s}}^T V_k|^2. \tag{2.26}$$

Let $S^{-1}(\cdot)$ be the inverse column stacking operation that converts the $N^2 \times 1$ column vector V_k into a $N \times N$ square matrix $S^{-1}(V_k)$. In particular,

$$\tilde{h}_k(p, q) = S^{-1}(V_k) = \begin{pmatrix} V_{k,1} & V_{k,N+1} & \cdots & V_{k,N(N-1)+1} \\ V_{k,2} & V_{k,N+2} & \cdots & V_{k,N(N-1)+2} \\ \vdots & \vdots & \ddots & \vdots \\ V_{k,N} & V_{k,2N} & \cdots & V_{k,N^2} \end{pmatrix}, \tag{2.27}$$

where $V_{k,i}$ is the ith entry of V_k. Taking the inverse FFT of $\tilde{h}_k(p, q)$ leads to the kth equivalent kernel of the SVD model,

$$h_k(m, n) = \text{IFFT}\{\tilde{h}_k(p, q)\}, \quad m, n = 1, 2, \ldots, N. \tag{2.28}$$

Substituting Eqs. (2.27) and (2.28) into Eq. (2.26),

$$I(m, n) = \sum_{k=1}^{N^2} \alpha_k |h_k(m, n) \otimes M(m, n)|^2. \tag{2.29}$$

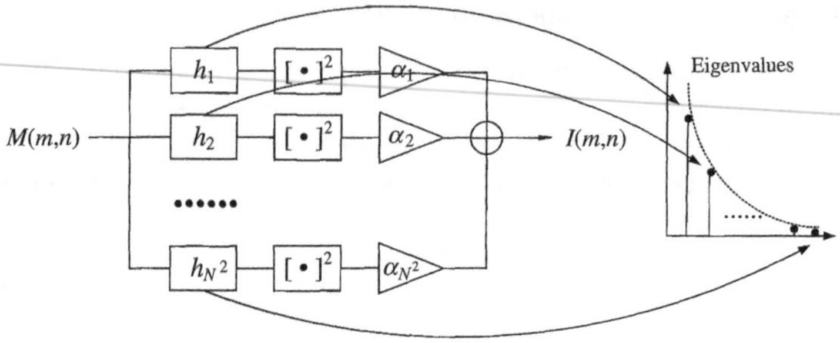

Figure 2.9 A partially coherent system represented by a SVD model as a sum of coherent systems.

Note that the partially coherent system is decomposed into the superposition of N^2 coherent systems. The scheme of the SOCS decomposition by SVD is depicted in Fig. 2.9, where each coherent system corresponds an eigenvalue of SVD decomposition. The ith order coherent approximation to the partially coherent system is defined as

$$I(m,n) \approx \sum_{k=1}^{i} \alpha_k |h_k(m,n) \otimes M(m,n)|^2, \quad i = 1, 2, \ldots, N^2. \tag{2.30}$$

An example of the first 50 eigenvalues of the SVD decomposition with small and medium partial coherence factors is illustrated in Fig. 2.10. In this simulation,

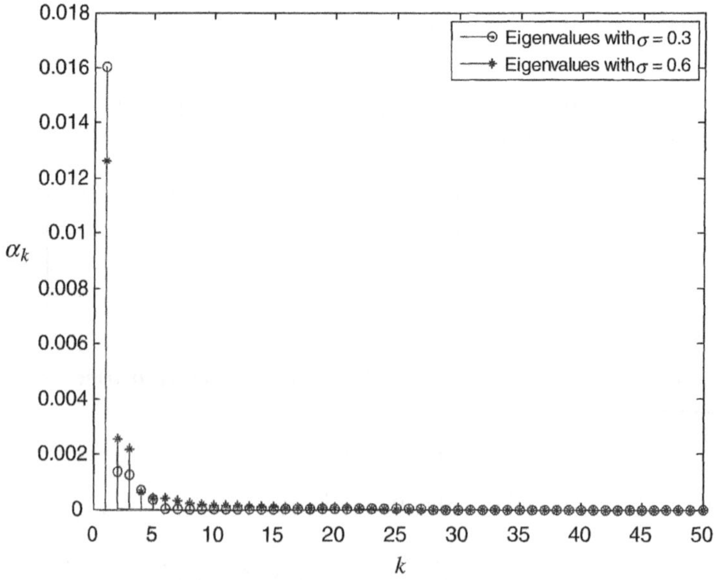

Figure 2.10 Eigenvalues α_k of sum of coherent systems decomposition by SVD.

the partially coherent illuminations are circular illuminations with partial coherence factors $\sigma = 0.3$ and $\sigma = 0.6$. The dimension of the discretization is $N = 51$. The pixel size is 11 nm × 11 nm. Thus, the effective source is

$$\tilde{\gamma}(f, g) = \frac{\lambda^2}{\pi(\sigma\mathrm{NA})^2}\mathrm{circ}\left(\frac{\lambda\sqrt{f^2+g^2}}{\sigma\mathrm{NA}}\right) = \begin{cases} \frac{\lambda^2}{\pi(\sigma\mathrm{NA})^2}, & \text{for } \sqrt{f^2+g^2} \le \frac{\sigma\mathrm{NA}}{\lambda}, \\ 0, & \text{elsewhere}, \end{cases} \tag{2.31}$$

where NA = 1.35 and $\lambda = 193$ nm. The amplitude impulse response is defined as the Fourier transform of the circular lens aperture with cutoff frequency NA/λ [7, 85]; therefore,

$$h(\mathbf{r}) = h(x, y) = \frac{J_1(2\pi r\mathrm{NA}/\lambda)}{2\pi r\mathrm{NA}/\lambda}. \tag{2.32}$$

The Fourier transform of $h(x, y)$ is

$$\tilde{h}(f, g) = \frac{\lambda^2}{\pi(\mathrm{NA})^2}\mathrm{circ}\left(\frac{\lambda\sqrt{f^2+g^2}}{\mathrm{NA}}\right) = \begin{cases} \frac{\lambda^2}{\pi(\mathrm{NA})^2}, & \text{for } \sqrt{f^2+g^2} \le \frac{\mathrm{NA}}{\lambda}, \\ 0, & \text{elsewhere}. \end{cases} \tag{2.33}$$

The amplitudes of the first and second equivalent kernels corresponding to the first and second largest eigenvalues with $\sigma = 0.3$ are illustrated in Fig. 2.11a and b, respectively. The amplitudes of the first and second equivalent kernels with $\sigma = 0.6$ are illustrated in Fig. 2.12a and b, respectively.

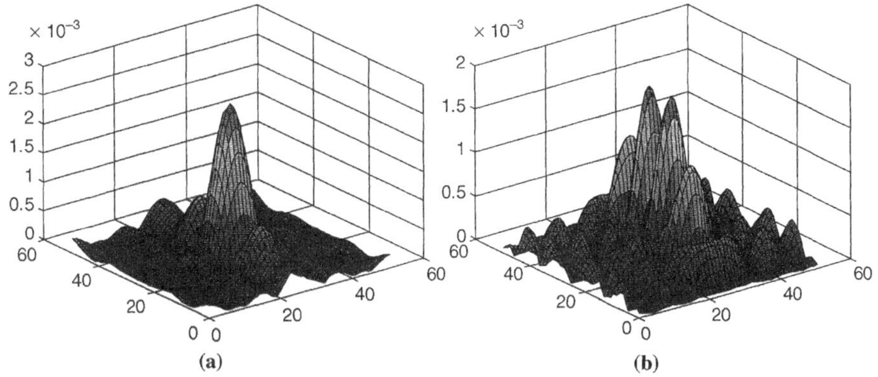

Figure 2.11 The amplitudes of (a) the first equivalent kernel corresponding to the largest eigenvalue $|\phi_1(x, y)|$ and (b) the second equivalent kernel corresponding to the second largest eigenvalue $|\phi_2(x, y)|$, with $\sigma = 0.3$.

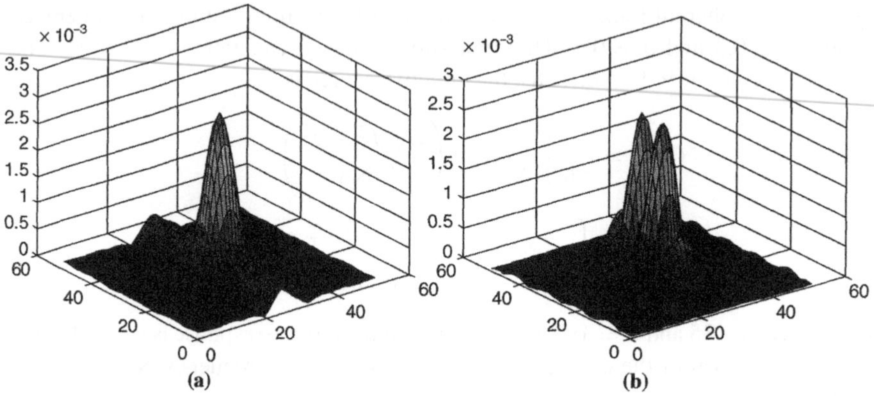

Figure 2.12 The amplitudes of (a) the first equivalent kernel corresponding to the largest eigenvalue $|\phi_1(x, y)|$ and (b) the second equivalent kernel corresponding to the second largest eigenvalue $|\phi_2(x, y)|$, with $\sigma = 0.6$.

It is noted that for the illuminations having small partial coherence factors, the eigenvalues decay very rapidly. It was proved that for partial coherence factors $\sigma \leq 0.5$, a partially coherent imaging system may be approximated to within 10% error by the first-order coherent approximation [58].

2.2.3 Average Coherent Approximation Model

The average coherent approximation model for the partially coherent imaging system was introduced by Salik et al. [75]. Different from the above SOCS models, the average approximation of PCI is to approximately decompose the contribution of the PCI into a coherent and an incoherent illumination component. Therefore,

$$\begin{aligned} I(\mathbf{r}) &= \iint M(\mathbf{r_1})M^*(\mathbf{r_2})\gamma(\mathbf{r_1}-\mathbf{r_2})h(\mathbf{r}-\mathbf{r_2})h^*(\mathbf{r}-\mathbf{r_2})\mathrm{d}\mathbf{r_1}\mathrm{d}\mathbf{r_2} \\ &\approx \left|\int M(\mathbf{r}')h_{\mathrm{C}}(\mathbf{r}',\mathbf{r})\mathrm{d}\mathbf{r}'\right|^2 + \int |M(\mathbf{r}')|^2|h_{\mathrm{I}}(\mathbf{r}',\mathbf{r})|^2\mathrm{d}\mathbf{r}', \end{aligned} \tag{2.34}$$

where $\mathbf{r}' = (x', y')$ and $h_{\mathrm{C}}(\mathbf{r}', \mathbf{r})$ and $h_{\mathrm{I}}(\mathbf{r}', \mathbf{r})$ are the equivalent amplitude impulse responses of the coherent and incoherent components, respectively. Furthermore,

$$h_{\mathrm{C}}(\mathbf{r}', \mathbf{r}) = f(\mathbf{r}', \mathbf{r})^{1/2}h(\mathbf{r}', \mathbf{r}) \tag{2.35}$$

and

$$h_{\mathrm{I}}(\mathbf{r}', \mathbf{r}) = [1 - f(\mathbf{r}', \mathbf{r})]^{1/2}h(\mathbf{r}', \mathbf{r}), \tag{2.36}$$

where

$$f(\mathbf{r}', \mathbf{r}) = \frac{\int |h(\mathbf{r}', \hat{\mathbf{r}})|^2\mu(\mathbf{r}, \tilde{\mathbf{r}})\mathrm{d}\tilde{\mathbf{r}}}{\int |h(\mathbf{r}', \hat{\mathbf{r}})|^2\mathrm{d}\hat{\mathbf{r}}} \tag{2.37}$$

and

$$\mu(\mathbf{r}, \hat{\mathbf{r}}) = \frac{\gamma(\mathbf{r}, \hat{\mathbf{r}})}{[\gamma(\mathbf{r}, \mathbf{r})\gamma(\hat{\mathbf{r}}, \hat{\mathbf{r}})]^{1/2}}. \tag{2.38}$$

In the equations above, $\hat{\mathbf{r}}$ and $\tilde{\mathbf{r}}$ are dummy variables. f is the fraction of coherent incident power with $0 \le f \le 1$. Taking the annular illuminating sources as an example, the function f is obtained by substituting Eqs. (2.20) and (2.22) into Eqs. (2.37) and (2.38), leading to

$$f(\mathbf{r}', \mathbf{r}) = \frac{1}{\int |(J_1(\pi \mathbf{r}/d))/(\pi \mathbf{r}/d)|^2 \, d\mathbf{r}} \int \left| \frac{(J_1(\pi \hat{\mathbf{r}}/d))}{(\pi \hat{\mathbf{r}}/d)} \right|^2 \times \left(\frac{(J_1(\pi(\hat{\mathbf{r}} - \dot{\mathbf{r}})/a_u))}{(\pi(\hat{\mathbf{r}} - \dot{\mathbf{r}})/a_u)} - \frac{a_u^2}{a_l^2} \frac{(J_1(\pi(\hat{\mathbf{r}} - \dot{\mathbf{r}})/a_l))}{(\pi(\hat{\mathbf{r}} - \dot{\mathbf{r}})/a_l)} \right) d\hat{\mathbf{r}}, \tag{2.39}$$

where $\dot{\mathbf{r}} = \mathbf{r} - \mathbf{r}'$, $d = \frac{\lambda}{2\mathrm{NA}}$, $a_u = 2D_{\mathrm{cu}}$, and $a_l = 2D_{\mathrm{cl}}$. Applying the Fourier transform, Eq. (2.39) becomes

$$f(\mathbf{r}', \mathbf{r}) = \frac{1}{\int |(J_1(\pi \mathbf{r}/d))/(\pi \mathbf{r}/d)|^2 \, d\mathbf{r}} \cdot \mathrm{IFFT}\{\mathrm{FFT}\{|h(\mathbf{r})|^2\} \cdot \mathrm{FFT}\{\gamma(\mathbf{r})\}\}, \tag{2.40}$$

where $\mathrm{FFT}\{\cdot\}$ and $\mathrm{IFFT}\{\cdot\}$ are the FFT and inverse FFT operations, respectively. It is noted that for some specific $h(\mathbf{r})$ and $\gamma(\mathbf{r})$, the condition $0 \le f \le 1$ may not be satisfied. According to Eq. (2.35), negative values of f will introduce complex pixel values in $h_C(\mathbf{r}', \mathbf{r})$. Similarly, Eq. (2.36) indicates that values of f larger than 1 will introduce complex pixel values in $h_I(\mathbf{r}', \mathbf{r})$. Nevertheless, in our extensive simulations, the average coherent approximation model leads to similar aerial image contours as those of the Fourier series expansion model. Substituting Eq. (2.40) into Eqs. (2.35) and (2.36), the equivalent amplitude impulse responses of the coherent and incoherent components can be found. Therefore, the partially coherent imaging system is approximately divided into a summation of a coherent system and an incoherent system. The scheme of the average coherent approximation model is depicted in Fig. 2.13.

The accuracy of the average coherence approximation model depends on the spatial coordinates, mask pattern, optical system kernel, and the complex degree of coherence [75]. Specifically, the error of the average coherence approximation model becomes smaller as the amplitude impulse response $h(\mathbf{r})$ becomes sharper or its energy is more concentrated. Figure. 2.14 illustrates the cross sections of the aerial imaging of two vertical bars based on the Fourier series expansion model and the average coherent approximation model. The mask dimension is 600 nm × 600 nm and $\lambda = 193$ nm. The source is a circular illumination with its Fourier series coefficients of the circular illumination being

$$\Gamma_{\mathbf{m}} = \begin{cases} 4D_c^2/\pi D^2, & \text{for } |\mathbf{m}| \le D/2D_c, \\ 0, & \text{elsewhere,} \end{cases} \tag{2.41}$$

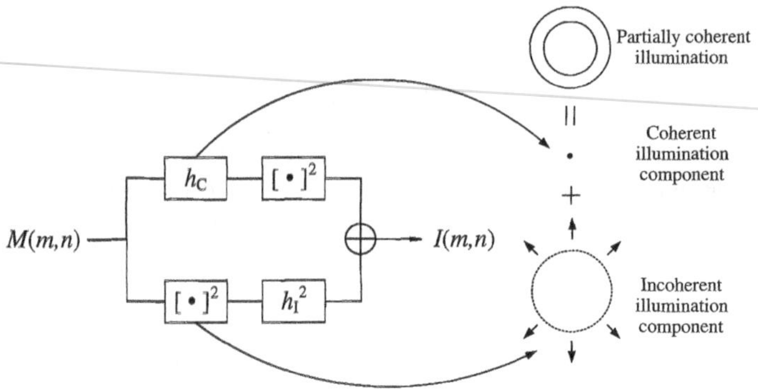

Figure 2.13 A partially coherent system represented by an average coherent approximation model as a sum of a coherent system and an incoherent system.

where $D = 600$ nm and $D_c = 8.6$ nm. $T = 3969$ is the number of Fourier series terms to represent the Fourier series expansion model. In Fig. 2.14, the solid lines and dashed–dotted lines represent the aerial imaging of the Fourier series expansion model and the average coherent approximation model, respectively. Since the accuracy of the Fourier series expansion model is the same as the discrete version of the Hopkins diffraction model, it is chosen as the criterion to measure the accuracy of the average coherent approximation model. The SNR is defined as the ratio between the energy of the accurate imaging and the error energy. In Fig. 2.14a, the numerical aperture NA $= 1.35$ and SNR $= 18.7$, while in Fig. 2.14b, NA $= 0.15$ and SNR $= 10.2$. It can be observed from Eq. (2.22) that larger NA corresponds to a sharper amplitude impulse response. The simulations show that the average coherent approximation model gives more accurate aerial imaging for sharper amplitude impulse response.

2.2.4 Discussion and Comparison

In this section, three kinds of approximation models of the partially coherent imaging system are discussed. Both the Fourier series expansion model and the SVD model belong to the SOCS models, which approximate the partially coherent system as a sum of coherent systems. The accuracy of both models is the same as the discrete version of the Hopkins diffraction model. However, they use different eigenvalue decomposition approaches. The Fourier series expansion model is based on 2D Fourier series expansion, while the SVD model applies the singular value decomposition. The advantage of the Fourier series expansion model is that the 2D Fourier series expansion of the complex degree of coherence of regular effective source shapes has close form solutions. Given the effective source shape, it is easy to calculate the 2D Fourier series expansion coefficients. However, the disadvantage is that all eigenvalues are the same; thus, all of them have to be taken into account in the evaluation of the aerial image. The computational cost is a polynomial of the number of the Fourier series terms used to represent the partially coherent imaging system, and in general, numerous terms are needed to attain an adequate representation.

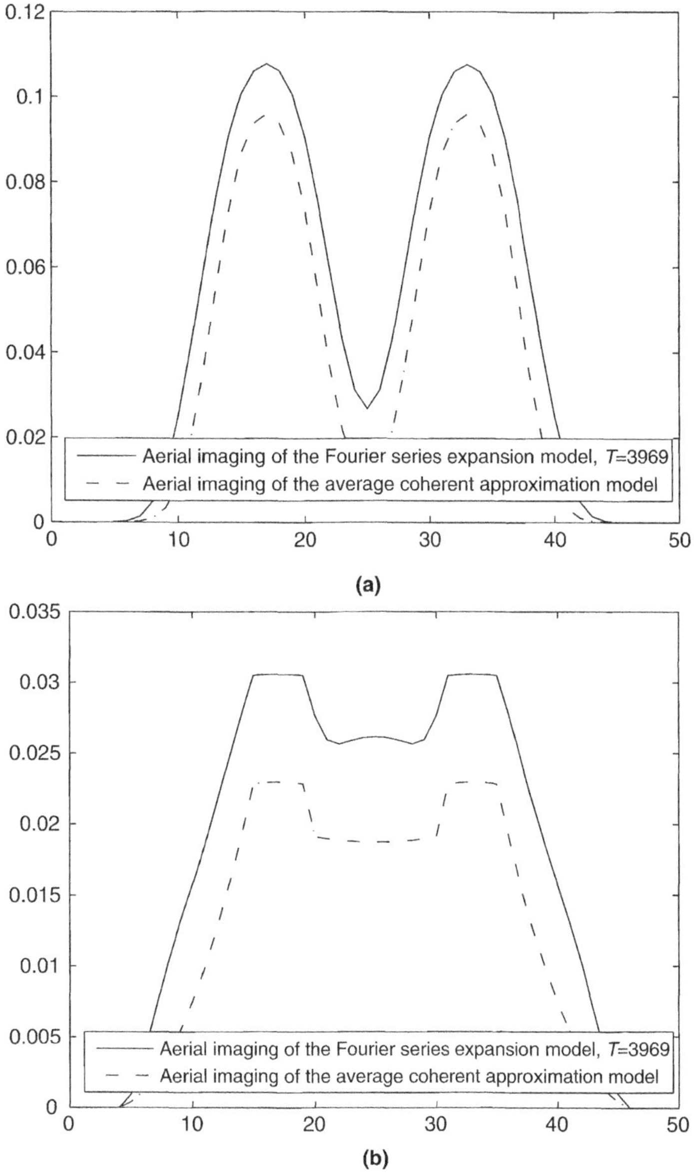

Figure 2.14 The average coherent approximation model gives more accurate aerial imaging for sharper amplitude impulse response. (a) NA = 1.35, corresponding to a sharper amplitude impulse response, SNR = 18.7; (b) NA = 0.15, corresponding to a smoother amplitude impulse response, SNR = 10.2.

On the other hand, the SVD model uses singular value decomposition to order the eigenvalue from large to small. The eigenvalues increasingly decay faster, as the partial coherence factor is decreased. Therefore, with small and medium partial coherence factors, the partially coherent systems can be approximated by several coherent components corresponding to the largest eigenvalues. However, the SVD model needs a singular value decomposition of the TCC matrix with a dimension of $N^2 \times N^2$, where N is the dimension of the mask. In general, N is very large, and the SVD is resource consuming.

Although accurate, the above SOCS models are computationally expensive. Different from the SOCS models, the average coherent approximation model decomposes the partially coherent imaging system into the superposition of a coherent and an incoherent illumination component. This model avoids the SVD and uses only two terms to represent the partially coherent imaging system, thus more computationally efficient than the SOCS models. However, the accuracy of the average coherence approximation model depends on the spatial coordinates, mask pattern, optical system kernel, and the complex degree of coherence.

2.3 SUMMARY

This chapter discussed the fundamentals of the optical lithography systems, where both Abbe's model and Hopkins diffraction model were summarized. In addition, this chapter discussed three approximation models to represent the partially coherent imaging systems, such as the Fourier series expansion model, the SVD model, and the average coherent approximation model.

3

Rule-Based Resolution Enhancement Techniques

3.1 RET TYPES

With Moore's law, rapid trend to reduce the critical dimension (CD) in optical lithography, imaging has become a low *k* process, where a sizable fraction of the transmitted light energy is concentrated in the high spatial frequency components of the mask spectrum. However, the low-pass filtering properties of the lens in the exposure system cuts off the high-frequency components leading to image distortion. To enhance the resolution and contrast of the circuit patterns, different types of RETs have been proposed, such as rule-based, model-based, and hybrid approaches. Conventionally, the terminologies of "rule-based," "model-based," and "hybrid" were only used for OPC methods. In this book, we generalize these terminologies to the entire scope of RETs, including OPC, PSM, and OAI. In the following, the concepts of these RET approaches are discussed. Subsequently, this chapter focuses on the discussion of the rule-based RETs. Model-based RETs are discussed at length in the following chapters.

3.1.1 Rule-Based RETs

In the rule-based approaches, adjustment strategies of the mask patterns are made up based on a set of locally restricted rules. The amount of correction applied to a feature or an edge is carried out in accordance with a predefined table [80, 92]. The specifics in the table can be derived from simulation, experiments, or their combination [92]. A flow chart of the rule-based RETs is shown in Fig. 3.1. Since the adjusted mask features are directly obtained by checking predefined tables, rule-based RETs are faster than other kinds of RETs. In addition, the rule-based RETs are more favorable than others because of the time-to-market requirement [57]. However, the rules are established based on the geometric information of the feature and its local environment. Therefore, the rule-based approaches, which are simple to implement, can just compensate for the warping in local areas.

Computational Lithography By Xu Ma and Gonzalo R. Arce

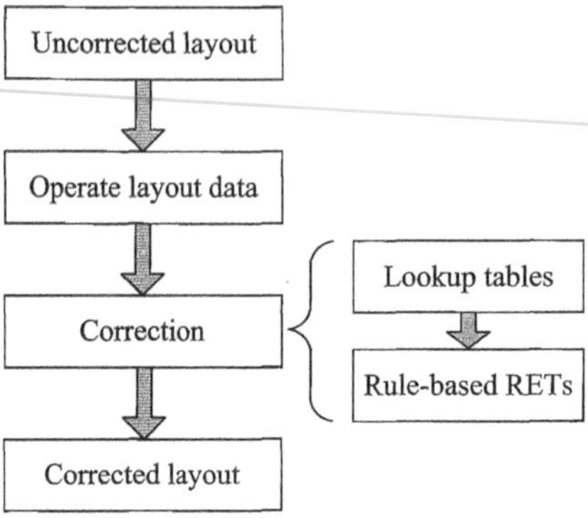

Figure 3.1 The flow chart of the rule-based RETs.

3.1.2 Model-Based RETs

In model-based RETs, the adjustment of the mask is calculated based on the mathematical models, which represent the image formation process of the optical lithography system. The effects of exposure in optical lithography may be represented by the various models described in Chapter 2. In addition, other effects should be taken into account in the models, such as photoresist acid diffusion, flare, and pattern loading in reactive ion etch. These effects can be modeled reliably on first principles, experiments, or their combination [92]. The model-based approaches include inverse and forward methods. The inverse methods, illustrated in Fig. 3.2, start from the desired output pattern and iteratively obtain the optimized layout. The forward methods, illustrated in Fig. 3.3, continuously modify the original layouts until the output patterns and the mask manufacturability properties are acceptable. Since a proper model will capture more factors influencing the optical imaging process, the model-based RETs result in more universal and aggressive strategies than the rule-based RETs. Thus, the model-based approaches may obtain global optimal solutions for the RETs. On the other hand, the formulation of accurate and efficient models is the primary bottleneck of the model-based RETs [92]. The model cannot be too complex to solve. However, simple models ignoring some variables and effects will introduce error. The second disadvantage is that model-based RETs are not as computationally efficient as the rule-based RETs. However, advances in computational platforms continue to alleviate this disadvantage.

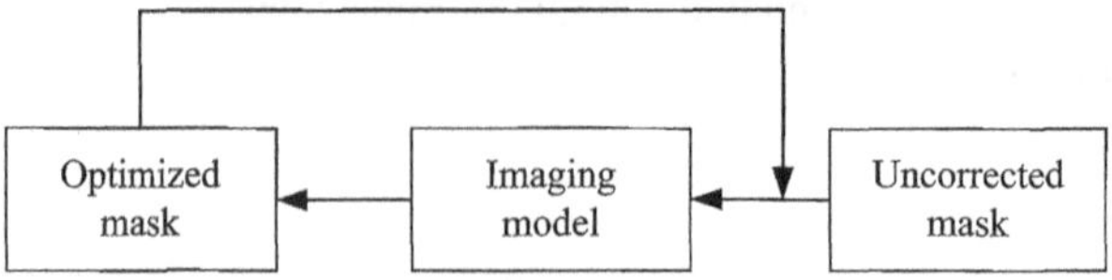

Figure 3.2 The flow chart of the inverse model-based RETs.

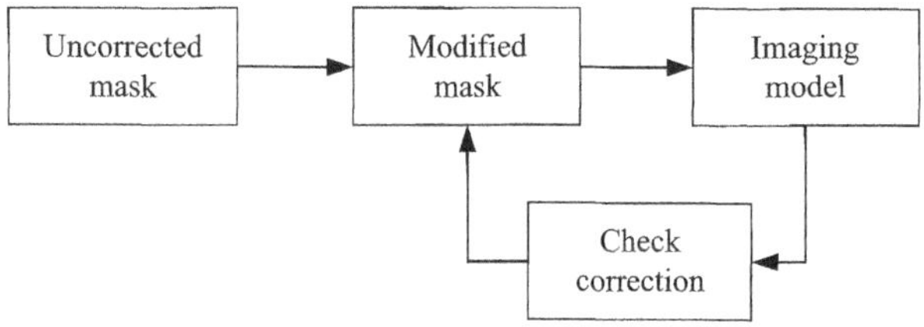

Figure 3.3 The flow chart of the forward model-based RETs.

3.1.3 Hybrid RETs

To overcome the limits of both rule-based and model-based RETs, hybrid RETs have been considered. Hybrid RET methods combine rule-based and model-based RETs, where their synergy is exploited to adjust the mask pattern. In most of current hybrid RETs, the image distortion is mainly compensated by model-based RETs. The residual error is subsequently reduced using the rule-based RETs [100]. The main challenges of hybrid RETs include decomposition of the design process into the rule-based and model-based realms and consistency between these two methods [92].

3.2 RULE-BASED OPC

The low-pass filtering properties of the lens, diffraction, and interference effects of optical system introduce the distortion to the printed image on the wafer. In general, there are four types of image distortion [92]. First, the variation of the printed image under different environments with the same nominal critical dimension (CD). Second, the distortion that occurs when changes of the nominal CD are not reflected linearly in the printed image. As the critical dimension decreases, this nonlinearity effect becomes pronounced and even leads to features not being printed. Third, the line shortening that is shown in Fig. 3.4a. Finally, the corner rounding as shown in Fig. 3.4b.

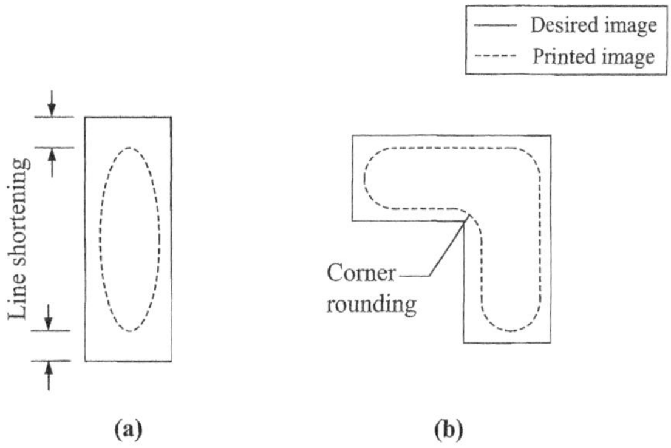

Figure 3.4 (a) Line-shortening and (b) corner-rounding artifacts.

In Fig. 3.4, the solid lines show the desired patterns, and the dashed lines show the printed patterns.

To compensate for these image distortions, the optical proximity correction methods modify the mask amplitude by the addition of subresolution features adjacent or separated from the original features such that the output patterns are as close to the desired pattern as possible. Basically, there are four rule-based OPC approaches. The first is to recover the nonprinting features, referred to as catastrophic OPC. Second, the one-dimensional OPC, which minimizes the line-width variation caused by the environment changing or nonlinearity effect. Third, the line-shortening reduction OPC. Finally, the corner-rounding correction OPC. In contrast to the one-dimensional OPC, the corner-rounding correction method modifies mask pattern in both x- and y-axes. Thus, this approach is also referred to as two-dimensional OPC. The details of these OPC approaches will be discussed next.

3.2.1 Catastrophic OPC

The goal of catastrophic OPC is only to guarantee that a feature on the mask can be printed on the wafer without feature size control. Thus, this method can be simply implemented by enlarging the target features on the mask. However, the main challenge is identification of the target patterns, which requires the understanding of the working of the circuit [92].

3.2.2 One-Dimensional OPC

3.2.2.1 Line Biasing The line biasing method simply uses the following rules to reduce the line-width variation: dense and sparse lines are made thinner, while lines with medium periods are made thicker. However, the resolution of the line-width adjustment is limited by the pixel size on the mask. An adjustment accuracy of $\frac{\delta}{2}$ requires a pixel resolution of δ [92]. Therefore, finer resolution can be obtained by the reduction of pixel size on the mask. However, the pixel size reduction dramatically increases the mask fabrication time.

To overcome this limitation, halftone techniques are applied to alleviate the conflict between resolution and mask fabrication time [19, 34, 51, 92]. According to Eq. (1.5), mask features with spatial period smaller than $\frac{\lambda}{\mathrm{NA}(1+\sigma)}$ cannot be resolved by the optical system, and only the average transmittance is imaged. Based on this concept, the halftone technique adds periodic blocks on the edges of lines, as illustrated in Fig. 3.5. In Fig. 3.5, the left pattern is the mask. The height of the segments is δ, and the width is $m\delta$. The period of the segments is

$$n\delta \leq \frac{\lambda}{\mathrm{NA}(1+\sigma)}, \quad n \in \mathbf{Z}. \tag{3.1}$$

The right pattern is the printed image, where the segments of the edge on the mask are not resolved. However, an average transmittance is imaged on the edge with a

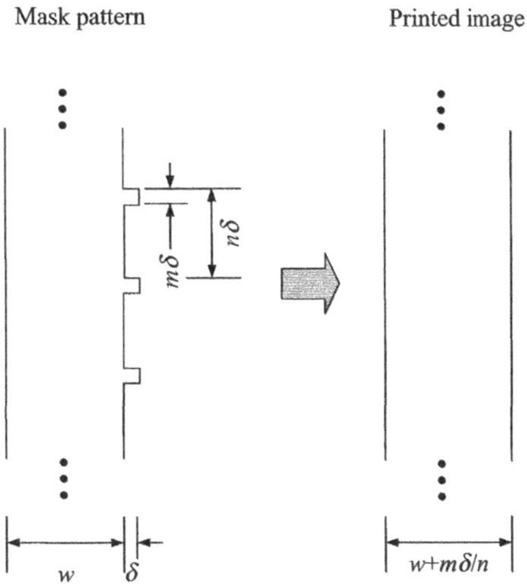

Figure 3.5 Halftone techniques applied in line biasing method.

width of

$$\delta_{\text{add}} = \frac{m\delta}{n}. \tag{3.2}$$

In addition, the resolution of the line-width adjustment can be enhanced by the asymmetric structures of periodic blocks applied on both sides of the lines [91]. The asymmetric halftone technique is illustrated in Fig. 3.6, where the two sets of segments

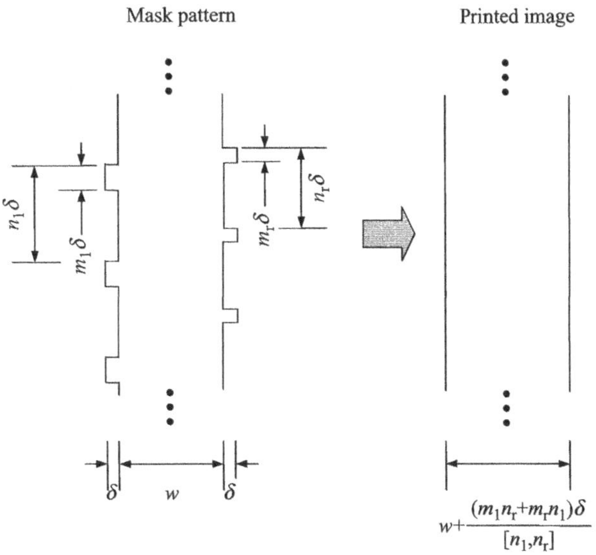

Figure 3.6 Asymmetric halftone technique applied on both sides of the line.

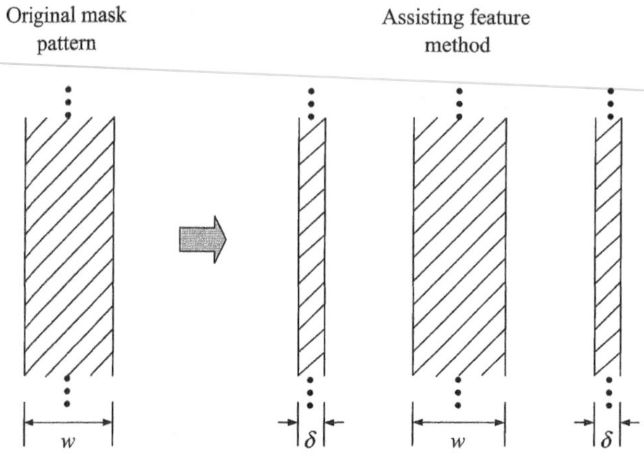

Figure 3.7 Assisting feature method applied on both sides of the line.

apply different periods of $n_l\delta$ and $n_r\delta$, respectively. On the left side, the height of the segments is δ and the width is $m_l\delta$. On the right side, the height of the segments is δ and the width is $m_r\delta$. Therefore, the additional thickness imaged on the edge of the line is

$$\delta_{\text{add}} = \frac{(m_l n_r + m_r n_l)\delta}{[n_l, n_r]}, \tag{3.3}$$

where $[\cdot, \cdot]$ denotes the least common multiple of the arguments.

3.2.2.2 Assisting Feature Although the line biasing method is simple to implement, it cannot improve the overall image quality [92]. While the off-axis illumination is used to improve the image quality of dense patterns, the image quality of sparse patterns may be enhanced by the assisting feature method, which is shown in Fig. 3.7. In Fig. 3.7, the left figure shows the original mask of a line pattern. The right figure shows the assisting feature method. This method adds on both sides of the line, a set of small assisting features, which are not printed on the wafer. These assisting features create an equivalent dense environment, whose image quality may be improved by the off-axis illumination. Thus, the assisting feature method is used to simultaneously adjust the line width and enhance the image quality.

The disadvantage of the assisting feature method is the complexity of implementation. To effectively compensate for the image distortion, several factors have to be considered, such as number, size, position of assisting features, and so on [50, 92].

3.2.3 Line-Shortening Reduction OPC

Line-shortening phenomenon is very common in optical lithography. The simplest and most effective method to reduce line shortening is lengthening of the line, which

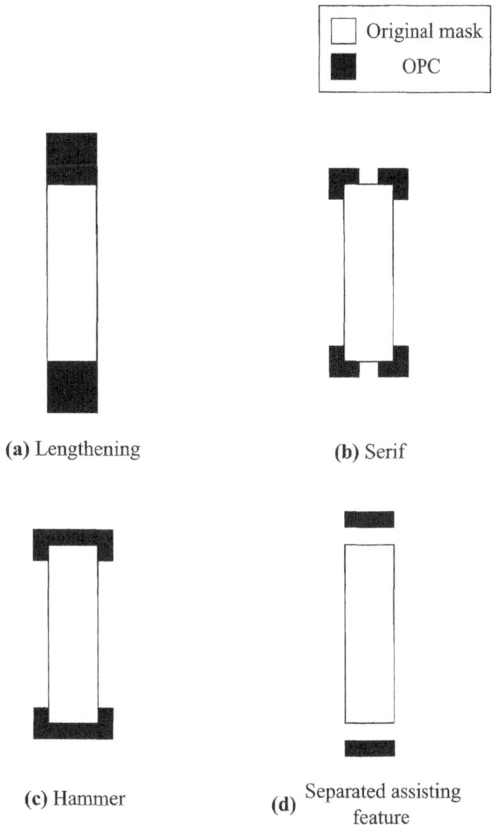

Figure 3.8 Line-shortening reduction OPC. (a) Lengthening, (b) serif, (c) hammer, and (d) separated assisting feature.

is shown in Fig. 3.8a. In Fig. 3.8, the white blocks represent the original mask patterns, and the black blocks represent mask adjustments of OPC. However, under a dense pattern environment, there is not enough space for line lengthening [92]. To solve this problem, approaches involving serif and hammer are used to emphasize the ends of the line. These two methods are shown in Fig. 3.8b and c, respectively. In addition, separated assisting features illustrated in Fig. 3.8d may also be applied to reduce the line shortening.

3.2.4 Two-Dimensional OPC

The goal of two-dimensional OPC is to correct the corner rounding. The prevalent method is to exploit both hammers and serifs, as shown in Fig. 3.9. In Fig. 3.9, the left figure shows the original mask pattern, and the right figure shows the corner-rounding correction method.

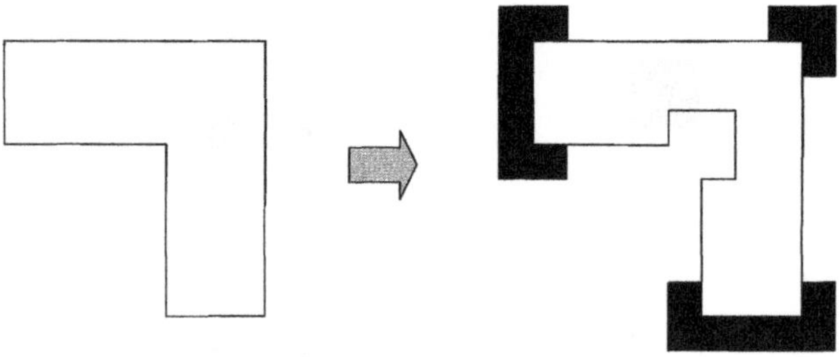

Figure 3.9 Two-dimensional OPC used to correct the corner rounding.

3.3 RULE-BASED PSM

The application of the OPC methods encounters energy diffusing problems, where unwanted energy in the opaque (chrome) regions appears due to the proximity of the neighboring transparent (quartz) features [69]. Phase-shifting masks induce phase shifts in the transmitted field that have a favorable constructive or destructive interference effect to remove the unwanted energy in the opaque regions. As described in Section 1.4.2, there are three types of PSMs: alternating phase-shifting masks, attenuated phase-shifting masks, and chromeless phase-shifting masks. In this section, we focus on the alternating phase-shifting masks, which modulate the phases of the adjacent features on the mask by 180° out of phase with each other.

There are two kinds of applications of alternating PSMs: dark-field application and light-field application [92]. In the dark-field application, the printed image is constructed by the exposed areas on the wafer. In the light-field application, the printed image is constructed by the unexposed areas, which are created by destructive interference. In the following, the design rules for both applications are discussed.

3.3.1 Dark-Field Application

Following are the design rules of dark-field application [92]:

1. When the distance between two features is smaller than the critical dimension, they must have different phases.
2. Each feature must have only one phase.

However, when single exposure is used to project the image on the wafer, the two rules above will conflict for some patterns. One of these patterns is shown in Fig. 3.10. In Fig. 3.10, the left figure shows the mask pattern, which cannot simultaneously meet both the rules above. The middle figure shows the conflict of phase assignment. Suppose phase 0° is assigned to features A and B, and phase 180° is assigned to

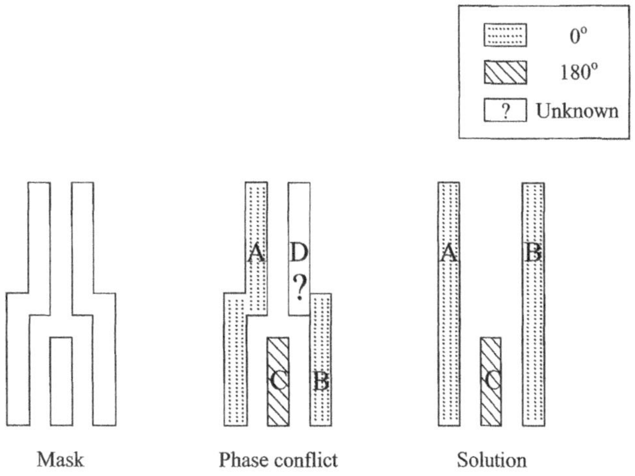

Figure 3.10 Conflict of phase assignment and its rule-based solution.

feature C. Thus, if feature D is assigned with 0°, then the first rule will be violated. If feature D is assigned with 180°, then the second rule will be violated. To satisfy both rules, the mask pattern may be modified as shown in the right figure in Fig. 3.10. The conflict of the phase assignment is solved by increasing the separation between features A and D.

3.3.2 Light-Field Application

In the light-field application, the printed image is constructed by the unexposed areas, which are created by destructive interference. An example of light-field application is illustrated in Fig. 3.11, where the desired pattern sought on the wafer is a horizontal line. In Fig. 3.11, the first figure is the light-field PSM, where the opaque line is between the transparent regions with phases 0° and 180°. The destructive interference effect of the transparent regions results in a dark horizontal line shape between the transparent regions. However, other unexpected dark regions are also introduced at the boundaries between the transparent regions, as shown in the second figure. To remove these unexpected dark regions, a dark-field trim mask shown in the third figure is exploited in a second exposure. The second exposure removes the unexpected dark regions and the final printed image is shown in the fourth figure.

The design rules of light-field application are more complicated than the dark-field application. To use the light-field mask, some significant issues have to be considered, including [92]:

1. Line width below which the phase shifting is needed, and beyond which the phase shifting is unnecessary.
2. Dimensions of the phase-shifting regions.
3. Distance between different phase-shifting regions.

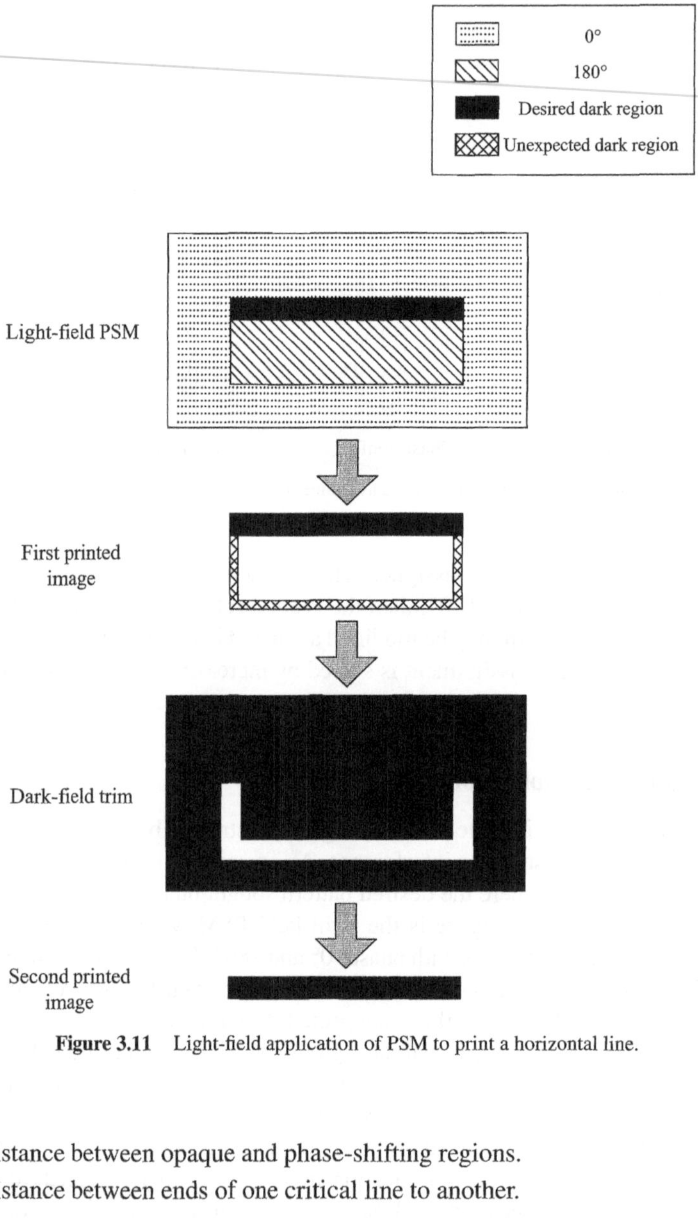

Figure 3.11 Light-field application of PSM to print a horizontal line.

4. Distance between opaque and phase-shifting regions.
5. Distance between ends of one critical line to another.

3.4 RULE-BASED OAI

As discussed in Section 1.4.3, a variety of off-axis illuminations includes dipole, quadrupole, annular, separated, fourfold, CQUEST, and QUASAR illuminations. Among them, the dipole, quadrupole, and annular illuminations have been extensively studied and are commonly used. The off-axis illumination improves image quality of the dense patterns compared to circular illumination having large σ [92]. To implement the off-axis illuminations, several issues have to be considered as described in Ref. [92]:

1. To print dense patterns, a circular illumination with large σ or off-axis illumination is preferred.
2. To print sparse patterns, a circular illumination with small σ is preferred, where $\sigma \in [0.3, 0.5]$.
3. To print both dense and sparse patterns, a circular illumination with medium σ is preferred, where $\sigma \in [0.4, 0.6]$.
4. The minimum size of the poles and annulus should be constrained. Extremely small effective source will result in excessive image ringing at the transition between exposed and unexposed regions. It will also increase the image sensitivity to aberrations and aggravate source intensity imbalance [9, 31, 32, 92]. In addition, focus and dose variation at the wafer plane can also be accentuated by extremely small effective source [30, 92].
5. Modified illumination configurations are sometimes used, which are implemented by blinding some parts of original illuminations. Since some radiation energy is blocked or redirected, the modified illuminations will reduce the effective energy exposed on the wafer, thus increasing the exposure time.
6. Changing illumination configurations are sometimes needed between exposure processes. However, the recalibration operations are necessary to maintain the uniformity of the illumination. These additional operations will severely increase the exposure time.

3.5 SUMMARY

This chapter first discussed the classification of the RET methods. Subsequently, it focused on the rule-based RETs, where the rule-based OPC, PSM, and OAI approaches were described in detail. As the counterpart of the rule-based RETs, the model-based RET methods are emphasized in the book hereafter.

4

Fundamentals of Optimization

4.1 DEFINITION AND CLASSIFICATION

4.1.1 Definitions in the Optimization Problem

Optimization is an important tool in decision science and in the analysis of physical systems [53]. In a mathematical framework, optimization is the minimization or maximization of a function subject to constraints on its variables. The optimization problem can be formulated as following [53]:

$$\min_{\underline{x}} = f(\underline{x}) \quad \text{subject to} \quad \begin{array}{l} c_i(\underline{x}) = 0, \ i \in \varepsilon, \\ c_i(\underline{x}) \geq 0, \ i \in \zeta, \end{array} \tag{4.1}$$

where $\underline{x}$ is the vector of variables, which depends on certain characteristics of the system. f is the objective function (or cost function), a scalar function of $\underline{x}$ that is to be maximized or minimized. c_i are scalar constraint functions, of $\underline{x}$ which define certain equations and inequalities that the variable vector must satisfy. The set of $\underline{x}$ satisfying all the constraints are referred to as feasible region.

The process of identifying the objective function, variable vector, and constraint functions is referred to as modeling. A good model should be complex enough to represent the useful characteristics of the system and simple enough to be easily solved. After modeling, a proper optimization algorithm can be selected and used to solve the problem described in Eq. (4.1). There is no universal optimization algorithm but rather a collection of algorithms, each of which is suitable for a particular type of optimization problem. The choice of the optimization algorithm is significant, since it determines whether the solution can be found. If it can be found, whether the problem can be solved rapidly [53].

Optimization algorithms are iterative. They begin with an initial guess $\underline{x}_0$ of the solution $\underline{x}^*$ and generate a sequence of improved estimates of $\underline{x}^*$. The algorithms terminate when some conditions are satisfied. Different algorithms use different strategies to update the estimates of $\underline{x}^*$, based on objective function f, constraint functions, and their first and second derivatives. Good algorithms should be robust, efficient, and accurate. Robust algorithms perform well on a wide variety of problems in a specific

Computational Lithography By Xu Ma and Gonzalo R. Arce

class. Efficient algorithms do not need excessive running time and memory. Accurate algorithms can obtain a solution close to the minimum or minimum without being overly sensitive to the errors. These properties of good algorithms are sometimes in conflict. Thus, a user should keep a balance among these properties.

4.1.2 Classification of Optimization Problems

According to the domain of the variable vector $\underline{x}$, the optimization problems can be classified into continuous optimization problems and discrete optimization problems. The continuous optimization problems allow $\underline{x}$ to be of any value in an infinite continuous set, for example, the real domain. The continuous optimization problems are normally easier to solve because the smoothness of the functions makes it possible to use objective and constraint information at a particular point $\underline{x}$ to deduce information about the functions' behavior at all points close to $\underline{x}$ [53]. In contrast, in the discrete optimization problem, $\underline{x}$ belongs to a finite discrete set [15, 52, 56, 88]. Specifically, if $\underline{x}_i \in \mathbf{Z}$ or $\underline{x}_i \in \{0, 1\}$, the problem is referred to as an integer programming problem, where $\underline{x}_i$ is the ith entry of $\underline{x}$ and $\mathbf{Z}$ is the integer set. If some of the variables are not constrained to be integer or binary variables, the problem is referred to as a mixed integer programming problem. It is obvious that integer programming problems belong to the class of discrete optimization problems.

According to the constraint functions in Eq. (4.1), the optimization problems can be classified into constrained and unconstrained optimization problems. For constrained optimization problems, $\varepsilon \neq \varnothing$ and $\zeta \neq \varnothing$ in Eq. (4.1), where $\varnothing$ is the null set. For unconstrained optimization problems, $\varepsilon = \varnothing$ and $\zeta = \varnothing$ in Eq. (4.1). In some cases, the constrained optimization problems can be transformed to the unconstrained optimization problems by some parameter transformation of variable $\underline{x}$, or through replacing the constraint functions by penalty terms added to the objective function.

According to the linearity of the objective function and constraint functions, the optimization problems can be classified into linear and nonlinear programming problems. If the objective function and all the constraint functions are linear functions, the problem is referred to as linear programming problem. If either the objective function or any of the constraint functions is a nonlinear function, the problem is referred to as nonlinear programming problem.

According to the certainty of the model, the optimization problems can be classified into deterministic and stochastic optimization problems. In deterministic optimization problems, the models are completely known. In the stochastic optimization problems, the models include some unknown quantities. Usually the stochastic optimization problem is solved based on some additional knowledge about these unknown quantities, from which the solution is produced by optimizing the mathematically expected performance of the models.

Based on the above classifications, the inverse lithography optimization problems discussed in the following chapters belong to the class of continuous, constrained, nonlinear, and deterministic optimization problems. To make the optimization problems analytically tractable, parameter transformations are exploited to translate constrained problems to unconstrained problems for the gradient-based inverse lithography optimization.

4.2 UNCONSTRAINED OPTIMIZATION

The gradient-based inverse lithography optimization discussed in the following chapters is formulated as an unconstrained problem. Thus, this section will address some important definitions, theorems, and descriptions on the solutions and algorithms of unconstrained optimization problems.

4.2.1 Solution of Unconstrained Optimization Problem

Global and local minimizers are defined in the following.

Definition 4.1 (Global minimizer). A solution $\underline{x}^*$ is a global minimizer if $f(\underline{x}^*) \leq f(\underline{x})$ for all $\underline{x} \in \mathcal{S}$, where $\mathcal{S}$ is the overall interested domain of $\underline{x}$.

Definition 4.2 (Weak local minimizer). A solution $\underline{x}^*$ is a weak local minimizer if there is a neighborhood $\mathcal{N} \subseteq \mathcal{S}$ of $\underline{x}^*$ such that $f(\underline{x}^*) \leq f(\underline{x})$ for all $\underline{x} \in \mathcal{N}$, where a neighborhood of $\underline{x}^*$ is an open set that contains $\underline{x}^*$.

Definition 4.3 (Strict local minimizer). A solution $\underline{x}^*$ is a strict local minimizer if there is a neighborhood $\mathcal{N} \subseteq \mathcal{S}$ of $\underline{x}^*$ such that $f(\underline{x}^*) < f(\underline{x})$ for all $\underline{x} \in \mathcal{N}$ with $\underline{x} \neq \underline{x}^*$.

Definition 4.4 (Isolated local minimizer). A solution $\underline{x}^*$ is an isolated local minimizer if there is a neighborhood $\mathcal{N} \subseteq \mathcal{S}$ of $\underline{x}^*$ such that $\underline{x}^*$ is the only local minimizer in $\mathcal{N}$. While strict local minimizers are not guaranteed to be isolated, any isolated local minimizer is strict.

The global minimizer is often difficult to find, because the knowledge of the objective function f is usually locally available. Thus, most optimization algorithms are able to find only local minimizers [53]. Figure 4.1 illustrates a function f with many local minimizers. The optimization algorithms applied to this kind of objective function tend to be trapped at local minimizers.

Several definitions and theorems to recognize local minimum and global minimum are listed next. The proofs of these theorems are skipped and can be found in Ref. [53].

Theorem 4.1 (First-order necessary condition). If $\underline{x}^*$ is a local minimizer and the objective function f is continuously differentiable in an open neighborhood of $\underline{x}^*$, then $\nabla f(\underline{x}^*) = 0$. In addition, $\underline{x}^*$ is referred to as the stationary point if $\nabla f(\underline{x}^*) = 0$.

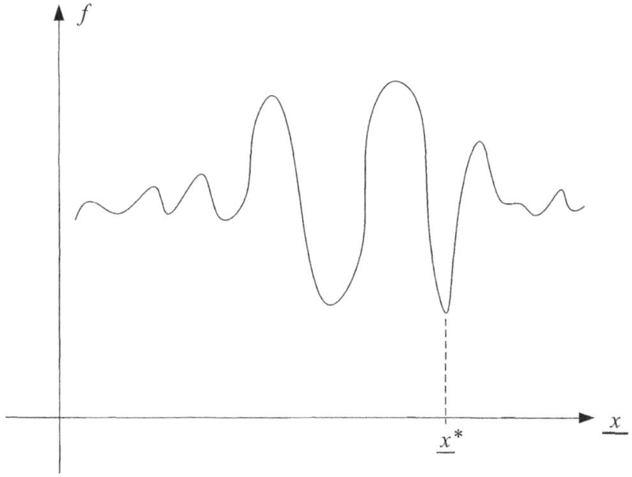

Figure 4.1 An objective function with many local minimizers.

Theorem 4.2 (Second-order necessary condition). If $\underline{x}^*$ is a local minimizer and the second derivative of the objective function $\nabla^2 f$ exists and is continuous in an open neighborhood of $\underline{x}^*$, then $\nabla f(\underline{x}^*) = 0$ and $\nabla^2 f(\underline{x}^*)$ is positive semidefinite. In addition, a function f is referred to as the smooth function if its second derivative exist and is continuous.

Theorem 4.3 (Second-order sufficient condition). Assuming that $\nabla^2 f$ is continuous in an open neighborhood of $\underline{x}^*$, $\nabla f(\underline{x}^*) = 0$ and $\nabla^2 f(\underline{x}^*)$ is positive definite. Then $\underline{x}^*$ is a strict local minimizer of f. Note that Theorem 4.3 is a sufficient, but not necessary, condition for the strict local minimizer.

Theorem 4.4. When f is convex, any local minimizer $\underline{x}^*$ is a global minimizer of f. In addition, if f is differentiable, then any stationary point $\underline{x}^*$ ($\nabla f(\underline{x}^*) = 0$) is a global minimizer of f. The convexity of function f is defined as following.

Definition 4.5 (Convex set). A set $\mathcal{S} \in \Re^n$ is a convex set if any two points $\underline{x}, \underline{y} \in \mathcal{S}$ satisfy $\alpha\underline{x} + (1-\alpha)\underline{y} \in \mathcal{S}$ for all $\alpha \in [0, 1]$.

Definition 4.6 (Convex function). A function f is a convex function if its domain $\mathcal{S} \in \Re^n$ is a convex set and if for any two points $\underline{x}, \underline{y} \in \mathcal{S}$, the following property is satisfied:

$$f(\alpha\underline{x} + (1-\alpha)\underline{y}) \leq \alpha f(\underline{x}) + (1-\alpha) f(\underline{y}), \quad \forall\, \alpha \in [0, 1]. \tag{4.2}$$

In addition, f is strictly convex if the inequality in Eq. (4.2) is replaced by

$$f(\alpha\underline{x} + (1-\alpha)\underline{y}) < \alpha f(\underline{x}) + (1-\alpha) f(\underline{y}), \quad \forall\, \alpha \in (0, 1) \text{ with } \underline{x} \neq \underline{y}. \tag{4.3}$$

The shape of a convex function is shown in Fig. 4.2.

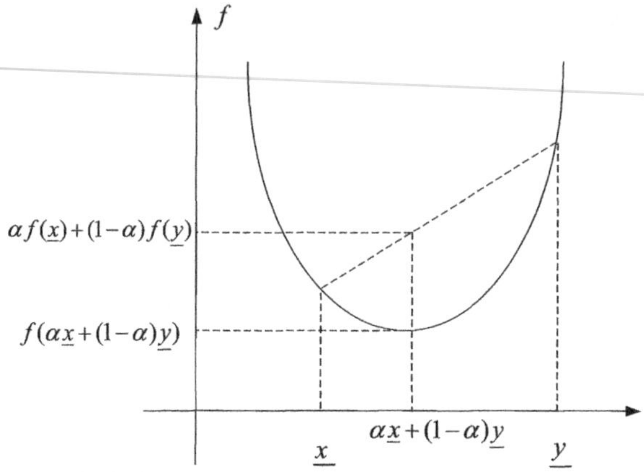

Figure 4.2 The shape of a convex function.

Definition 4.7 (Concave function). A function f is a concave function if $-f$ is a convex function.

Definition 4.8 (Convex programming). Convex programming is a constrained optimization problem described in Eq. (4.1), where f is convex, equality constraints $c_i(\underline{x}),\ i \in \varepsilon$ are linear, and $c_i(\underline{x}),\ i \in \zeta$ are concave.

4.2.2 Unconstrained Optimization Algorithms

A variety of optimization algorithms for unconstrained optimization of smooth functions have been developed during the past 40 years [53]. Generally speaking, these algorithms begin at a starting point $\underline{x}_0$, which is a reasonable estimate of the solution based on the prior knowledge of the system. Otherwise, the starting point may be chosen by some systematic algorithms. After the initialization, the optimization algorithms update the estimate of the solution at each iteration, generating a series $\{\underline{x}_k\}_{k=0}^{\infty}$. At each iteration, the optimization algorithms update the estimate from $\underline{x}_k$ to $\underline{x}_{k+1}$ based on the knowledge of the objective function f at $\underline{x}_k$, and, if possible, also based on the knowledge of the previous iteration $\underline{x}_0, \underline{x}_1, \ldots, \underline{x}_{k-1}$. Hopefully, this series can generally approach to the exact solution $\underline{x}^*$. The optimization algorithms terminate when no more progress can be made or when it seems that the current estimate $\underline{x}_k$ is enough close to solution $\underline{x}^*$. Different algorithms are mainly distinguished by the strategies of updating the estimate from $\underline{x}_k$ to $\underline{x}_{k+1}$. There are two typical strategies: line search strategy and trust region strategy. The derivations of both strategies rely on the Taylor's theorem, which is described as following.

Theorem 4.5 (Taylor's theorem). Suppose that $f : \Re^n \rightarrow \Re$ is continuously differentiable and that $\underline{p} \in \Re^n$. Then,

$$f(\underline{x} + \underline{p}) = f(\underline{x}) + \nabla f(\underline{x} + t\underline{p})^T \underline{p}, \tag{4.4}$$

for some $t \in (0, 1)$. Moreover, if f is twice continuously differentiable, then

$$\nabla f(\underline{x} + \underline{p}) = \nabla f(\underline{x}) + \int_0^1 \nabla^2 f(\underline{x} + t\underline{p})^T \underline{p} \mathrm{d}t \tag{4.5}$$

and that

$$f(\underline{x} + \underline{p}) = f(\underline{x}) + \nabla f(\underline{x})^T \underline{p} + \frac{1}{2}\underline{p}^T \nabla^2 f(\underline{x} + t\underline{p})\underline{p}, \tag{4.6}$$

for some $t \in (0, 1)$. Figure 4.3 illustrates the approximation of a function using the Taylor's theorem. In Fig. 4.3, let $\underline{x} = x$ and $\underline{p} = p$ be one-dimensional variables. Consider a one-dimensional function

$$f(x + p) = \mathrm{e}^{x+p}. \tag{4.7}$$

In Eq. (4.6),

$$f(x) + \nabla f(x)^T p = \mathrm{e}^x + \mathrm{e}^x p. \tag{4.8}$$

Assuming $\underline{x} = 0$, Eqs. (4.7) and (4.8) are modified as

$$f(x + p)|_{x=0} = \mathrm{e}^p, \tag{4.9}$$

which is shown by the solid line in Fig. 4.3, and

$$f(x) + \nabla f(x)^T p|_{x=0} = 1 + p, \tag{4.10}$$

which is shown by the dashed line in Fig. 4.3. Note that around the point $x = 0$, Eq. (4.9) may be approximated by Eq. (4.10). The residue of the approximation is

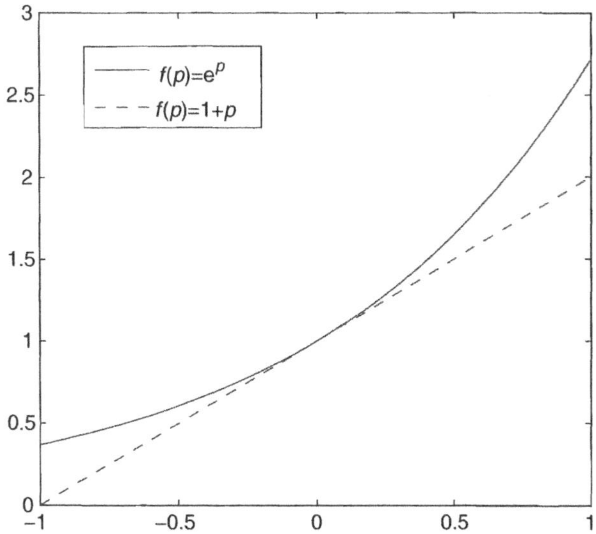

Figure 4.3 The Taylor expansion.

attributed to the third term on the right-hand side in Eq. (4.6). The line search and trust region strategies are respectively summarized as following.

4.2.2.1 Line Search Strategy In the line search strategy, the optimization algorithms first choose a direction $\underline{p}_k$, along which the current estimate $\underline{x}_k$ is moved toward the new estimate $\underline{x}_{k+1}$. The second step is to find an optimal or suboptimal step length, a distance to move along $\underline{p}_k$, by solving the following minimization problem:

$$\min_{\alpha_k>0} f(\underline{x}_k + \alpha_k \underline{p}_k). \tag{4.11}$$

The exact minimization of Eq. (4.11) may be computationally complex and unnecessary. Instead, the line search algorithms generate a limited number of trial step lengths until it finds one that loosely approximates the minimum of Eq. (4.11) [53]. Sometimes, the step length can also be assigned heuristically and empirically. In the following, four typical line search optimization algorithms are summarized. The details of these algorithms can be found in Ref. [53].

Steepest Descent Method According to Theorem 4.5, we have

$$f(\underline{x}_k + \alpha_k \underline{p}_k) = f(\underline{x}_k) + \alpha_k \nabla f(\underline{x}_k)^T \underline{p}_k + \frac{1}{2}\alpha_k^2 \underline{p}_k^T \nabla^2 f(\underline{x}_k + t\underline{p}_k)\underline{p}_k, \tag{4.12}$$

for some $t \in (0, \alpha_k)$. The rate of change in f along the direction $\underline{p}_k$ at $\underline{x}_k$ is $\nabla f(\underline{x}_k)^T \underline{p}_k$. Hence, the unit direction $\underline{p}_k$ of most rapid decrease is the solution to the problem

$$\min_{\underline{p}_k} \nabla f(\underline{x}_k)^T \underline{p}_k, \quad \text{subject to } \|\underline{p}_k\| = 1. \tag{4.13}$$

The solution of Eq. (4.13) is

$$\underline{p}_k^* = -\frac{\nabla f_k}{\|\nabla f_k\|}, \tag{4.14}$$

where $\nabla f_k = \nabla f(\underline{x}_k)$. The steepest descent method updates the estimate along $\underline{p}_k^*$ in Eq. (4.14). A variety of methods have been developed to select the step length α_k [53]. The advantage of the steepest descent method is that only the first derivative ∇f_k is required to be calculated. However, it may converge slowly for some optimization problems. The steepest descent direction for a two-dimensional function is shown in Fig. 4.4. In the following chapters, the steepest descent method is applied to the gradient-based inverse lithography optimization.

Newton Method According to Theorem 4.5, we have

$$f(\underline{x}_k + \underline{p}_k) \approx f_k + \nabla f_k^T \underline{p}_k + \frac{1}{2}\underline{p}_k^T \nabla^2 f_k \underline{p}_k, \tag{4.15}$$

where $f_k = f(\underline{x}_k)$, $\nabla f_k = \nabla f(\underline{x}_k)$, and $\nabla^2 f_k = \nabla^2 f(\underline{x}_k)$. If $\nabla^2 f$ is sufficiently smooth, the error introduced by the approximation in Eq. (4.15) is only $O(\|\underline{p}_k\|^3)$.

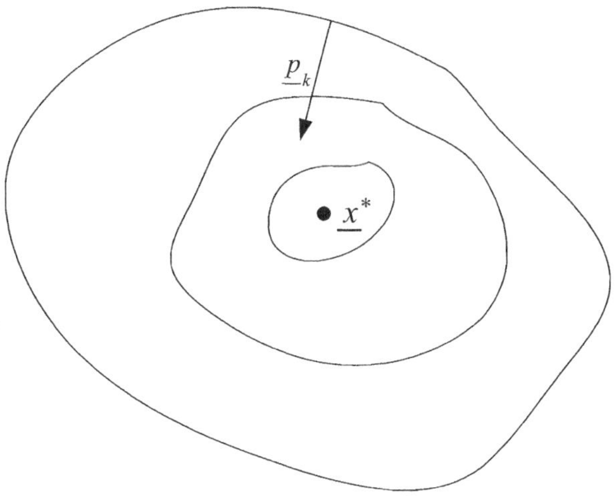

Figure 4.4 The steepest descent direction for a two-dimensional function.

Thus, if $\|\underline{p}_k\|$ is small, the approximation is very accurate. The Newton direction $\underline{p}_k$ is the solution to the following problem:

$$\min_{\underline{p}_k} f_k + \nabla f_k^T \underline{p}_k + \frac{1}{2} \underline{p}_k^T \nabla^2 f_k \underline{p}_k. \tag{4.16}$$

Assuming that $\nabla^2 f_k$ is positive definite, the solution of Eq. (4.16) is

$$\underline{p}_k^* = -(\nabla^2 f_k)^{-1} \nabla f_k. \tag{4.17}$$

Most line search implementations of Newton method use the unit step length $\alpha_k = 1$ where possible and adjust α_k only when it does not produce a satisfactory reduction in the value of f [53]. Note that the calculation of the Newton direction requires that $\nabla^2 f_k$ is positive definite. When $\nabla^2 f_k$ is not positive definite, a set of modification algorithms have been developed to adjust $\nabla^2 f_k$ as a positive definite matrix [53]. The advantage of the Newton method is that its local convergence rate is quadratic. From an estimate $\underline{x}_k$ close to the exact solution $\underline{x}^*$, the Newton algorithm may converge very fast with high accuracy. However, the drawback of the Newton method is that the second derivative $\nabla^2 f_k$ is needed.

Quasi-Newton Method The quasi-Newton method is a modification of Newton method that does not require the calculation of $\nabla^2 f_k$ and attains a superlinear rate of convergence. The quasi-Newton method approximates $\nabla^2 f_{k+1}$ as B_{k+1}, which satisfies

$$B_{k+1} \underline{s}_k = \underline{y}_k, \tag{4.18}$$

where $\underline{s}_k = \underline{x}_{k+1} - \underline{x}_k$ and $\underline{y}_k = \nabla f_{k+1} - \nabla f_k$. The approximation B_{k+1} can be successively updated based on the previous B_k. Two most popular formulas for updating

B_{k+1} are the symmetric rank one (SR1) formula, defined by [53]

$$B_{k+1} = B_k + \frac{(\underline{y}_k - B_k\underline{s}_k)(\underline{y}_k - B_k\underline{s}_k)^T}{(\underline{y}_k - B_k\underline{s}_k)^T \underline{s}_k}, \tag{4.19}$$

and the BFGS formula, defined by [53]

$$B_{k+1} = B_k - \frac{B_k\underline{s}_k\underline{s}_k^T B_k}{\underline{s}_k^T B_k \underline{s}_k} + \frac{\underline{y}_k\underline{y}_k^T}{\underline{y}_k^T\underline{s}_k}. \tag{4.20}$$

Given the B_k, the search direction of quasi-Newton method is formulated as

$$\underline{p}_k = -B_k^{-1}\nabla f_k. \tag{4.21}$$

To avoid calculating the inverse matrix B_k^{-1}, some practical implementations of the quasi-Newton method successively update B_{k+1}^{-1} instead of B_{k+1}. Assuming $H_{k+1} = B_{k+1}^{-1}$, the update is

$$H_{k+1} = (\mathbf{I} - \rho_k\underline{s}_k\underline{y}_k^T)H_k(\mathbf{I} - \rho_k\underline{y}_k\underline{s}_k^T) + \rho_k\underline{s}_k\underline{s}_k^T, \tag{4.22}$$

where $\mathbf{I}$ is the identity matrix and $\rho_k = \frac{1}{\underline{y}_k^T\underline{s}_k}$. Given the H_k, the search direction of quasi-Newton method can be reformulated as

$$\underline{p}_k = -H_k\nabla f_k. \tag{4.23}$$

Conjugate Gradient Method The conjugate gradient method initializes the search direction as $\underline{p}_0 = -\nabla f_0$. At each iteration, the search direction is updated as

$$\underline{p}_{k+1} = -\nabla f_{k+1} + \beta_{k+1}\underline{p}_k, \tag{4.24}$$

where β_{k+1} is a scalar that ensures that $\underline{p}_{k+1}$ and $\underline{p}_k$ are conjugate with respect to some matrices [16, 23, 27, 53]. Two vectors $\underline{p}_i$ and $\underline{p}_j$ are conjugate with respect to a symmetric positive definite matrix A, if

$$\underline{p}_i^T A \underline{p}_j = 0, \quad \text{for all } i \neq j. \tag{4.25}$$

The step length α_k is calculated by line search method that identifies an approximate minimum of the objective function f along $\underline{p}_k$. The conjugate gradient search direction $\underline{p}_k$ is more effective than the steepest descent direction, but has similar computational complexity. However, its convergence rate is not as fast as that of the Newton or quasi-Newton methods.

4.2.2.2 Trust Region Strategy The trust region strategy constructs a model function m_k to approximate the behavior of objective function f in a neighborhood $\mathcal{N}$ of the current estimate $\underline{x}_k$. Assuming the next estimate $\underline{x}_{k+1} = \underline{x}_k + \underline{p}_k$, the trust region strategy solves the following subproblem:

$$\min_{\underline{p}_k} m_k(\underline{x}_k + \underline{p}_k), \tag{4.26}$$

where $\underline{x}_k + \underline{p}_k$ lies inside the trust region.

The quadratic model function is usually used in trust region strategy, which is formulated as

$$m_k(\underline{x}_k + \underline{p}_k) = f_k + \nabla f_k^T \underline{p}_k + \frac{1}{2}\underline{p}_k^T B_k \underline{p}_k, \tag{4.27}$$

where B_k is the Hessian $\nabla^2 f_k$ or some approximation to it. Various kinds of trust regions have been investigated. The ball-shaped trust region defined by $\|\underline{p}_k\|_2 \leq \Delta_k$ is extensively used, where $\Delta_k > 0$ is called the trust region radius. In addition, the elliptical and box-shaped trust regions may also be used [53]. It is important to modify the trust region radius Δ_k at each iteration, according to the agreement between the model function m_k and the objective function f in previous trust region. If Δ_k is too large, m_k may diverge from f at some points far from current estimate $\underline{x}_k$, thus leading to inaccurate approximation of f. If Δ_k is too small, the trust region strategy cannot attain sufficient reduction in the value of f.

Basically, the line search and trust region strategies differ in the order in which they choose the search direction $\underline{p}_k$ and the step length α_k [53]. The line search strategy first determines the search direction $\underline{p}_k$. Subsequently, the step length α_k is identified to obtain sufficient reduction in the value of f. In contrast, the trust region strategy first determines the trust region radius Δ_k. Then, the search direction $\underline{p}_k$ and the step length α_k are chosen simultaneously by solving the subproblem in Eq. (4.26). If the objective function f cannot be successfully reduced, the trust region strategy reduces Δ_k and tries again.

Several different types of trust region algorithms have been developed, such as the dogleg method and the two-dimensional subspace minimization method among others. The following chapters apply the steepest descent algorithm to solve the inverse lithography optimization problem. The details of trust region strategies are skipped here, and may be found in Ref. [53].

4.3 SUMMARY

This chapter summarized the definition and the classification of different optimization problems. The unconstrained optimization was then discussed at length. As one of the most important unconstrained optimization algorithms, the steepest descent method will be applied to solve the RET optimization problems in the following chapters.

5

Computational Lithography with Coherent Illumination

Due to resolution limits of optical lithographic systems, the electronics industry has relied on RET to compensate and minimize mask distortions as they are projected onto semiconductor wafers [92]. Resolution in optical lithography obeys the Rayleigh criterion resolution$(R) = k\frac{\lambda}{\text{NA}}$, where λ is the wavelength, NA is the numerical aperture taking on values around 0.9 for most lithography systems used today, and k is the process constant that can be minimized through RET methods [11, 37, 76, 77]. OPC methods modify the mask amplitude by the addition of subresolution features to the mask pattern. PSMs, commonly attributed to Levenson et al. [35], induce phase shifts in the transmitted field that have a favorable constructive or destructive interference effect. Thus, a suitable modulation of both the phase and the intensity of the incident light can be used to effectively compensate for some of the resolution-limiting phenomena in optical diffraction.

Several approaches of OPC and PSM optimizations have been proposed in the literature. These range from heuristic and empirically based design rules to computationally expensive optimization-based inverse algorithms. Sherif et al. derived an iterative approach to generate binary masks [79]. Liu and Zakhor developed a binary and phase-shifting mask design strategy based on the branch and bound algorithm and simulated annealing [38]. Pati and Kailath exploited a class of approximations for partially coherent imaging systems to develop suboptimal projections onto convex sets for PSM designs [58]. In addition, Erdmann et al. proposed an automatic optimization of the mask and the illumination parameters with a genetic algorithm [21]. Pang et al. gave an overview of inverse lithography techniques (ILTs) and provided some simulations to demonstrate the benefit of ILT [55]. Granik described and compared the solutions of inverse mask problems [26]. All the methods mentioned above, however, are not based on gradient-type optimization and thus the searching process for a suitable solution is either computationally expensive or not efficient. Therefore, the challenge is to develop computationally efficient methods to design precompensated masks, also referred to as inverse lithography technology [68].

Computational Lithography By Xu Ma and Gonzalo R. Arce

This chapter focuses on gradient-based OPC and PSM optimization algorithms in coherent imaging systems. As discussed in Chapter 2, the intensity distribution on the wafer in coherent imaging system is formulated as

$$I(\mathbf{r}) = |M(\mathbf{r}) \otimes h(\mathbf{r})|^2. \tag{5.1}$$

Two classes of PSM are discussed: First, the two-phase PSM, where the transmission coefficients on the mask of different polarizations have 180° out of phase with respect to each other. While this approach is very effective in some cases, the end result is that the search generally fails to generate adequate PSM for mask patterns having arbitrary Manhattan geometries [58]. To overcome this limitation, a generalized PSM optimization is introduced here, capable of generating arbitrary number of phase levels on the mask [40, 41].

Let $M(x, y)$ be the input mask to an optical lithography system $T\{\cdot\}$, approximated as a low-pass spatial filter followed by a soft threshold operation, which accounts for the photoresist effect. The output pattern is denoted as $Z(x, y) = T\{M(x, y)\}$. Given a $N \times N$ desired output pattern $\tilde{Z}(x, y)$, the goal of OPC and PSM optimizations is to find the optimized $M(x, y)$ called $\hat{M}(x, y)$ such that the distance

$$D = d(Z(x, y), \tilde{Z}(x, y)) = d(T\{M(x, y)\}, \tilde{Z}(x, y)) \tag{5.2}$$

is minimized, where $d(\cdot, \cdot)$ is the square of the l^2 norm criterion. The OPC and PSM inverse lithography optimization problem can thus be formulated as the search of $\hat{M}(x, y)$ such that

$$\hat{M}(x, y) = \arg \min_{M(x,y)} d(T\{M(x, y)\}, \tilde{Z}(x, y)). \tag{5.3}$$

For the OPC and two-phase PSM optimization, $\hat{M}(x, y)$ is searched in the $N \times N$ real space $\Re^{N\times N}$. For the generalized PSM optimization, $\hat{M}(x, y)$ is searched in the $N \times N$ complex space $C^{N\times N}$.

5.1 PROBLEM FORMULATION

Recently, Poonawala and Milanfar introduced a powerful optimization framework for inverse lithography based on a pixel-based, continuous function formulation, well suited for gradient-based search [68]. Based on a steepest descent search, their approach exploits the rich theory of regularized iterative optimization [83]. In Poonawala and Milanfar's work, an approximated forward process model is exploited to represent the optical lithography system [67, 68]. Figure 5.1 illustrates the scheme of the approximated forward process model. In Fig. 5.1, the mask is the input of the system. The amplitude pattern of the propagating light is modified by the assisting features on the binary mask. The phase pattern of the propagating light is modified by the phase-shifting material overcoating the mask. Light propagating through the mask pattern is affected by diffraction and mutual interference—a phenomenon described by the Hopkins diffraction model [7, 14]. The light that is transmitted through the mask and

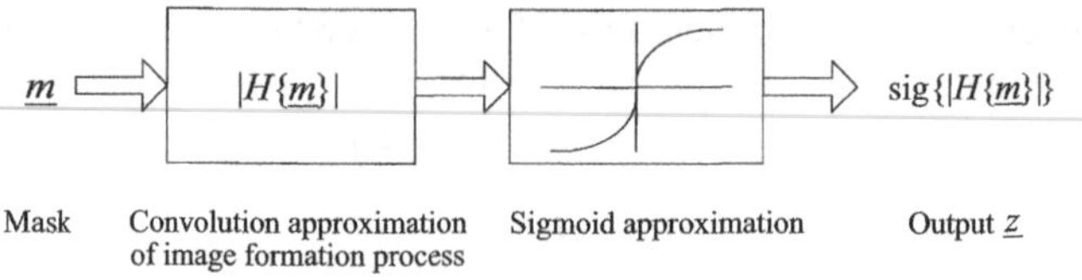

Figure 5.1 Approximated forward process model of coherent imaging system.

lenses reaches a light-sensitive photoresist, which is subsequently developed through the use of solvents. The thickness of the remaining photoresist after development is proportional to the exposure dose exceeding a given threshold intensity. Assume that positive photoresist is used in this optical lithography system. In a positive photoresist process, almost all the photoresist material remains in the low-exposure area on the wafer and is removed in the high-exposure area. Between these two extremes is the transition region. For mathematical simplicity, it is assumed that when the light field exceeds a threshold, the exposed area becomes a high-exposure area, otherwise, a low-exposure area. Thus, a hard threshold operation, which is a shifted unit step function $U(x - t_r)$, can adequately represent the photoresist effect described above and the output pattern of the optical system is binary. The hard threshold function to represent the photoresist effect is defined as

$$\Lambda(x) = \begin{cases} 0 & : \quad x \leq t_r, \\ 1 & : \quad x > t_r, \end{cases} \tag{5.4}$$

where t_r is the threshold.

In Fig. 5.1, $|\cdot|$ is the element-by-element absolute operation. For coherent imaging systems, the aerial image formation process can be approximated by a convolution between the mask pattern and a Gaussian low-pass filter h. In the pixel-based algorithm, pixel size = resolution(R) = $k\frac{\lambda}{\mathrm{NA}}$. The standard deviation of the Gaussian low-pass filter h is $\sigma = \frac{R \times \mathrm{NA}}{\lambda} = k$. The output of the convolution and the absolute operation model is the electric field amplitude of the aerial image. Further, since the derivative of the sigmoid function exists, it is used to approximate the hard threshold function. The sigmoid function is defined as

$$\mathrm{sig}(x) = \frac{1}{1 + \exp[-a(x - t_r)]}, \tag{5.5}$$

where t_r is the process threshold and a dictates the steepness of the sigmoid function. Figure 5.2 illustrates the curves of the hard threshold function and sigmoid functions with $t_r = 0.5$. In Fig. 5.2, solid line represents the hard threshold function $\Lambda(x)$. Dashed line, dashed–dotted line, and dotted line represent the sigmoid functions with a =10, 30, and 110, respectively.

In the OPC and two-phase PSM optimization, the mask is represented by a matrix $M \in \Re^{N \times N}$. In the generalized PSM optimization, the mask is represented by a matrix $M \in C^{N \times N}$. $\underline{m}_{N^2 \times 1}$ is the $N^2 \times 1$ equivalent raster-scanned vector representation of

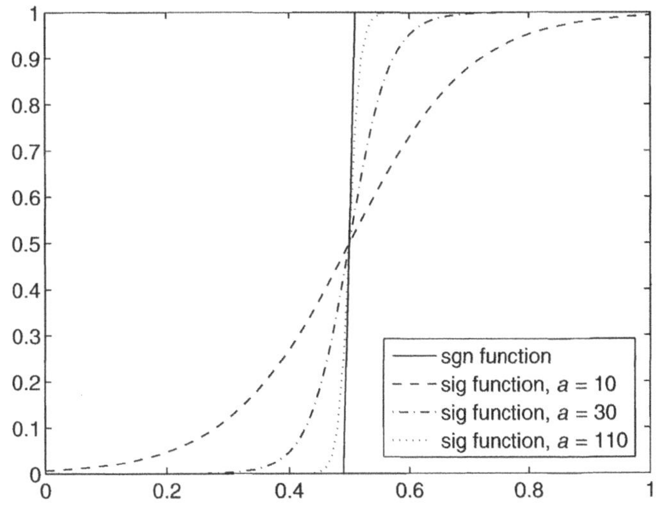

Figure 5.2 Hard threshold function and sigmoid functions.

the mask pattern M and is denoted as $\underline{m}$ for short notation. Following the definitions above, the notations used are as under:

1. A convolution matrix H is a $N^2 \times N^2$ matrix with an equivalent two-dimensional low-pass filter h.
2. The desired $N \times N$ binary output pattern is denoted as $\tilde{Z}$. It is the desired light distribution sought on the wafer. Its vector representation is denoted as $\underline{\tilde{z}}$.
3. The output of the sigmoid function is the $N \times N$ real-valued image denoted as

$$Z = \text{sig}(|H\{M\}|). \tag{5.6}$$

   The equivalent vector is denoted as $\underline{z} \in \Re^{N^2 \times 1}$.
4. The hard threshold version of Z is the binary output pattern denoted as

$$Z_{\text{b}} = \Lambda(|H\{M\}|). \tag{5.7}$$

   Its equivalent vector is denoted as $\underline{z}_{\text{b}} \in \Re^{N^2 \times 1}$, with all entries constrained to 0 or 1.
5. In the OPC optimization, the hard threshold version of M is the binary mask M_{b}. Its equivalent vector is denoted as $\underline{m}_{\text{b}} \in \Re^{N^2 \times 1}$, with all entries constrained to 0 or 1. In the PSM optimization, the discrete version of M is the pole-level mask M_{p}. Its equivalent vector is denoted as $\underline{m}_{\text{p}}$. For the two-phase PSM, all entries of $\underline{m}_{\text{p}}$ are constrained to -1, 0, or 1. For the generalized PSM, the amplitudes of all entries of $\underline{m}_{\text{p}}$ are constrained to 0 or 1. The phases are constrained to be several discrete phase levels.

6. The optimized $N \times N$ mask denoted as $\hat{M}$ minimizes the distance between Z and $\tilde{Z}$, that is,

$$\hat{M} = \arg\min_{M} d(\text{sig}\{|H\{M\}|\}, \tilde{Z}). \tag{5.8}$$

   For the optimized binary mask and two-phase PSM, its equivalent vector is denoted as $\underline{\hat{m}} \in \Re^{N^2 \times 1}$. For the optimized generalized PSM, $\underline{\hat{m}} \in C^{N^2 \times 1}$.
7. In the OPC optimization, the binary optimized mask $\hat{M}_\text{b}$ is the hard threshold version of $\hat{M}$. Its equivalent vector is denoted as $\underline{\hat{m}}_\text{b} \in \Re^{N^2 \times 1}$, with all entries constrained to 0 or 1. In the PSM optimization, the pole-level optimized mask $\hat{M}_\text{p}$ is the quantization of $\hat{M}$. Its equivalent vector is denoted as $\underline{\hat{m}}_\text{p}$. For the two-phase PSM, all entries of $\underline{\hat{m}}_\text{p}$ are constrained to -1, 0, or 1. For the generalized PSM, the amplitudes of all entries of $\underline{\hat{m}}_\text{p}$ are constrained to 0 or 1. The phases are constrained to be of several discrete phase levels.

Given the gray-level pattern $\underline{z} = \text{sig}\{|H\ \underline{m}|\}$, the ith entry in this vector can be represented as

$$\underline{z}_i = \frac{1}{1 + \exp\left[-a\left|\sum_{j=1}^{N^2} h_{ij}\underline{m}_j\right| + at_r\right]}, \quad i = 1, \ldots, N^2, \tag{5.9}$$

where h_{ij} is the i, jth entry of the filter. In the optimization process, $\underline{\hat{m}}$ is searched to minimize the square of the l^2 norm of the difference between $\underline{z}$ and $\underline{\tilde{z}}$. Therefore,

$$\underline{\hat{m}} = \arg\min_{\underline{m}}\{F(\underline{m})\}, \tag{5.10}$$

where the cost function $F(\cdot)$ is defined as

$$F(\underline{m}) = \|\underline{\tilde{z}} - \underline{z}\|_2^2 = \sum_{i=1}^{N^2}(\tilde{z}_i - z_i)^2$$
$$= \sum_{i=1}^{N^2}\left(\tilde{z}_i - \frac{1}{1 + \exp\left[-a\left|\sum_{j=1}^{N^2} h_{ij}\underline{m}_j\right| + at_r\right]}\right)^2. \tag{5.11}$$

5.2 OPC OPTIMIZATION

5.2.1 OPC Design Algorithm

For the OPC optimization, $\underline{m}$ can only have entry values of 0 or 1, leading to a discrete combinatorial optimization problem. To make the optimization problem analytically tractable, the entry values of $\underline{m}$ are relaxed to lie in the range of [0, 1]. Thus, the optimization problem given in Eq. (5.10) is constrained by the following inequality:

$$0 \leq \underline{m}_k \leq 1, \quad k = 1, \ldots, N^2. \tag{5.12}$$

The bound-constrained optimization is then reduced to an unconstrained optimization problem using the following parameter transformation,

$$\underline{m}_k = \frac{1 + \cos\underline{\theta}_k}{2}, \quad k = 1, \ldots, N^2, \tag{5.13}$$

where $\underline{\theta}_k \in (-\infty, \infty)$. Defining the unconstrained parameter vector $\underline{\theta} = [\theta_1, \ldots, \theta_{N^2}]^T$, the optimization problem is formulated as

$$\hat{\underline{\theta}} = \arg\min_{\underline{\theta}}\{F(\underline{\theta})\}. \tag{5.14}$$

Since $0 \leq \underline{m}_i \leq 1$, $|H\underline{m}| = H\underline{m}$ and thus the cost function is

$$\begin{aligned} F(\underline{\theta}) &= \|\tilde{\underline{z}} - \underline{z}\|_2^2 = \sum_{i=1}^{N^2} (\tilde{z}_i - z_i)^2 \\ &= \sum_{i=1}^{N^2} \left(\tilde{\underline{z}}_i - \frac{1}{1 + \exp\left[-a\left(\sum_{k=1}^{N^2} h_{ik} \frac{1 + \cos\underline{\theta}_k}{2}\right)\right] + at_r} \right)^2. \end{aligned} \tag{5.15}$$

The steepest descent method is used to optimize the above problem. The gradient $\nabla F(\underline{\theta})$ derived in Ref. [68] can be calculated as follows:

$$\nabla F(\underline{\theta}) = \underline{d}_\theta = 2a(H^T[(\tilde{\underline{z}} - \underline{z}) \odot \underline{z} \odot (\underline{1} - \underline{z})]) \odot \sin(\underline{\theta}), \tag{5.16}$$

where $\nabla F(\underline{\theta}) \in \Re^{N^2 \times 1}$, $\odot$ is the element-by-element multiplication operator, and $\underline{1} = [1, \ldots, 1]^T \in \Re^{N^2 \times 1}$. Assuming that $\underline{\theta}^k$ is the kth iteration result, at the $k + 1th$ iteration

$$\underline{\theta}^{k+1} = \underline{\theta}^k - s\underline{d}_\theta^k, \tag{5.17}$$

where s is the step size. It is noted that Eq. (5.16) can be quickly and directly carried out on the 2D image array with no need for the raster scanning operation [68]. This feature reduces the computational complexity of the described algorithm and also simplifies its implementation. In addition, since the optimized binary mask $\hat{M}$ should be a perturbation of the desired output pattern $\tilde{Z}$, the initial mask pattern is set to be $\tilde{Z}$. This initialization leads to a quick convergence of the parameter $\underline{\theta}$. Subsequently, the optimized binary mask $\hat{M}$ can be obtained from $\underline{\theta}$ by Eq. (5.13).

The iterative optimization above, in general, leads to real-valued solutions that are not constrained to a binary mask. Therefore, a post-processing step is needed to obtain the binary-optimized mask $\hat{\underline{m}}_{\rm b}$. In the post-processing step, the real-valued optimized mask $\hat{\underline{m}}$ is quantized by a global threshold t_m as

$$\hat{\underline{m}}_{{\rm b}k} = U(\hat{\underline{m}}_k - t_m) = \begin{cases} 0 & : \quad \hat{\underline{m}}_k \leq t_m, \\ 1 & : \quad \hat{\underline{m}}_k > t_m, \end{cases} \qquad k = 1, \ldots, N^2. \tag{5.18}$$

The pattern error E is defined as the square of the l^2 norm of the difference between the desired output pattern $\tilde{Z}$ and the actual binary output pattern $Z_{\rm b}$, that is,

$$E = \sum_{i=1}^{N^2} |\tilde{\underline{z}}_i - \underline{z}_{{\rm b}i}|^2 = \sum_{i=1}^{N^2} |\tilde{\underline{z}}_i - \Lambda_i(|H\underline{m}_{\rm b}|)|^2. \tag{5.19}$$

When the pattern error is reduced to a tolerable level, the steepest descent iteration is stopped.

5.2.2 Simulations

In this section, simulation results of the OPC optimization in coherent imaging systems are presented. Figure 5.3 shows the OPC optimization with a desired pattern of two

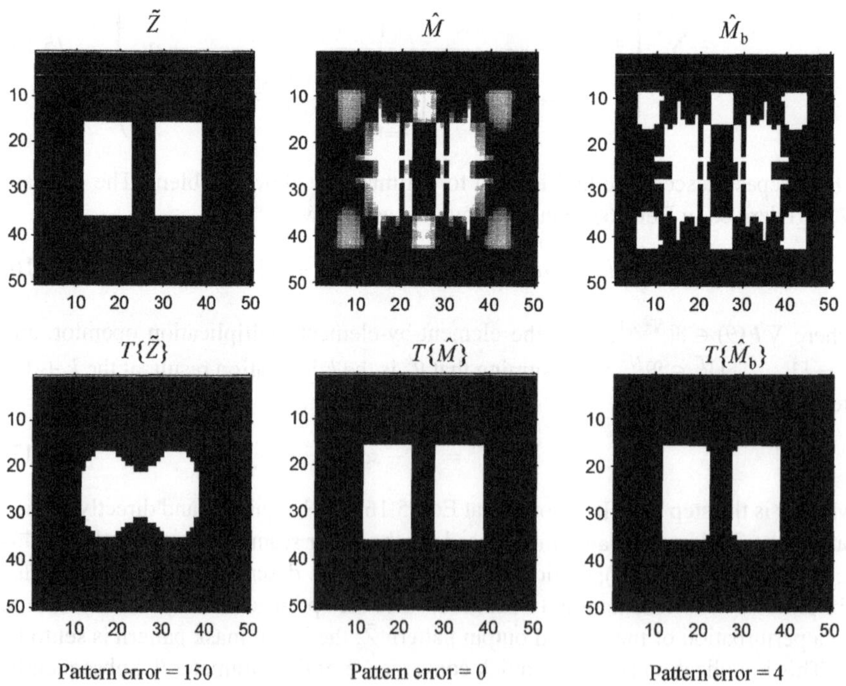

Figure 5.3 OPC optimization with a desired pattern of vertical bars. *Top row* (input masks) (left to right): Desired pattern, optimized real-valued mask, and optimized binary mask obtained using a threshold t_m. *Bottom row*: Indicates the corresponding binary output patterns. Black and white represent 0 and 1, respectively.

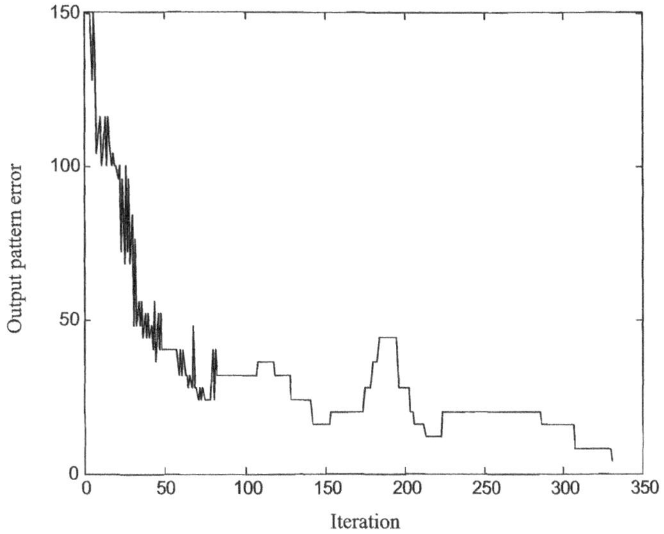

Figure 5.4 Convergence of the OPC optimization algorithm versus steepest descent iterations for Fig. 5.3.

vertical bars. From left to right, the top row shows the desired pattern, optimized real-valued mask, and optimized binary mask obtained using a threshold t_m. The bottom row indicates the corresponding binary output patterns. Initially, we assign $\underline{\theta}_i = \frac{\pi}{5}$ for the transparent regions, and $\underline{\theta}_i = \frac{4\pi}{5}$ for the opaque regions. The parameters used in the simulation are $a = 90$, $t_r = 0.5$, $t_m = 0.5$, and 15×15 Gaussian low-pass filter with $k = 5$ and $s = 0.2$. Black and white represent 0 and 1, respectively. If the desired pattern is used as the input mask, the output pattern error is 150. In addition, the two vertical bars cannot be distinguished. The optimized real-valued mask leads to a zero output pattern error. After the post-processing step, the optimized binary mask leads to an output pattern error of 4. The OPC optimization algorithm effectively reduces the pattern errors. Figure 5.4 shows the convergence of the OPC optimization algorithm versus steepest descent iterations. It is noted that the cost function decays fast and converges to a low pattern error.

Figure 5.5 shows another simulation with a desired pattern of four horizontal bars. The parameters used in the simulation are $a = 80$, $t_r = 0.5$, $t_m = 0.5$, and 11×11 Gaussian low pass filter with $k = 14$ and $s = 0.5$. Black and white represent 0 and 1, respectively. The described algorithm effectively reduces the output pattern errors from 642 to 4. Figure 5.6 shows the successful convergence of the OPC optimization algorithm.

5.3 TWO-PHASE PSM OPTIMIZATION

5.3.1 Two-Phase PSM Design Algorithm

This section focuses on the alternating PSM optimization [67]. The pixel values on the alternating phase-shifting mask can take values of 0, 1, or -1, where 0 represents

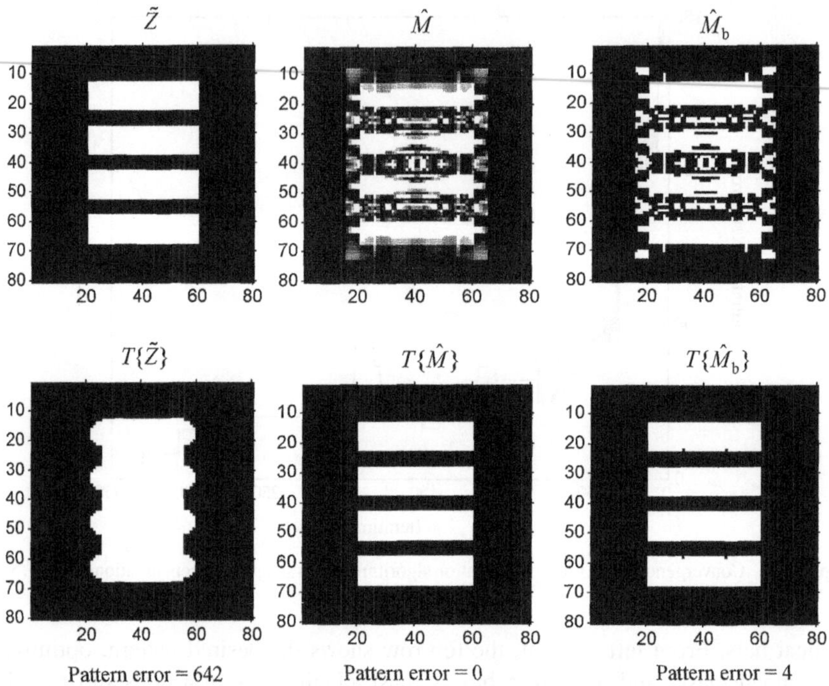

Figure 5.5 OPC optimization with a desired pattern of horizontal bars. *Top row* (input masks) (left to right): Desired pattern, optimized real-valued mask, and optimized binary mask obtained using a threshold t_m. *Bottom row:* Indicates the corresponding binary output patterns. Black and white represent 0 and 1, respectively.

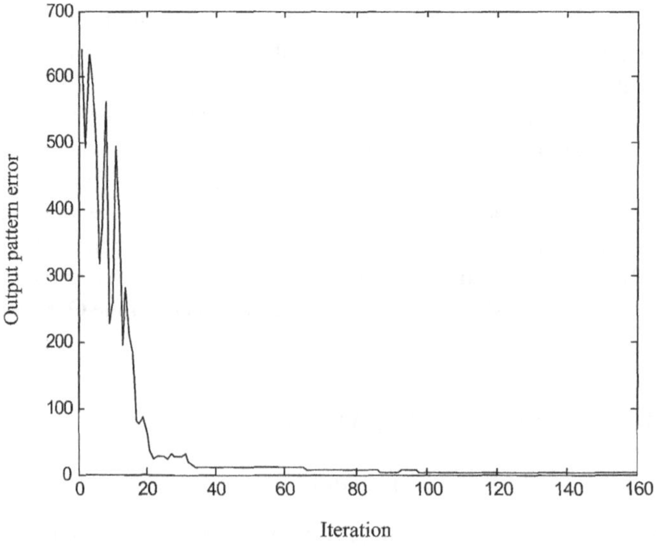

Figure 5.6 Convergence of the OPC optimization algorithm versus steepest descent iterations for Fig. 5.5.

the opaque region and 1 and -1 represent the transparent regions with 0 and π phases, respectively. To make the problem analytically tractable, the following constraint is imposed on the two-phase PSM optimization problem,

$$-1 \leq \underline{m}_k \leq 1, \quad k = 1, \ldots, N^2. \tag{5.20}$$

To convert the bound-constrained optimization problem to an unconstrained optimization problem, the following parameter transformation is used:

$$\underline{m}_k = \cos(\theta_k), \quad k = 1, \ldots, N^2, \tag{5.21}$$

where $\underline{\theta}_k \in (-\infty, \infty)$ and $\underline{m}_i \in [-1, 1]$. $\underline{\theta} = [\theta_1, \ldots, \theta_{N^2}]^T$ is defined as the unconstrained parameter vector. The two-phase PSM optimization problem can be formulated as

$$\hat{\underline{\theta}} = \arg\min_{\underline{\theta}} \{F(\underline{\theta})\}, \tag{5.22}$$

where the cost function $F(\cdot)$ is defined as

$$F(\underline{\theta}) = \sum_{i=1}^{N^2} \left(\underline{\tilde{z}}_i - \frac{1}{1 + \exp\left[-a \left| \sum_{j=1}^{N^2} h_{ij} \cos\underline{\theta}_j \right| + a t_r \right]} \right)^2$$

$$= \sum_{i=1}^{N^2} \left(\underline{\tilde{z}}_i - \frac{1}{1 + \exp\left[-a \sqrt{\left(\sum_{j=1}^{N^2} h_{ij} \cos\underline{\theta}_j \right)^2} + a t_r \right]} \right)^2 . \tag{5.23}$$

The steepest descent algorithm is used to optimize the above problem. According to Appendix A, the gradient of the cost function is

$$\nabla F(\underline{\theta}) = \underline{d} = 2a(H^T [(\underline{\tilde{z}} - \underline{z}) \odot \underline{z} \odot (\underline{1} - \underline{z}) \odot \underline{\text{sig}}]) \odot \sin \underline{\theta}, \tag{5.24}$$

where $\nabla F(\underline{\theta}) \in \Re^{N^2 \times 1}$ and $\odot$ is element-by-element multiplication operator. The $N^2 \times 1$ vector $\underline{\text{sig}}$ is defined as

$$\underline{\text{sig}} = (\text{sig}_1, \text{sig}_2, \ldots, \text{sig}_{N^2}), \tag{5.25}$$

where

$$\text{sig}_i = \begin{cases} 0 \; : & \sum_{j=1}^{N^2} h_{ij} \cos \leq 0, \\ 1 \; : & \sum_{j=1}^{N^2} h_{ij} \cos > 0, \end{cases} \quad i = 1, \ldots, N^2. \tag{5.26}$$

It is noted that Eq. (5.24) is not the same as Eq. (17) in Ref. [68], since we threshold the amplitude of the electric field $|H\underline{m}|$ in the approximated forward process model shown in Fig. 5.1. On the other hand, the intensity distribution $|H\underline{m}|^2$ is thresholded in Ref. [68]. Similar to the OPC optimization algorithm, a post-processing step is used to obtain the optimized two-phase PSM. Let the threshold version of $\hat{\underline{m}}$ is $\hat{\underline{m}}_{\text{p}}$, which is defined as

$$\hat{\underline{m}}_{\text{p}k} = \begin{cases} 1 & : \quad |\hat{\underline{m}}_k| > t_m, \\ 0 & : \quad -t_m \leq |\hat{\underline{m}}_k| \leq t_m, \\ -1 & : \quad |\hat{\underline{m}}_k| < -t_m, \end{cases} \qquad k = 1, \ldots, N^2. \tag{5.27}$$

The pattern error E is defined as the square of the l^2 norm of the difference between the desired output image $\tilde{Z}$ and the actual binary output pattern Z_{b}, that is

$$E = \sum_{i=1}^{N^2} |\tilde{\underline{z}}_i - \underline{z}_{\text{b}i}|^2 = \sum_{i=1}^{N^2} |\tilde{\underline{z}}_i - \Lambda_i(|H\underline{m}_{\text{p}}|)|^2. \tag{5.28}$$

When the pattern error is reduced to a tolerable level, the steepest descent iteration is stopped.

5.3.2 Simulations

In this section, simulation results of the two-phase PSM optimization in coherent imaging systems are presented. Figure 5.7 shows the two-phase PSM optimization with a desired pattern of two vertical bars. From left to right, the top row shows the desired pattern, optimized real-valued PSM, and optimized two-phase PSM obtained using a threshold t_m. The bottom row indicates the corresponding binary output patterns. The amplitude of the initial mask is the same as that of the desired output pattern. The phase assignment of the initial mask is done *a priori* and phases in neighboring blocks are assigned alternately. Specifically, we assign $\underline{\theta}_i = 0$ for the transparent regions with 0 phase, $\underline{\theta}_i = \pi$ for the transparent regions with π phase, and $\underline{\theta}_i = \frac{\pi}{2}$ for the opaque regions. The parameters used in the simulation are $a = 90$, $t_r = 0.5$, $t_m = 0.5$, and 15×15 Gaussian low-pass filter with $k = 5$ and $s = 1$. Black, gray, and white represent -1, 0 and 1, respectively. If the desired pattern is used as the input mask, the output pattern error is 150. In addition, the two vertical bars cannot be distinguished. The optimized real-valued mask leads to an output pattern error of 12. After the post-processing step, the optimized two-phase PSM leads to a output pattern error of 10. The described two-phase PSM optimization algorithm effectively reduces the pattern errors. Figure 5.8 shows the convergence of the two-phase PSM optimization algorithm versus steepest descent iterations.

Figure 5.9 shows another simulation with a desired pattern of four horizontal bars. The parameters used in the simulation are $a = 80$, $t_r = 0.5$, $t_m = 0.5$, and 11×11 Gaussian low-pass filter with $k = 14$ and $s = 0.5$. Black, gray, and white represent -1, 0, and 1, respectively. The described algorithm effectively reduces the output

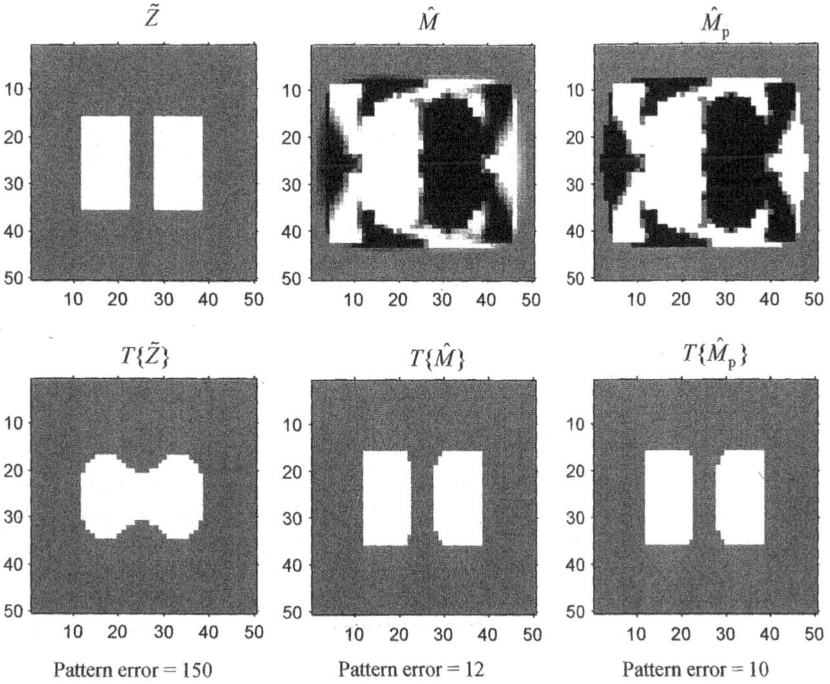

Figure 5.7 Two-phase PSM optimization with a desired pattern of vertical bars. *Top row* (input masks) (left to right): Desired pattern, optimized real-valued PSM, and optimized two-phase PSM obtained using a threshold t_m. *Bottom row:* Indicates the corresponding binary output patterns. Black, gray, and white represent −1, 0, and 1, respectively.

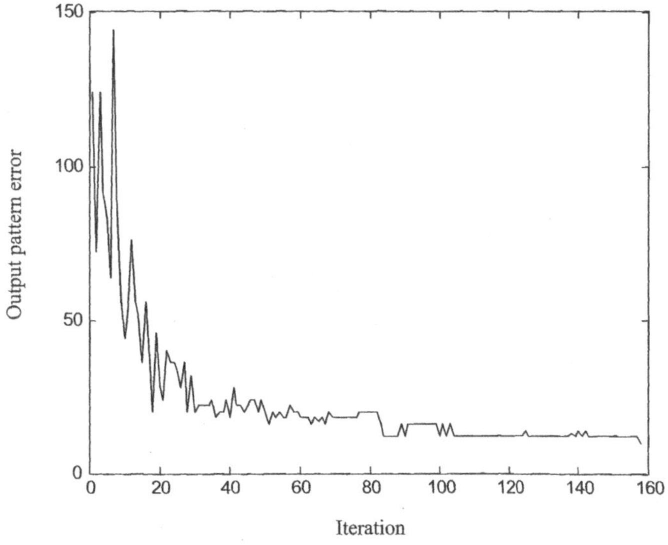

Figure 5.8 Convergence of the two-phase PSM optimization algorithm versus steepest descent iterations for Fig. 5.7.

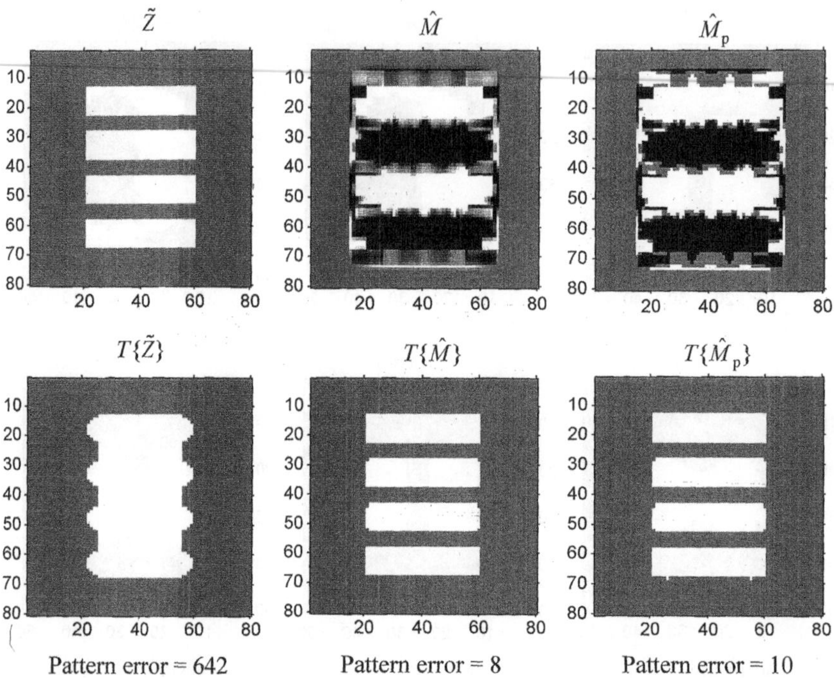

Figure 5.9 Two-phase PSM optimization with a desired pattern of horizontal bars. *Top row* (input masks) (left to right): Desired pattern, optimized real-valued PSM, and optimized two-phase PSM obtained using a threshold t_m. *Bottom row:* Indicates the corresponding binary output patterns. Black, gray, and white represent -1, 0, and 1, respectively.

pattern errors from 642 to 10. Figure 5.10 shows the successful convergence of the two-phase PSM optimization algorithm.

Comparing Figs. 5.3, 5.5, 5.7, and 5.9, it is noted that optimized binary masks even result in smaller output pattern errors than the optimized two-phase PSMs. However, the major advantage of the PSMs is the improvement in the contrast of the aerial images. Contrast is a measurement of image quality, which is defined as

$$C = \frac{I_{\max} - I_{\min}}{I_{\max} + I_{\min}} \times 100\% = \frac{E_{\max}^2 - E_{\min}^2}{E_{\max}^2 + E_{\min}^2} \times 100\%, \tag{5.29}$$

where $I_{\max}$ and $I_{\min}$ are the maximum and minimum intensities. $E_{\max}$ and $E_{\min}$ are the maximum and minimum amplitudes of the electric fields. For the simulations of two vertical bars, Fig. 5.11 shows the comparison of the electric field amplitudes generated in Figs. 5.3 and 5.7 in the 25th row. Solid and dashed lines show the electric field amplitudes generated by the optimized binary masks and two-phase PSMs, respectively. The contrast generated by the optimized binary mask is $C = 51.8\%$, while $C = 99.9\%$ for the optimized two-phase PSM. Similarly, for the simulations of four horizontal bars, Fig. 5.12 shows the comparison of the electric field amplitudes

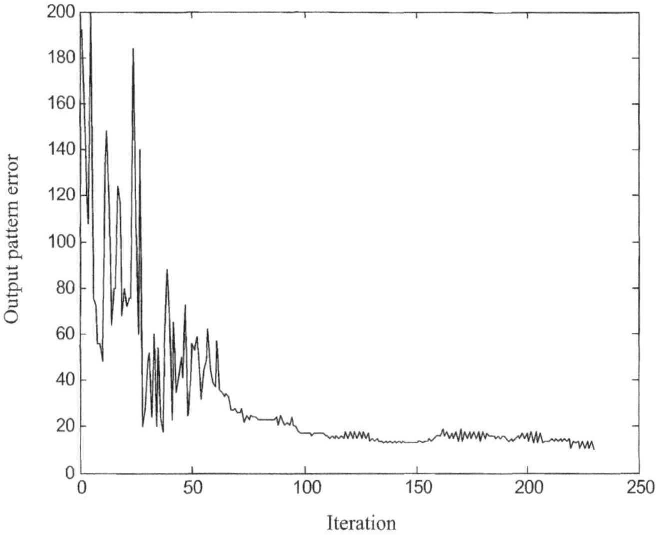

Figure 5.10 Convergence of the two-phase PSM optimization algorithm versus steepest descent iterations for Fig. 5.9.

generated in Figs. 5.5 and 5.9 in the 40th column. The contrast generated by the optimized binary mask is $C = 56.8\%$, while $C = 100\%$ for the optimized two-phase PSM. It is obvious that the optimized two-phase PSM effectively improves the contrast of the aerial image.

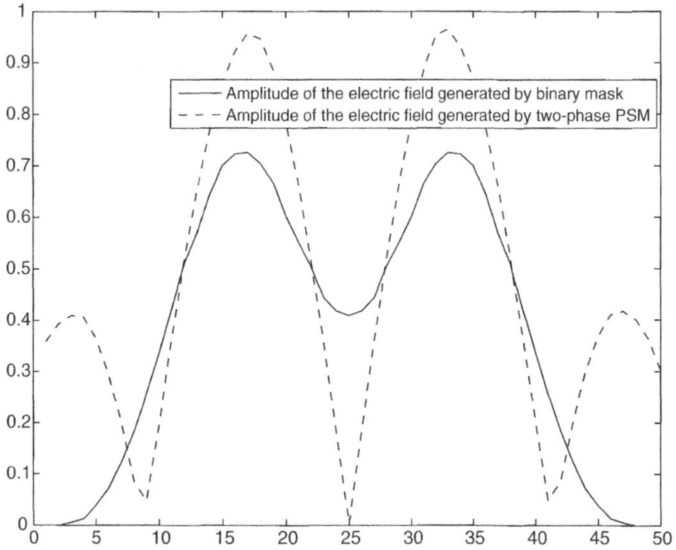

Figure 5.11 Comparison of the electric field amplitudes generated in Figs. 5.3 and 5.7 in the 25th row.

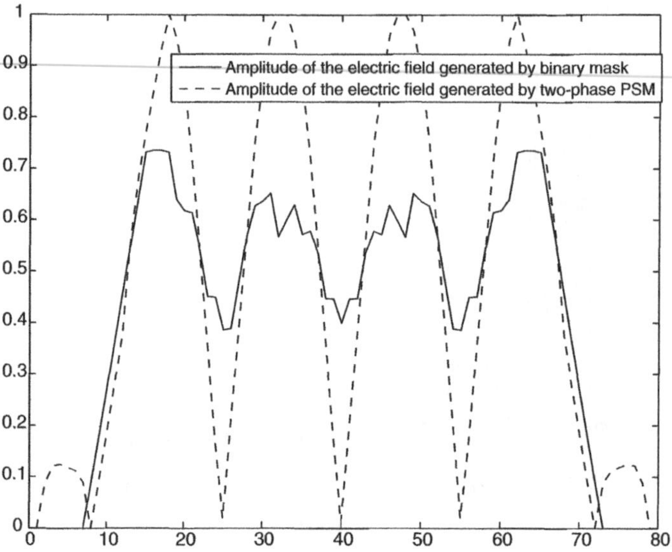

Figure 5.12 Comparison of the electric field amplitudes generated in Figs. 5.5 and 5.9 in the 40th column.

5.4 GENERALIZED PSM OPTIMIZATION

5.4.1 Generalized PSM Design Algorithm

While the two-phase PSM approach is very effective in some cases, the end result is that the search generally fails to generate adequate PSM for mask patterns having arbitrary Manhattan geometries and phase conflicts are likely to arise [40, 41]. A detailed description of Manhattan geometries is described in Appendix B. According to the "Four-Phase Theorem" described in Ref. [58], given an arbitrary pattern with a Manhattan geometry, a phase-shifting mask used to synthesize the image pattern must use a minimum of four distinct phase levels to avoid conflicts and ambiguities in the assignment of phases. The gradient-based two-phase PSM algorithm is not well suited for optimization of masks having more than two-phase levels (0 and π). This drawback motivates us to develop a generalized algorithm, admitting an arbitrary number of discrete phase levels, which overcomes this limitation.

The main goal of the generalized PSM is to obtain a generalized synthesis algorithm capable of generating arbitrary mask patterns. This is accomplished as follows: First, the iterative optimization framework is reformulated where the search trajectory is unconstrained in the complex plane. The optimization problem is formulated as

$$\hat{M}(x, y) = \arg \min_{M(x,y) \in C^{N \times N}} d\left(T\{M(x, y)\}, \tilde{Z}(x, y)\right), \tag{5.30}$$

where $C^{N \times N}$ is the $N \times N$ complex space. As expected, the resultant mask patterns obtained by Eq. (5.30) have arbitrary complex pixel values, and consequently a post-processing step is used to quantize the patterns into the desired four-phase level, shifting mask patterns. Denote the electrical field of the input mask as the complex-valued $N \times N$ matrix M and equivalently represent as $\underline{m} \in C^{N^2 \times 1}$. The pole-constrained mask M_p is defined as the quantization of M. The pixel magnitudes of M_p are quantized to 0 or 1. The pixel phases are quantized to several discrete phase levels. Its equivalent vector is denoted as $\underline{m}_\mathrm{p} \in C^{N^2 \times 1}$.

Let $\underline{r}$ and $\underline{\theta}$ be the magnitude and phase components of the complex-valued mask

$$\underline{m}_k = \underline{r}_k \mathrm{e}^{j\underline{\theta}_k}, \quad k = 1, \ldots, N^2, \tag{5.31}$$

where $j = \sqrt{-1}$, $\underline{\theta}_k \in (-\infty, \infty)$ and $\underline{r}_k \in [0, 1]$. The bound-constrained optimization is then reduced to an unconstrained optimization problem using the following parameter transformation:

$$\underline{r}_k = \frac{1 + \cos\underline{\phi}_k}{2}, \quad k = 1, \ldots, N^2, \tag{5.32}$$

where $\underline{\phi}_k \in (-\infty, \infty)$. Substituting $\underline{r}_k$ in Eq. (5.31) with Eq. (5.32), we have

$$\underline{m}_k = \frac{1 + \cos\underline{\phi}_k}{2} \mathrm{e}^{j\underline{\theta}_k}, \quad k = 1, \ldots, N^2. \tag{5.33}$$

Defining the vector $\underline{\theta} = [\theta_1, \ldots, \theta_{N^2}]^T$ and $\underline{\phi} = [\phi_1, \ldots, \phi_{N^2}]^T$, the optimization problem is formulated as

$$(\hat{\underline{\theta}}, \hat{\underline{\phi}}) = \arg \min_{(\underline{\theta}, \underline{\phi})} \{F(\underline{\theta}, \underline{\phi})\}, \tag{5.34}$$

where the cost function is

$$F(\underline{\theta}, \underline{\phi}) = \sum_{i=1}^{N^2} \left(\tilde{\underline{z}}_i - \frac{1}{1 + \exp\left[-a \sqrt{\left(\sum_{k=1}^{N^2} h_{ik} \frac{1 + \cos\underline{\phi}_k}{2} \mathrm{e}^{j\underline{\theta}_k} \right)^2} + at_r \right]} \right)^2. \tag{5.35}$$

The steepest descent method is used to optimize the above problem. The gradients $\nabla F(\underline{\theta}, \underline{\phi})_{\underline{\theta}}$ and $\nabla F(\underline{\theta}, \underline{\phi})_{\underline{\phi}}$ derived in Appendix A can be calculated as follows:

$$\begin{aligned} \nabla F(\underline{\theta}, \underline{\phi})_{\underline{\theta}} = \underline{d}_\theta = 2a \times \frac{\underline{1} + \cos\underline{\phi}}{2} \odot \sin\underline{\theta} \odot \{H^T [(\tilde{\underline{z}} - \underline{z}) \odot \underline{z} \odot (\underline{1} - \underline{z}) \\ \odot H(\underline{m}_\mathrm{R}) \odot T(\underline{m})]\} - 2a \times \frac{\underline{1} + \cos\underline{\phi}}{2} \odot \cos\underline{\theta} \\ \odot \{H^T [(\tilde{\underline{z}} - \underline{z}) \odot \underline{z} \odot (\underline{1} - \underline{z}) \odot H(\underline{m}_\mathrm{I}) \odot T(\underline{m})]\}, \end{aligned} \tag{5.36}$$

$$\nabla F(\underline{\theta}, \underline{\phi})_{\underline{\phi}} = \underline{d}_{\underline{\phi}} = a \times \sin\underline{\phi} \odot \cos\underline{\theta} \odot \{H^T[(\underline{\tilde{z}} - \underline{z}) \odot \underline{z} \odot (\underline{1} - \underline{z}) \odot H(\underline{m}_{\mathrm{R}}) \odot T(\underline{m})]\} + a \times \sin\underline{\phi} \odot \sin\underline{\theta} \odot \{H^T[(\underline{\tilde{z}} - \underline{z}) \odot \underline{z} \odot (\underline{1} - \underline{z}) \odot H(\underline{m}_{\mathrm{I}}) \odot T(\underline{m})]\}, \tag{5.37}$$

where $\nabla F(\underline{\theta}, \underline{\phi})_{\underline{\theta}}$, $F(\underline{\theta}, \underline{\phi})_{\underline{\phi}} \in \Re^{N^2 \times 1}$, $\odot$ is the element-by-element multiplication operator, and $T(\underline{m}) = [(H\underline{m}_{\mathrm{R}})^2 + (H\underline{m}_{\mathrm{I}})^2]^{-\frac{1}{2}}$. $\underline{m}_{\mathrm{R}}$ and $\underline{m}_{\mathrm{I}}$ are the real part and the imaginary part of $\underline{m}$. $\underline{1} = [1, \ldots, 1]^T \in \Re^{N^2 \times 1}$. Assuming $\underline{\theta}^k$ and $\underline{\phi}^k$ are the kth iteration results, at the $k+1$th iteration,

$$\underline{\theta}^{k+1} = \underline{\theta}^k - s_{\underline{\theta}} \underline{d}_{\underline{\theta}}^k, \tag{5.38}$$

$$\underline{\phi}^{k+1} = \underline{\phi}^k - s_{\underline{\phi}} \underline{d}_{\underline{\phi}}^k, \tag{5.39}$$

where $s_{\underline{\theta}}$ and $s_{\underline{\phi}}$ are the step sizes.

The iterative optimization above, in general, leads to complex-valued solutions that are not constrained to a discrete number of magnitudes and phase levels. Therefore, a post-processing step is needed to obtain the pole-level optimized mask $\underline{\hat{m}}_{\mathrm{p}}$. Since $\underline{\hat{m}}$ is complex valued, a two-step quantization process is followed. First, the magnitudes are quantized by a global threshold t_m as

$$|\hat{\underline{m}}_{\mathrm{p}k}| = U(|\hat{\underline{m}}_k| - t_m) = \begin{cases} 0 & : \quad |\hat{\underline{m}}_k| \le t_m, \\ 1 & : \quad |\hat{\underline{m}}_k| > t_m, \end{cases} \qquad k = 1, \ldots, N^2. \tag{5.40}$$

The phases are subsequently quantized to the nearest prescribed discrete phase level. In the following simulations of four-phase PSM optimization, the quantization of the phases is formulated as

$$\arg\{\hat{\underline{m}}_{\mathrm{p}k}\} = \begin{cases} \frac{\pi}{4} & : \quad 0 \le \arg\{\hat{\underline{m}}_k\} < \frac{\pi}{2}, \\ \frac{3\pi}{4} & : \quad \frac{\pi}{2} \le \arg\{\hat{\underline{m}}_k\} < \pi, \\ \frac{5\pi}{4} & : \quad \pi \le \arg\{\hat{\underline{m}}_k\} < \frac{3\pi}{2}, \\ \frac{7\pi}{4} & : \quad \frac{3\pi}{2} \le \arg\{\hat{\underline{m}}_k\} < 2\pi, \end{cases} \qquad k = 1, \ldots, N^2. \tag{5.41}$$

If two phase levels are used in the generalized PSM optimization, the quantization of the phases is formulated as

$$\arg\{\hat{\underline{m}}_{\mathrm{p}k}\} = \begin{cases} 0 & : \quad 0 \le \arg\{\hat{\underline{m}}_k\} < \frac{\pi}{2} \ \text{ or } \ \frac{3\pi}{2} \le \arg\{\hat{\underline{m}}_k\} < 2\pi, \\ \pi & : \quad \frac{\pi}{2} \le \arg\{\hat{\underline{m}}_k\} < \frac{3\pi}{2}, \end{cases} \tag{5.42}$$

where $k = 1, \ldots, N^2$. The pattern error E is defined as the square of the l^2 norm of the difference between the desired output image $\tilde{Z}$ and the actual binary output pattern Z_{b}, that is,

$$E = \sum_{i=1}^{N^2} |\underline{\tilde{z}}_i - \underline{z}_{\mathrm{b}i}|^2 = \sum_{i=1}^{N^2} |\underline{\tilde{z}}_i - \Lambda_i(|H\underline{m}_{\mathrm{p}}|)|^2. \tag{5.43}$$

When the pattern error is reduced to a tolerable level, the steepest descent iteration is stopped.

5.4.2 Simulations

To demonstrate the effect of the number of phases used in the generalized PSM design, consider the desired pattern shown in Fig. 5.13. The top horizontal block cross-connects two vertical parallel blocks. The phases assigned to the two parallel adjacent blocks cannot be the same, so as to exploit the PSM principle. Since the horizontal top block connects the parallel blocks, and since it would be desirable that there are no gaps introduced in the image, the horizontal block must be assigned an intermediate phase value that is distinct from the phases of the two parallel blocks.

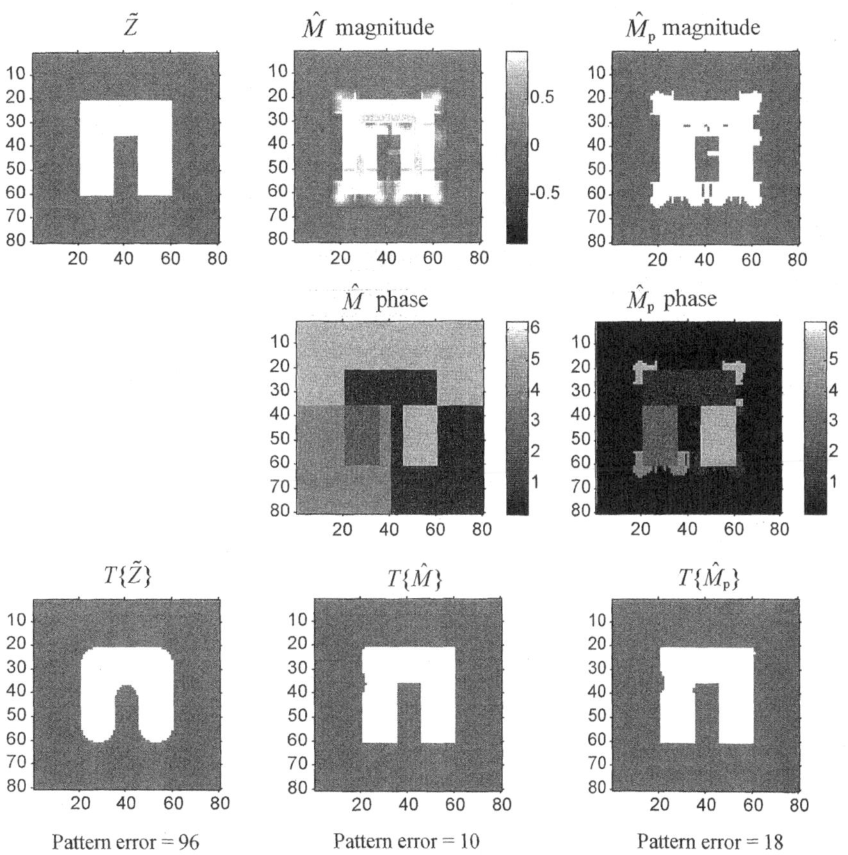

Figure 5.13 Generalized PSM optimization with a desired pattern of U-junction, where four phase levels are used. *Top row* (input masks) (left to right): Desired pattern, magnitude of the optimized complex-valued mask, and optimized pole-level mask obtained using a threshold t_m. *Middle row:* Indicates the phases of the optimized complex-valued mask. *Bottom row:* Indicates the corresponding binary output patterns. Gray, white and black represent 0, 1, and -1, respectively, in the top and bottom rows. Dark black, light black, dark gray, and light gray represent $\frac{\pi}{4}$, $\frac{3\pi}{4}$, $\frac{5\pi}{4}$, and $\frac{7\pi}{4}$, respectively, in the middle row.

Therefore, two-phase levels are not sufficient to attain this goal. The conflict can be eliminated with the use of four phases. In Fig. 5.13, the patterns in the top row and the middle row illustrate the input patterns. The output patterns are shown in the bottom row. The three images in the bottom row show the output patterns corresponding to the inputs of (left) the desired pattern ($\tilde{Z}$), (center) the complex level optimized mask ($\hat{M}$), and (right) the pole-level optimized mask ($\hat{M}_{\mathrm{p}}$). The aerial image formation process is approximated by a 11×11 Gaussian low-pass filter with $k = 14$, and in the sigmoid function, we assign parameters $a = 80$ and $t_r = 0.5$. The global threshold is $t_m = 0.5$. The step sizes are $s_{\underline{\phi}} = 2$ and $s_{\underline{\theta}} = 0.01$. The shape of the image used to initialize the iterative algorithm is the same as that of the desired binary output pattern $\tilde{Z}$. For $\underline{\phi}_k$, we assign the phase of $\frac{\pi}{5}$ corresponding to the areas having a magnitude of 1 and a phase of $\frac{4\pi}{5}$ for the areas having a magnitude of 0. The phase assignment must be done *a priori* and phases in neighboring blocks are assigned alternately. Because of the numerous *sine* and *cosine* functions in Eqs. (5.36) and (5.37), we intentionally avoid assigning $\underline{\theta}_k$ and $\underline{\phi}_k$ the values of $0, \frac{\pi}{2}, \pi$, or $\frac{3\pi}{2}$. Otherwise, Eq. (5.36) or (5.37) may reduce to a zero update, terminating the iteration. For a four-phase level mask design, empirical observations show that an efficient assignment of phase values to $\underline{\theta}_k$ is to select phases in the set of $\frac{\pi}{4}, \frac{3\pi}{4}, \frac{5\pi}{4}$, and $\frac{7\pi}{4}$. Further, regions around a block should be assigned a phase value that is $\frac{\pi}{2}$ different from that of the block and is on the same side of the imaginary axis. In Fig. 5.13, gray, white, and black represent 0, 1 and -1, respectively, in the top and bottom rows. Dark black, light black, dark gray, and light gray represent $\frac{\pi}{4}, \frac{3\pi}{4}, \frac{5\pi}{4}$, and $\frac{7\pi}{4}$, respectively, in the middle row. If the desired output pattern is used as the mask, the output pattern error is 96. The optimized four-phase PSM reduces the output pattern error to 18 and obtains a better fidelity of the output pattern. Figure 5.14 shows the successful convergence of the generalized

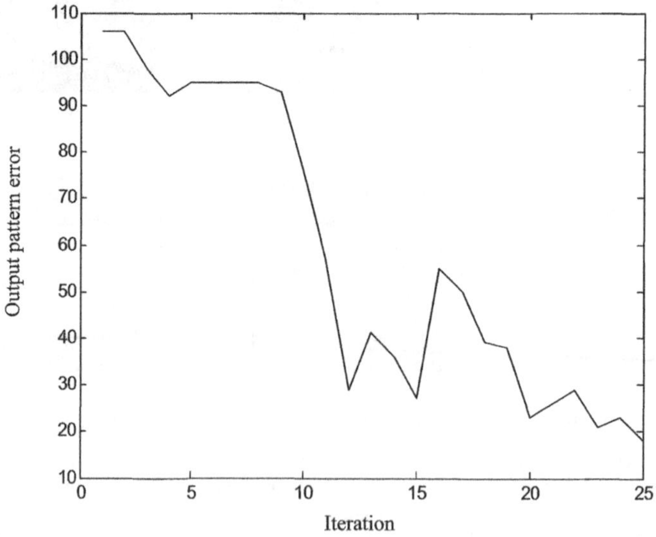

Figure 5.14 Convergence of the generalized PSM optimization algorithm versus steepest descent iterations for Fig. 5.13.

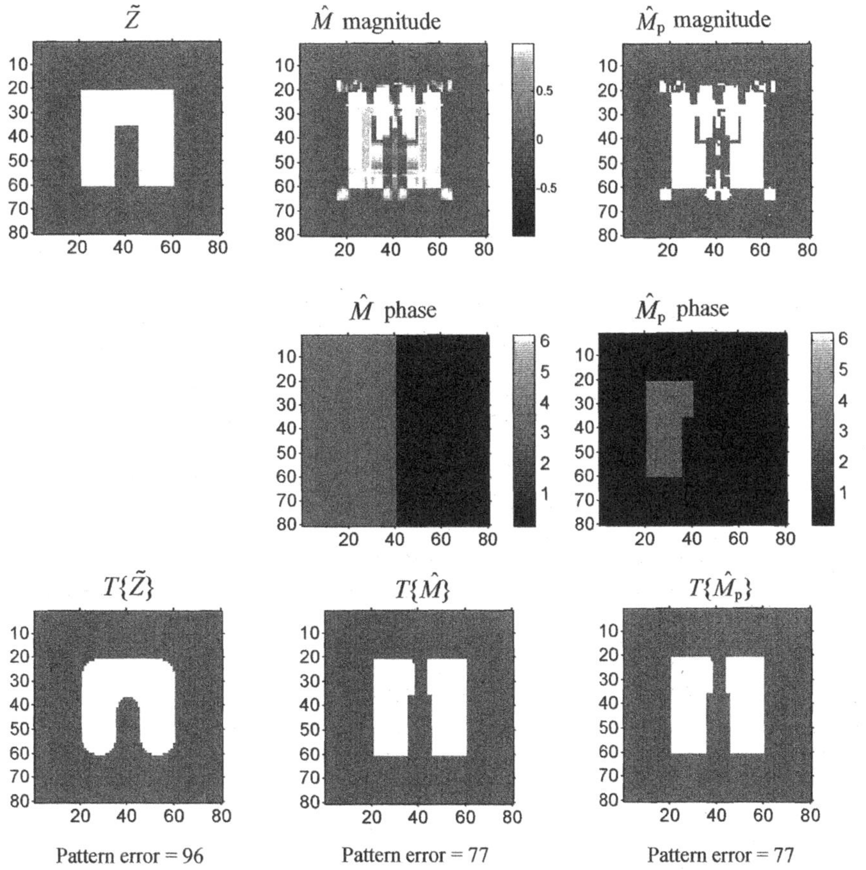

Figure 5.15 Generalized PSM optimization with a desired pattern of U-junction, where two phase levels are used. *Top row* (input masks) (left to right): Desired pattern, magnitude of the optimized complex-valued mask, and optimized two-phase mask obtained using a threshold t_m. *Middle row:* Indicates the phases of the optimized complex-valued mask. *Bottom row:* Indicates the corresponding binary output patterns. Gray, white, and black represent 0, 1, and -1, respectively, in the top and bottom rows. Gray and black represent π and 0, respectively, in the middle row.

PSM optimization algorithm versus steepest descent iterations for Fig. 5.13. This generalized PSM approach has proved efficient in our extensive simulation analysis. As a comparison to the four-phase PSM design, the experiment using just two phases is illustrated in Fig. 5.15, where all the parameters are the same as those in Fig. 5.13. In Fig. 5.15, gray, white, and black represent 0, 1 and -1, respectively, in the top and bottom rows. Gray and black represent π and 0, respectively, in the middle row. Note that a gap appears on the top connection of the output pattern, as expected.

It should be noted that for some special patterns, the four-phase levels are not necessary to avoid the phase assignment conflict. For instance, the parallel bar pattern in Fig. 5.16 can be attained by a two-phase mask. As shown in Fig. 5.16, the generalized

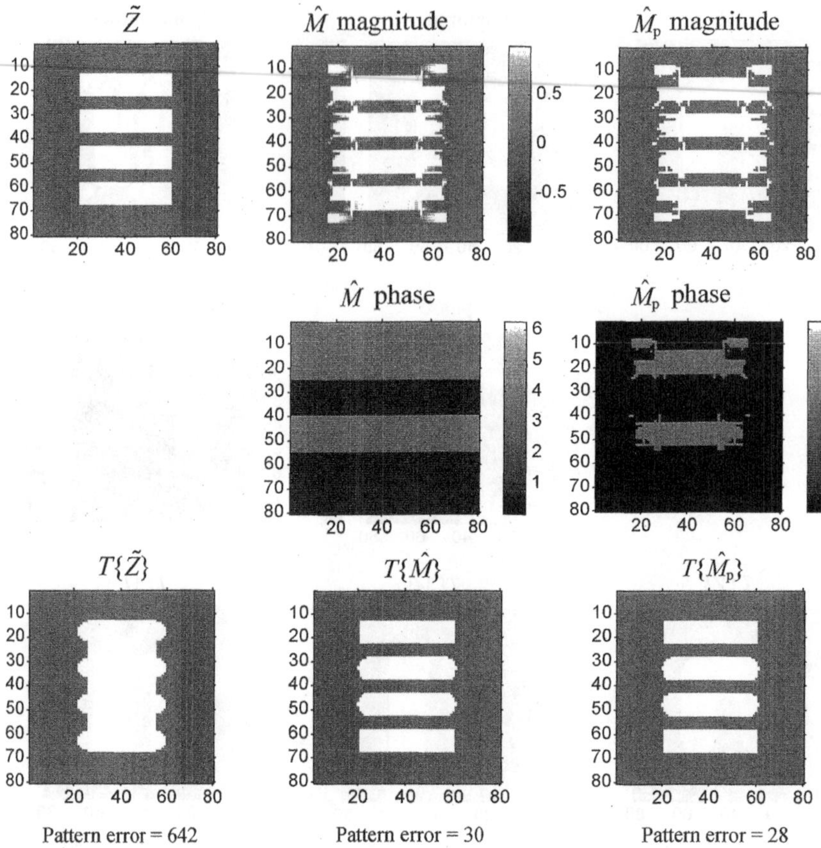

Figure 5.16 Generalized PSM optimization with a desired pattern of four horizontal bars, where two phase levels are used. *Top row* (input masks) (left to right): Desired pattern, magnitude of the optimized complex-valued mask, and optimized two-phase mask obtained using a threshold t_m. *Middle row:* Indicates the phases of the optimized complex-valued mask. *Bottom row:* Indicates the corresponding binary output patterns. Gray, white, and black represent 0, 1, and -1, respectively, in the top and bottom rows. Gray and black represent π and 0, respectively, in the middle row.

PSM algorithm is also capable of designing two-phase masks. In Fig. 5.16, the shape of the image used to initialize the iterative algorithm is the same as the desired binary output pattern $\tilde{Z}$. We assign $\underline{\phi}_k = \frac{\pi}{5}$ corresponding to the areas having a magnitude of 1 and $\underline{\phi}_k = \frac{4\pi}{5}$ for the areas having magnitude of 0. For the phase assignment, $\underline{\theta}_k = \frac{6\pi}{5}$ for the first and the third bars and their surrounding areas. $\underline{\theta}_k = \frac{\pi}{5}$ for the second and the fourth bars and their surrounding areas. In Fig. 5.16, gray, white, and black represent 0, 1, and -1, respectively, in the top and bottom rows. Gray and black represent π and 0, respectively, in the middle row. Other parameters used in Fig. 5.16 are the same as those in Fig. 5.13. If the desired output pattern is used as the mask, the output pattern error is 642, and the four bars cannot be distinguished. The optimized pole-level PSM reduces the output pattern error to 28 and obtains a much

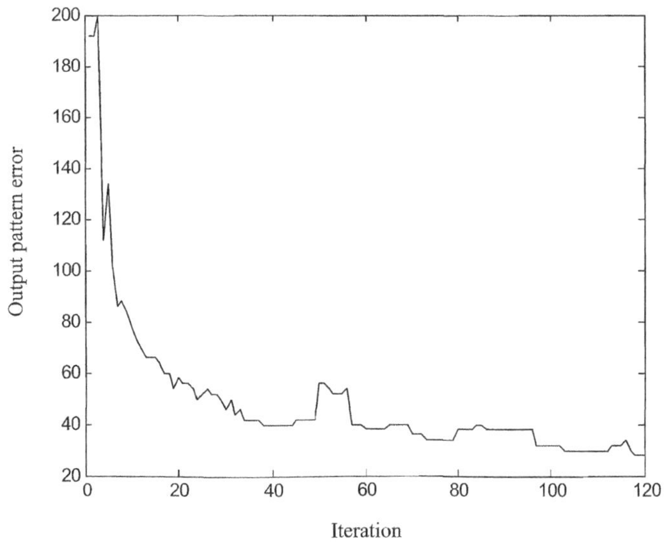

Figure 5.17 Convergence of the generalized PSM optimization algorithm versus steepest descent iterations for Fig. 5.16.

better fidelity of the output pattern. Figure 5.17 shows the successful convergence of the generalized PSM optimization algorithm versus steepest descent iterations for Fig. 5.16. To summarize, the generalized PSM optimization algorithm is effective to generate arbitrary number of phase levels, thus, eliminating the conflict of phase assignment.

5.5 RESIST MODELING EFFECTS

In Section 5.1, the photoresist effect is modeled by the hard threshold function. In the optimization algorithms, the hard threshold function is replaced by the sigmoid function to make the cost function differentiable. However, as described in Section 1.3, the normalized remaining thickness of the photoresist is a nonlinear function of the exposure dose. It is observed from Fig. 1.7 that the actual behavior of the photoresist development is more accurately modeled by the sigmoid function defined in Eq. (5.5).

Assume that photoresist development behavior obeys a sigmoid function $\widetilde{\mathrm{sig}}(x)$ with parameters $\widetilde{a} = 10$ and $\widetilde{t_r} = 0.5$. In the optimization algorithms, the photoresist effect is approximated by another sigmoid function $\mathrm{sig}(x)$ with parameters a and t_r. The process to select the parameters a and t_r is referred to as resist modeling. Thus, one task of the lithographers is to select the proper parameter values a and t_r to find the best approximation of the photoresist behavior, which they use in the optical lithography processes. If the selected parameters a and t_r are equal or close to $\widetilde{a} = 10$

and $\widetilde{t_r} = 0.5$, then the resist model sig(x) will give an adequate representation of the photoresist behavior $\widetilde{\mathrm{sig}}(x)$. Thus, the optimized masks will lead to smaller output pattern errors. In contrast, if the parameters a and t_r are poorly selected and are far from $\widetilde{a} = 10$ and $\widetilde{t_r} = 0.5$, then the optimized masks will lead to larger output pattern errors.

The resist modeling effects of the parameter a on the OPC optimization in coherent imaging system are illustrated in Fig. 5.18. From left to right, the top row shows the optimized binary mask and corresponding output pattern obtained using parameters

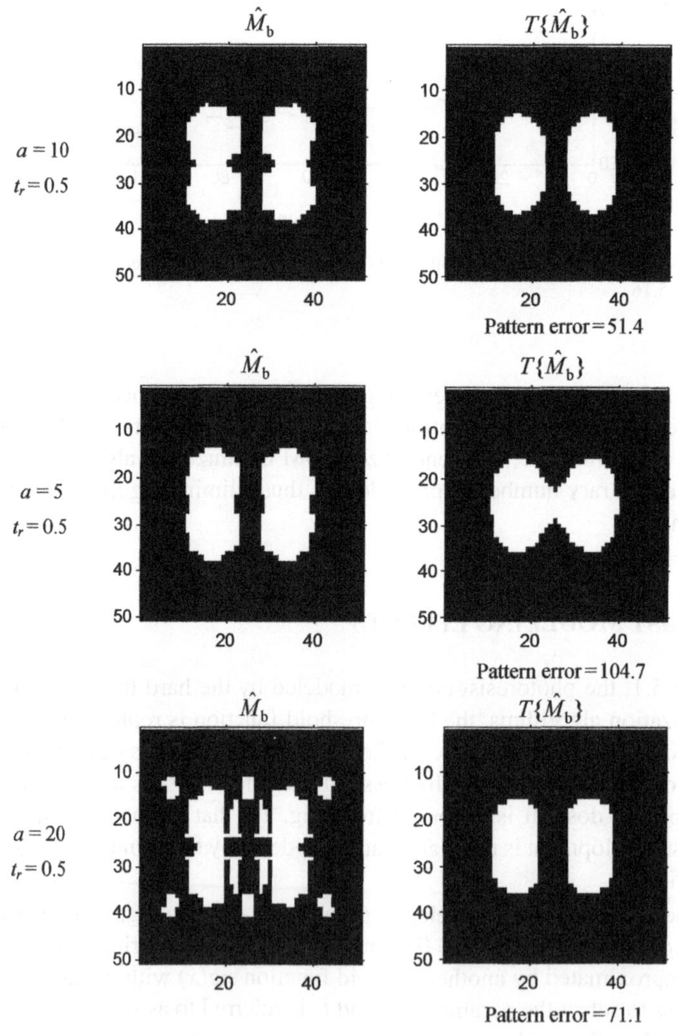

Figure 5.18 The resist modeling effects of parameter a on the OPC optimization in coherent imaging system. Left column depicts the optimized masks using the various selected resist parameters. The right column depicts the corresponding output patterns.

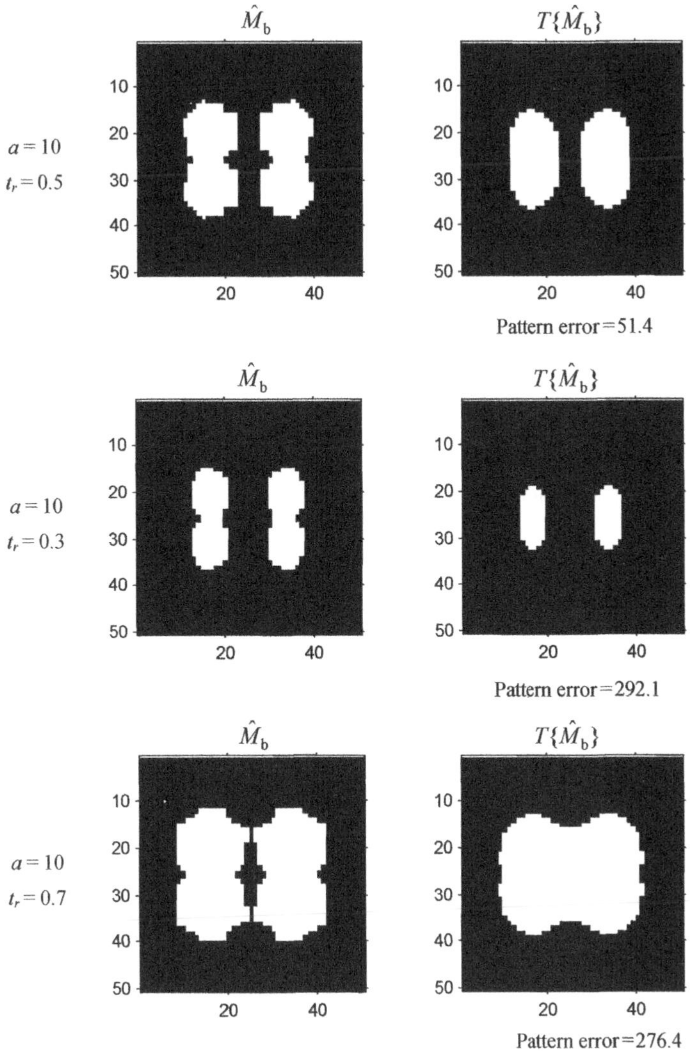

Figure 5.19 The resist modeling effects of parameter t_r on the OPC optimization in coherent imaging system. Left column depicts the optimized masks using the various selected resist parameters. The right column depicts the corresponding output patterns.

$a = \widetilde{a} = 10$ and $t_r = \widetilde{t_r} = 0.5$. Thus, in the top row, the selected parameters are consistent with the actual photoresist behavior. The output pattern error of the optimized binary mask is 51.4. In the middle row, the parameter t_r is selected equal to $\widetilde{t_r} = 0.5$. However, the parameter a is poorly selected as $a = 5 < \widetilde{a}$. The optimized mask using the erroneous parameter is shown in the left middle row image. The output pattern error is increased to 104.7. In the bottom row, $t_r = \widetilde{t_r} = 0.5$. However, $a = 20 > \widetilde{a}$. The output pattern error is increased to 71.1.

The resist modeling effects of the parameter t_r on the OPC optimization in coherent imaging system are illustrated in Fig. 5.19. Again, from left to right, the top row shows the optimized binary mask and corresponding output pattern with parameters $a = \widetilde{a} = 10$ and $t_r = \widetilde{t_r} = 0.5$, which are consistent with the actual photoresist behavior. The output pattern error is 51.4. In the middle row, the parameter a is selected equal to $\widetilde{a} = 10$. However, the parameter t_r is poorly selected as $t_r = 0.3 < \widetilde{t_r}$. The output pattern error is increased to 292.1. In the bottom row, $a = \widetilde{a} = 10$. However, $t_r = 0.7 > \widetilde{t_r}$. The output pattern error is increased to 276.4.

The above simulation results explain the importance and the necessity to properly choose the parameters of the photoresist model according to the specific photoresist used in practice. In a practical scenario, the curves representing the actual photoresist behavior may be obtained from physical experiments. According to these curves, a variety of differentiable functions may be used to approximate the photoresist characteristics. If functions other than the sigmoid function are chosen to model the resist behavior, the derivations developed in this book would need to be modified according to the selected resist function approximations.

5.6 SUMMARY

This chapter first reviewed the prior works on OPC and two-phase PSM optimization algorithms based on coherent imaging system. Then, this chapter focused on the development of the generalized PSM optimization approaches for inverse lithography. This generalized approach may solve the PSM optimization problem with arbitrary Manhattan geometries. Finally, this chapter discussed the resist modeling effects on the optimization of the mask patterns.

6

Regularization Framework

Inverse lithography is an ill-posed problem where numerous input patterns can lead to the same binary output pattern. Regularization in ILT seeks to bias the solution space to sample solutions that have some favorable properties [40, 41, 68]. According to Chapter 5, the described OPC and PSM optimization algorithms result in the optimized mask with continuous amplitude and phase, referred to as the continuous mask, which is not physically realizable. To overcome this limitation, the post-processing steps are used to quantize the amplitude and phase of the optimized mask to several discrete levels, resulting in the pole-level mask. However, these post-processing steps are suboptimal with no guarantee that the pattern error is under the goal [42, 43, 68]. To reduce the error increase contributed by the post-processing step, it is desired to obtain an optimized continuous mask, which is close to the optimized pole-level mask. Furthermore, the OPC and PSM optimization algorithms in Chapter 5 lead to optimized mask patterns containing numerous details, which may bring difficulty to mask fabrication. Most of the details consist of singular transmission and opaque pixels. To control the manufacturing cost, we would like to reduce the complexity of the optimized masks.

One approach to obtain the optimized solution with prescribed properties is through regularization during the optimization process [83]. Regularization framework is formulated as follows:

$$\underline{\hat{m}} = \arg\min_{\underline{\hat{m}}}\{F(\underline{m}) + \gamma R(\underline{m})\}, \tag{6.1}$$

where $F(\underline{m})$ is the cost function described in Chapter 5, referred to as the data fidelity term. $R(\underline{m})$ is the regularization term, which is used to reduce the solution space and constrain the optimized results. γ is the user-defined parameter to reveal the weight of the regularization. The regularization framework was first employed in the context of lithography by Peckerar and Marrian [59, 60] to solve the proximity effect problem arising in e-beam lithography. In Section 6.1, the discretization penalties are discussed, which are used to direct the optimized continuous mask toward the optimized pole-level mask. In Section 6.2, several complexity penalties are utilized to control the manufacturing cost of the optimized masks. All the regularization frameworks are tailored for both OPC and PSM optimizations.

6.1 DISCRETIZATION PENALTY

6.1.1 Discretization Penalty for OPC Optimization

From Figs. 5.3 and 5.5, it is noted that the post-processing step may increase the output pattern errors of the optimized binary masks. For the OPC optimization, the discretization penalty is exploited to attain near-binary gray optimized mask, whose pixel values are close to 0 or 1 [68]. The discretization penalty term for OPC optimization is

$$R_D(\underline{m}) = \sum_{k=1}^{N^2}[1 - (2\underline{m}_k - 1)^2] = 4\underline{m}^T(\underline{1} - \underline{m}), \tag{6.2}$$

where $\underline{1} = [1, \ldots, 1]^T \in \Re^{N^2 \times 1}$ Thus, for each pixel $\underline{m}_k$,

$$r_D(\underline{m}_k) = 1 - (2\underline{m}_k - 1)^2. \tag{6.3}$$

The curve of Eq. (6.3) is shown in Fig. 6.1. The incurred penalty is zero for pixel value 0 or 1, and is increased as the pixel value moves away from these two limits. The penalty is maximum for the pixel value 0.5. Thus, this discretization penalty term directs the optimized gray mask toward a binary one. To apply the steepest descent algorithm, the gradient of the discretization penalty term is calculated as

$$\nabla R_D(\underline{m}) = -8\underline{m}_k + 4. \tag{6.4}$$

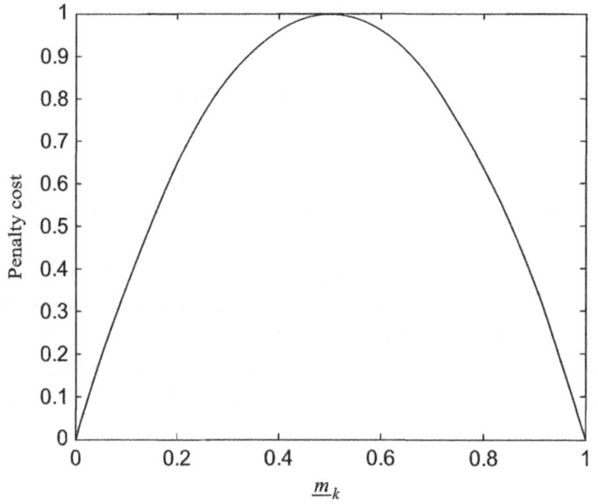

Figure 6.1 Discretization penalty cost function for OPC optimization.

According to Eq. (6.1), the cost function is adjusted as

$$J(\underline{m}) = F(\underline{m}) + \gamma_D R_D(\underline{m}). \tag{6.5}$$

Thus, the gradient of the overall cost function is adjusted as

$$\nabla J(\underline{m}) = \nabla F(\underline{m}) + \gamma_D \nabla R_D(\underline{m}). \tag{6.6}$$

Using the discretization penalty term in Eq. (6.2), the simulation of Fig. 5.3 can be repeated. The result is illustrated in Fig. 6.2. The parameters used in Fig. 6.2 are the same as those in Fig. 5.3 and $\gamma_D = 0.025$. Compared with Fig. 5.3, the optimized continuous mask in Fig. 6.2 is more close to the optimized binary mask. As expected, the discretization penalty effectively reduces the error accumulation from the post-processing step and leads to a zero output pattern error. Similarly, the simulation in Fig. 5.5 is repeated in Fig. 6.3. The parameters used in Fig. 6.3 are the same as those in Fig. 5.5 and $\gamma_D = 0.01$. The output pattern error is also reduced from 4 to 0.

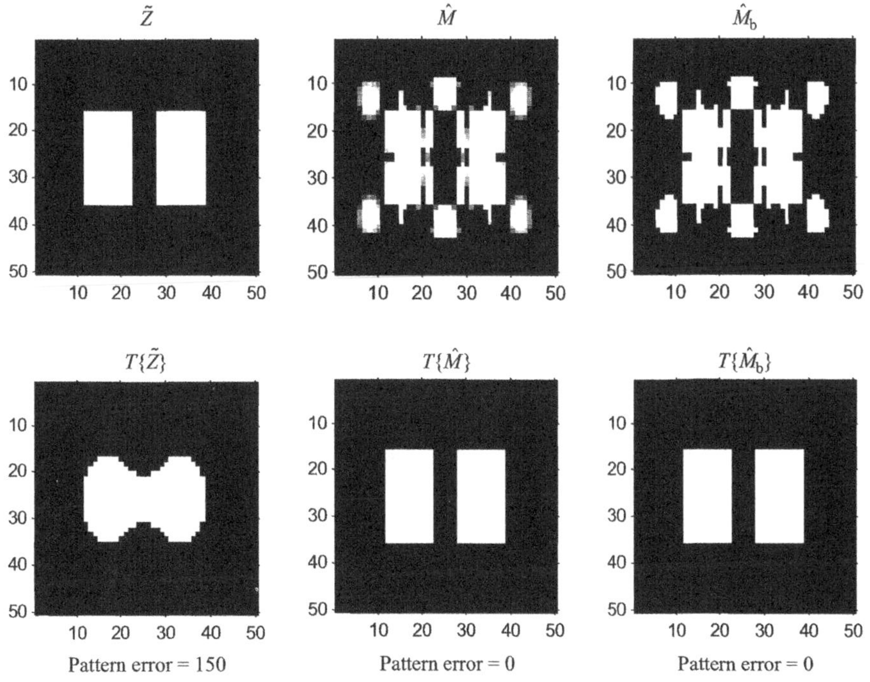

Figure 6.2 OPC optimization using discretization penalty with a desired pattern of vertical bars. *Top row* (input masks) (left to right): Desired pattern, optimized real-valued mask, and optimized binary mask obtained using a threshold t_m. *Bottom row:* Indicates the corresponding binary output patterns. $\gamma_D = 0.025$. Black and white represent 0 and 1, respectively.

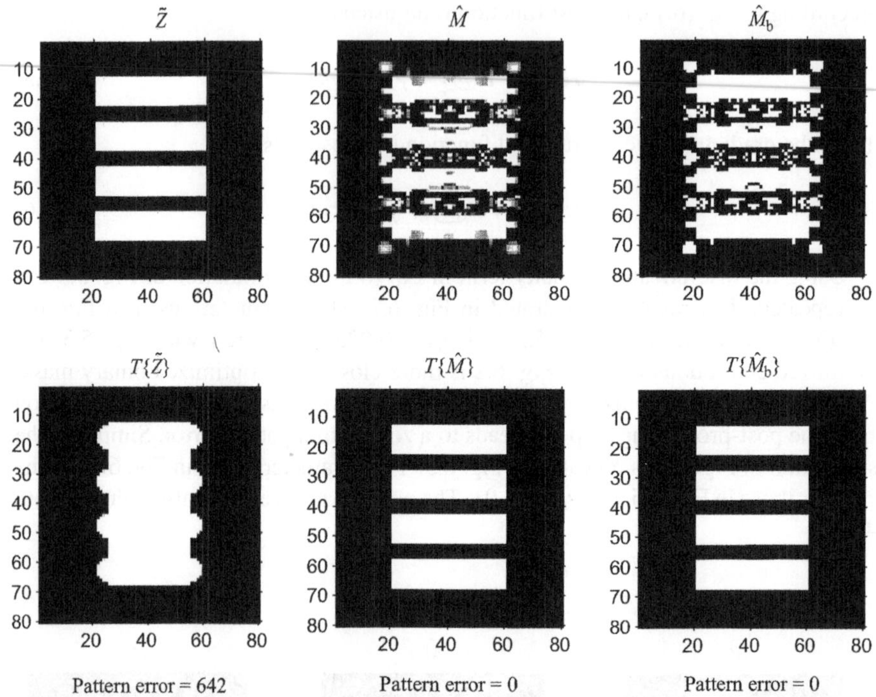

Figure 6.3 OPC optimization using discretization penalty with a desired pattern of four horizontal bars. *Top row* (input masks) (left to right): Desired pattern, optimized real-valued mask, and optimized binary mask obtained using a threshold t_m. *Bottom row:* Indicates the corresponding binary output patterns. $\gamma_D = 0.01$. Black and white represent 0 and 1, respectively.

6.1.2 Discretization Penalty for Two-Phase PSM Optimization

In Ref. [67], Poonawala and Milanfar introduced a discretization penalty for the alternating PSM optimization, where the pixel values were constrained around -1, 0, or 1. In this section, we present another discretization penalty term for the two-phase PSM optimization. The goal is not only to constrain the pixel value toward discrete levels but also to reduce the complexity of the mask pattern. In Figs. 5.7 and 5.9, the optimized two-phase PSMs contain numerous details composed of singular transmission or opaque pixels. In addition, since the transmission regions are too close to each other, the opaque regions between them become discontinuous. To remove these singular pixels, different transmission regions should be separated at a distance. A discretization penalty to achieve this goal is

$$R_D(\underline{m}) = \sum_{k=1}^{N^2} \underline{m}_k^2 = \underline{m}^T \underline{m}. \tag{6.7}$$

For each pixel $\underline{m}_k$,

$$r_D(\underline{m}_k) = \underline{m}_k^2. \tag{6.8}$$

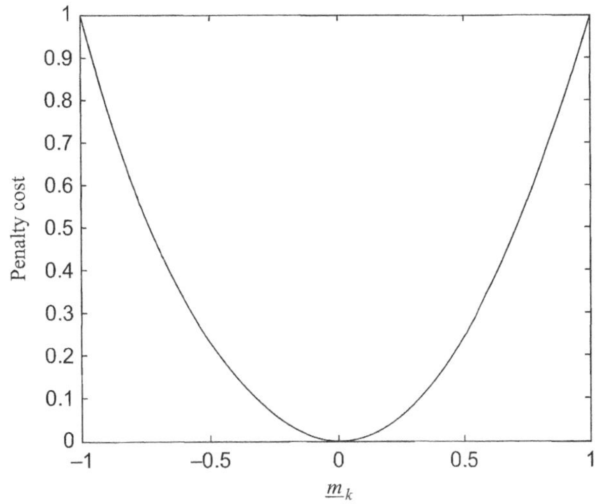

Figure 6.4 Discretization penalty cost function for the two-phase PSM optimization.

The curve of Eq. (6.8) is shown in Fig. 6.4. The zero penalty is assigned to the opaque pixels and maximum penalty is assigned to the transmission pixels. The gradient of the discretization penalty term is calculated as

$$\nabla R_D(\underline{m}) = 2\underline{m}. \tag{6.9}$$

Using the described discretization penalty in Eq. (6.7), the simulations of Figs. 5.7 and 5.9 are repeated in Figs. 6.5 and 6.6. The parameters used in Figs. 6.5 and 6.6 are the same as those in Figs. 5.7 and 5.9, respectively. In addition, $\gamma_D = 0.0175$ in Fig. 6.5 and $\gamma_D = 0.0025$ in Fig. 6.6. It is noted that the discretization penalty successfully separates different transmission regions and remove some singular pixels, thus reducing the complexity of the optimized mask patterns. However, the discretization penalty term introduces a small error increase.

6.1.3 Discretization Penalty for Generalized PSM Optimization

The discretization penalty for the generalized PSM optimization is used to reduce the complexity of the mask patterns, where both magnitude and phase of the optimized mask are constrained around several prescribed discrete levels. Thus, the penalty term is divided into amplitude discretization penalty and phase discretization penalty. According to Eq. (6.1), the overall cost function is adjusted as

$$J(\underline{m}) = F(\underline{m}) + \gamma_{\mathrm{A}} R_{\mathrm{A}}(\underline{\phi}) + \gamma_{\mathrm{P}} R_{\mathrm{P}}(\underline{\theta}), \tag{6.10}$$

where $R_{\mathrm{A}}(\underline{\phi})$ is the amplitude discretization penalty term and $R_{\mathrm{P}}(\underline{\theta})$ is the phase discretization penalty term.

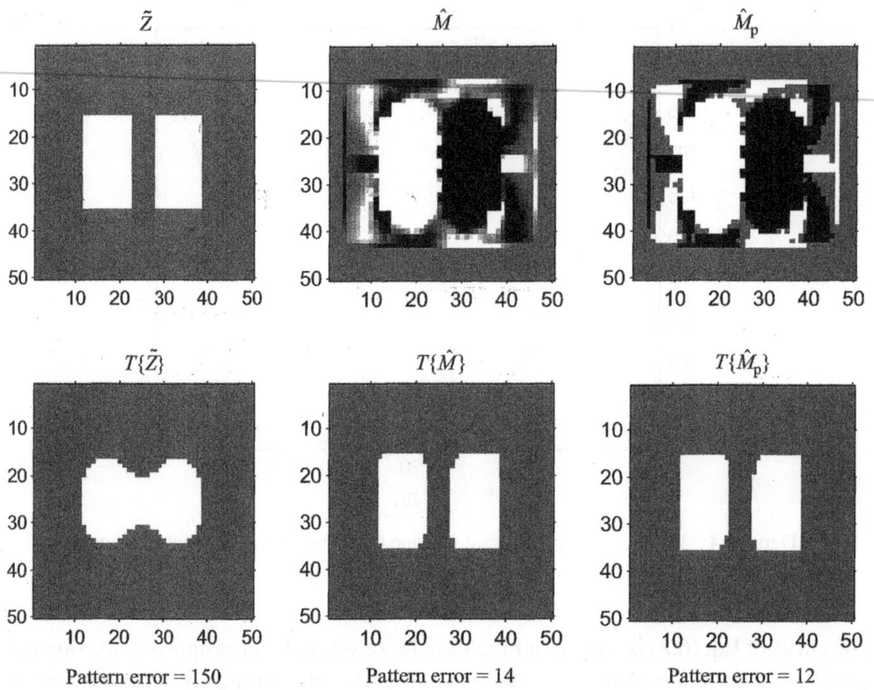

Figure 6.5 Two-phase PSM optimization using discretization penalty with a desired pattern of vertical bars. *Top row* (input masks) (left to right): Desired pattern, optimized real-valued mask, and optimized two-phase PSM obtained using a threshold t_m. *Bottom row:* Indicates the corresponding binary output patterns. $\gamma_D = 0.0175$. Black, gray, and white represent -1, 0, and 1, respectively.

The amplitude discretization penalty term is

$$R_A(\underline{\phi}) = R_A(|\underline{m}|) = \sum_{k=1}^{N^2} |\underline{m}|_k^2 = |\underline{m}|^T |\underline{m}|. \tag{6.11}$$

For each pixel value, the corresponding penalty is the quadratic function

$$r_A(|\underline{m}|_k) = |\underline{m}|_k^2 = \left(\frac{1+\cos\underline{\phi}_k}{2}\right)^2, \quad k = 1, \ldots, N^2. \tag{6.12}$$

The curve of Eq. (6.12) is shown in Fig. 6.7. According to Eq. (6.11), the gradient of $R_A(\underline{\phi})$ is

$$\nabla R_A(\underline{\phi}) = 2|\underline{m}| = (\underline{1} + \cos\underline{\phi}) \odot \left(-\frac{1}{2}\sin\underline{\phi}\right), \tag{6.13}$$

where $\underline{1} = [1, \ldots, 1]^T$.

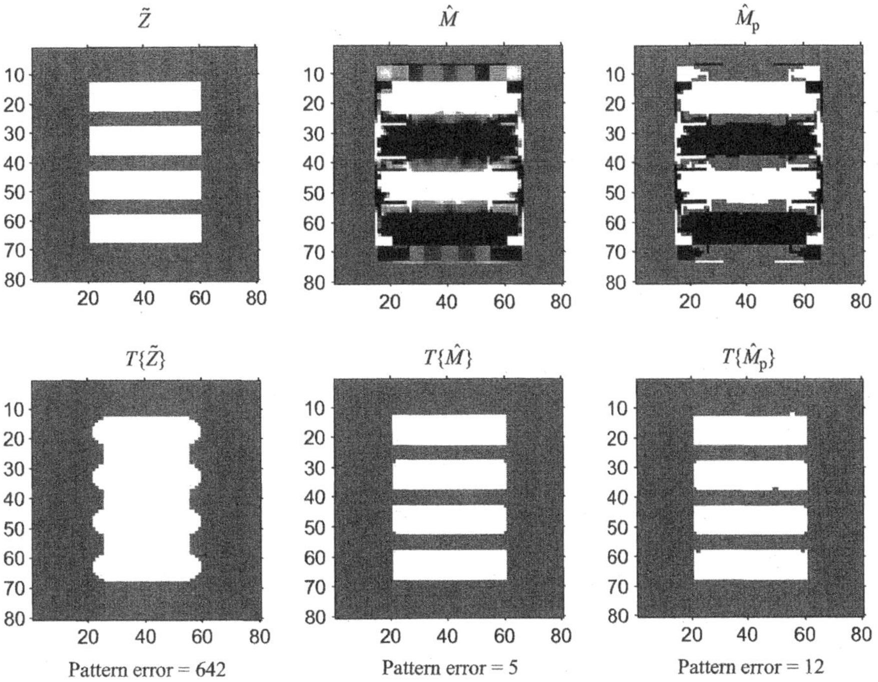

Figure 6.6 Two-phase PSM optimization using discretization penalty with a desired pattern of four horizontal bars. *Top row* (input masks) (left to right): Desired pattern, optimized real-valued mask, and optimized two-phase PSM obtained using a threshold t_m. *Bottom row:* Indicates the corresponding binary output patterns. $\gamma_D = 0.0025$. Black, gray, and white represent -1, 0, and 1, respectively.

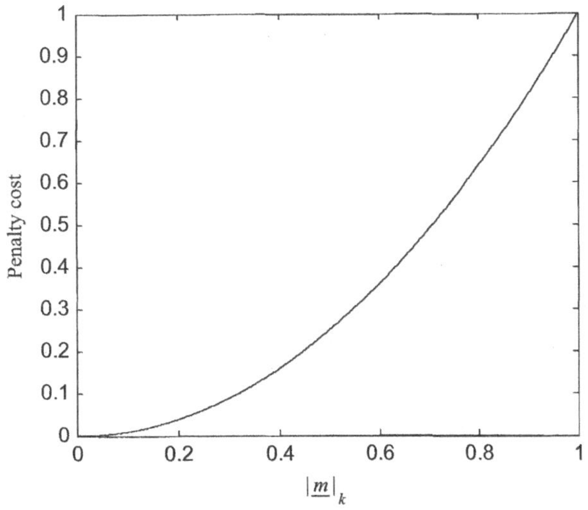

Figure 6.7 Amplitude discretization penalty cost function for the generalized PSM optimization.

In the phase penalty, the phases are constrained to the closest phase levels. If we use two-phase levels, the regularization term is

$$R_P(\underline{\theta}) = \sum_{k=1}^{N^2} \left[\sin\left(2\underline{\theta}_k - \frac{\pi}{2}\right) + 1\right]^2$$
$$= \left[\sin\left(2\underline{\theta} - \frac{\pi}{2}\right) + \underline{1}\right]^T \left[\sin\left(2\underline{\theta} - \frac{\pi}{2}\right) + \underline{1}\right]. \quad (6.14)$$

For each pixel value, the corresponding penalty depicted by solid line in Fig. 6.8 is

$$r_P(\underline{\theta}_k) = \left[\sin\left(2\underline{\theta}_k - \frac{\pi}{2}\right) + 1\right]^2, \quad k = 1, \ldots, N^2. \quad (6.15)$$

According to Eq. (6.14), the gradient of $R_P(\underline{\theta})$ is

$$\nabla R_P(\underline{\theta}) = 4\left[\sin\left(2\underline{\theta} - \frac{\pi}{2}\right) + \underline{1}\right]^T \cos\left(2\underline{\theta} - \frac{\pi}{2}\right). \quad (6.16)$$

If the four-phase levels are considered, the regularization term is obtained as

$$R_P(\underline{\theta}) = \sum_{k=1}^{N^2} \left[\sin\left(4\underline{\theta}_k - \frac{3\pi}{2}\right) + 1\right]^2$$
$$= \left[\sin\left(4\underline{\theta}_k - \frac{3\pi}{2}\right) + \underline{1}\right]^T \left[\sin\left(4\underline{\theta}_k - \frac{3\pi}{2}\right) + \underline{1}\right]. \quad (6.17)$$

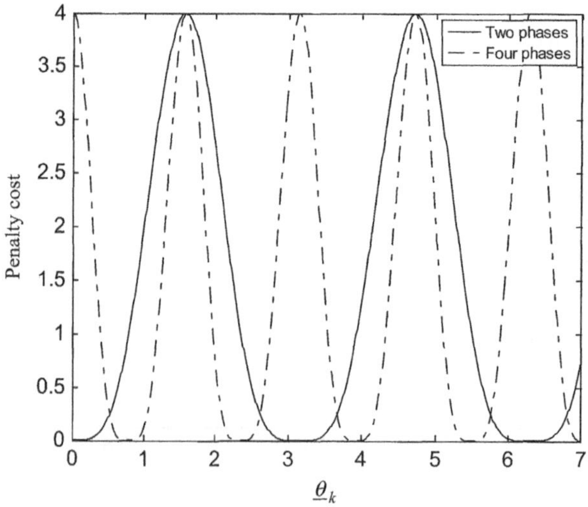

Figure 6.8 Phase discretization penalty cost function for the generalized PSM optimization.

For each pixel value, the corresponding penalty depicted by the dashed–dotted line in Fig. 6.8 is

$$r_{\mathrm{P}}(\underline{\theta}_k) = \left[\sin\left(4\underline{\theta}_k - \frac{3\pi}{2}\right) + 1\right]^2, \quad k = 1, \ldots, N^2, \tag{6.18}$$

and the gradient of $R_{\mathrm{P}}(\underline{\theta})$ is

$$\nabla R_{\mathrm{P}}(\underline{\theta}) = 8\left[\sin\left(4\underline{\theta} - \frac{3\pi}{2}\right) + \underline{1}\right]^T \cos\left(4\underline{\theta} - \frac{3\pi}{2}\right). \tag{6.19}$$

Using the discretization penalty described in Eqs. (6.11), (6.14) and (6.17), the experiment shown in Figs. 5.13 and 5.16 can be repeated. The result is illustrated in

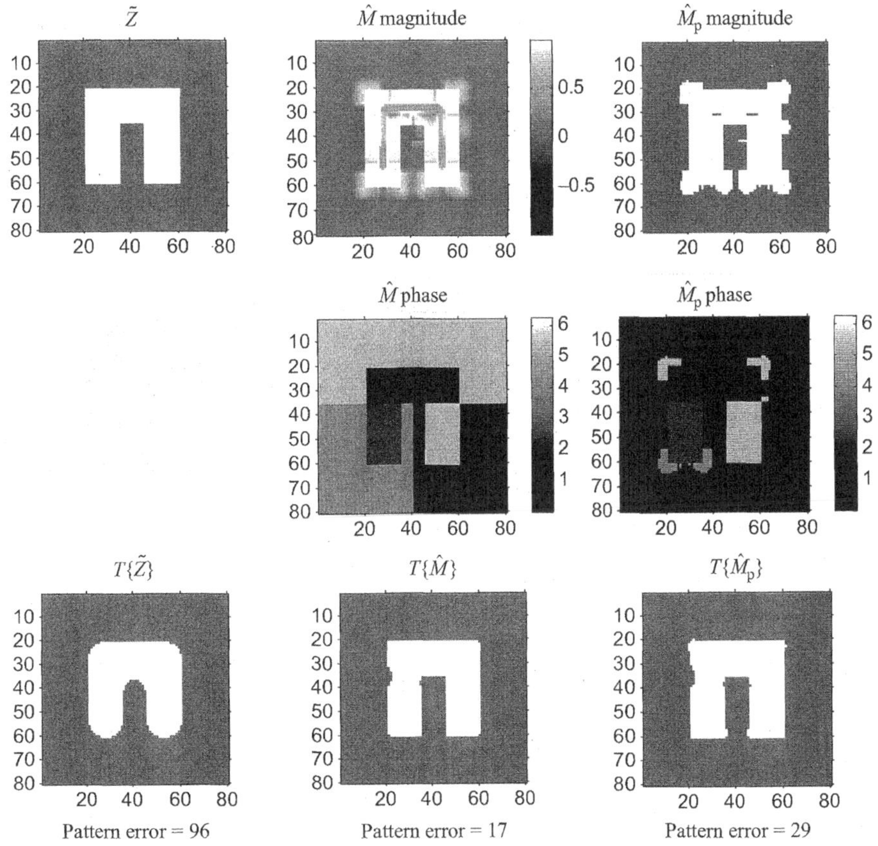

Figure 6.9 Generalized PSM optimization using discretization penalty with a desired pattern of U-junction. *Top row* (input masks) (left to right): Desired pattern, magnitude of the optimized complex-valued mask, and optimized two-phase mask obtained using a threshold t_m. *Middle row:* Indicates the phases of the optimized complex-valued mask. *Bottom row:* Indicates the corresponding binary output patterns. $\gamma_{\mathrm{A}} = 0.045$ and $\gamma_{\mathrm{P}} = 0.001$. Gray, white, and black represent 0, 1, and -1, respectively, in the top and bottom rows. Dark black, light black, dark gray, and light gray represent $\frac{\pi}{4}$, $\frac{3\pi}{4}$, $\frac{5\pi}{4}$, and $\frac{7\pi}{4}$, respectively, in the middle row.

Figs. 6.9 and 6.10. The parameters used in Figs. 6.9 and 6.10 are the same as those in Figs. 5.13 and 5.16, respectively. In addition, $\gamma_A = 0.045$ and $\gamma_P = 0.001$ in Fig. 6.9. $\gamma_A = 0.001$ and $\gamma_P = 0.0001$ in Fig. 6.10. It is shown that the discretization penalty leads to fewer transmission pixels and thus fewer details in the attained mask patterns. However, some details still remain in the mask patterns. To remove those details, an additional penalty can be considered. The complexity penalty regularization method is introduced next for this goal.

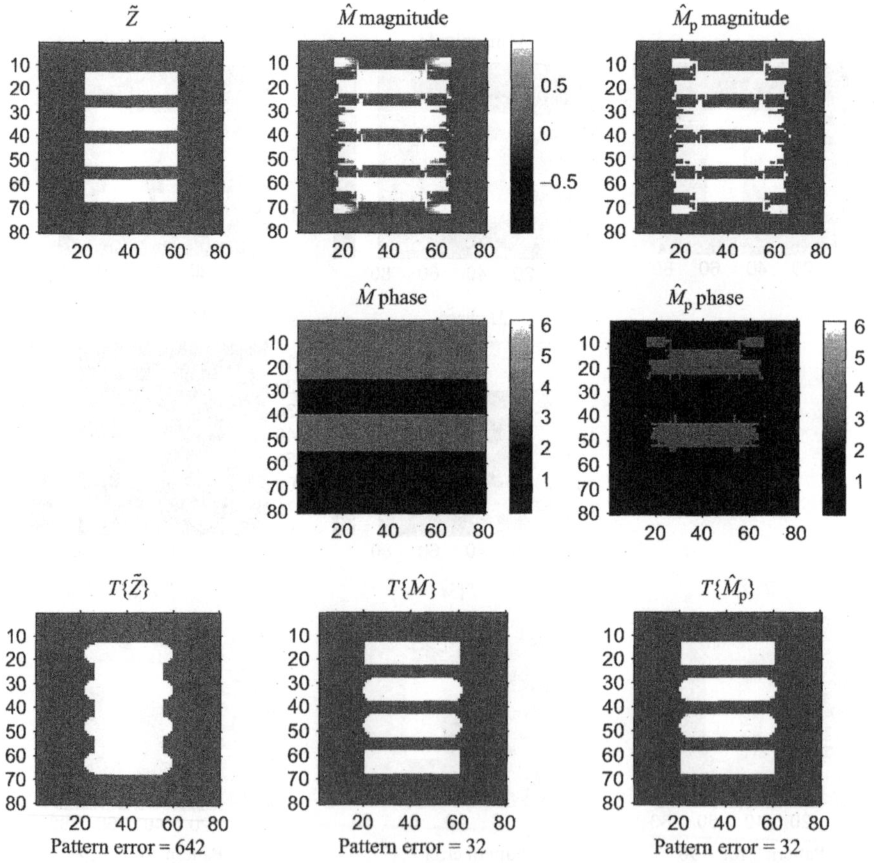

Figure 6.10 Generalized PSM optimization using discretization penalty with a desired pattern of four horizontal bars, where two phase levels are used. *Top row* (input masks) (left to right): Desired pattern, magnitude of the optimized complex-valued mask, and optimized two-phase mask obtained using a threshold t_m. *Middle row:* Indicates the phases of the optimized complex-valued mask. *Bottom row:* Indicates the corresponding binary output patterns. $\gamma_A = 0.001$ and $\gamma_P = 0.0001$. Gray, white, and black represent 0, 1, and -1, respectively, in the top and bottom rows. Gray and black represent π and 0, respectively, in the middle row.

6.2 COMPLEXITY PENALTY

6.2.1 Total Variation Penalty

The pixel-based OPC and PSM optimization approaches allow tremendous flexibility in representing the mask patterns, however, they usually result in complex optimized masks, which are difficult to manufacture [69]. A well-known penalty to remove details is total variation penalty, which was used in Ref. [68] and applied to the OPC optimization. In this section, we extend this total variation penalty to the two-phase PSM optimization. First, the activation pattern $\underline{f}$, indicating the complexity of the mask pattern, is defined as

$$\underline{f}_k = ||\underline{m}|_k - \underline{\tilde{z}}_k|, \quad k = 1, \ldots, N^2, \tag{6.20}$$

where each pixel of $\underline{f}$ denotes the l^1 norm between the mask magnitude and the desired output pattern. The total variation penalty R_{TV} is chosen as the local variation of the activation pattern [22, 68],

$$R_{\mathrm{TV}}(\underline{m}) = \|\nabla \underline{f}\|_1 = \|\mathbf{Q_x}\underline{f}\|_1 + \|\mathbf{Q_y}\underline{f}\|_1, \tag{6.21}$$

where $\|\cdot\|_1$ is the l^1 norm of the argument. $\mathbf{Q_x}, \mathbf{Q_y} \in \Re^{N^2 \times N^2}$ represent the first (directional) derivatives and are defined as $\mathbf{Q_x} = \mathbf{I} - \mathbf{S_x}$ and $\mathbf{Q_y} = \mathbf{I} - \mathbf{S_y}$, where $\mathbf{S_x}$ and $\mathbf{S_y}$ shift a $N \times N$ matrix along horizontal (right) and vertical (up) direction by one pixel, respectively. The gradient $\nabla R_{\mathrm{TV}}(\underline{m}) \in \Re^{N^2 \times 1}$ is calculated as

$$\nabla R_{\mathrm{TV}}(\underline{m}) = [\mathbf{Q_x}^T \mathrm{sign}(\mathbf{Q_x}\underline{f}) + \mathbf{Q_y}^T \mathrm{sign}(\mathbf{Q_y}\underline{f})] \odot \mathrm{sign}(|\underline{m}| - \underline{\tilde{z}}) \odot \mathrm{sign}(\underline{m}). \tag{6.22}$$

For the OPC optimization, where $\underline{m}_k \geq 0$, Eq. (6.22) reduces to [68]

$$\nabla R_{\mathrm{TV}}(\underline{m}) = [\mathbf{Q_x}^T \mathrm{sign}(\mathbf{Q_x}\underline{f}) + \mathbf{Q_y}^T \mathrm{sign}(\mathbf{Q_y}\underline{f})] \odot \mathrm{sign}(\underline{m} - \underline{\tilde{z}}), \tag{6.23}$$

where

$$\underline{f}_k = |\underline{m}_k - \underline{\tilde{z}}_k|, \quad k = 1, \ldots, N^2. \tag{6.24}$$

The extended total variation penalty is tailored to the OPC and two-phase PSM optimization. The cost function is adjusted as

$$J(\underline{m}) = F(\underline{m}) + \gamma_{\mathrm{TV}} R_{\mathrm{TV}}(\underline{m}). \tag{6.25}$$

Using the total penalty described in Eq. (6.21), the experiments shown in Figs. 6.3 and 6.6 can be repeated in Figs. 6.11 and 6.12. The parameters used in Figs. 6.11 and 6.12 are the same as those in Figs. 6.3 and 6.6, respectively. In addition, $\gamma_{\mathrm{TV}} = 0.025$ in Fig. 6.11 and $\gamma_{\mathrm{TV}} = 0.008$ in Fig. 6.12. For both of these simulations, the total variation penalty effectively reduces the complexity of the optimized binary and phase-shifting masks. However, the total variation penalty will have a trade-off reducing the pattern complexity while increasing the pattern errors. Since the total

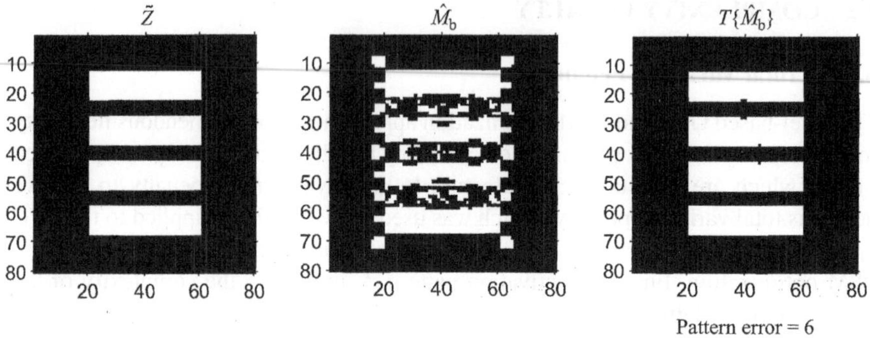

Figure 6.11 OPC optimization using discretization penalty and total variation penalty. *Left to right:* Desired pattern, optimized binary mask, and binary output pattern. $\gamma_D = 0.01$ and $\gamma_{\text{TV}} = 0.025$. Black and white represent 0 and 1, respectively.

variation penalty removes some small assisting features on the masks, the distortions of the output images are increased.

6.2.2 Global Wavelet Penalty

While total variation regularization is utilized to attain better manufacturability of mask patterns in the OPC and two-phase PSM optimization, it is not adequate to reduce the complexity of the generalized PSMs. Generalized to the complex domain, the gradients of the total variation penalty are given as

$$\nabla R_{\text{TV}}(\underline{\phi}) = [\mathbf{Q_x}^T \text{sign}(\mathbf{Q_x}\underline{f}) + \mathbf{Q_y}^T \text{sign}(\mathbf{Q_y}\underline{f})] \odot -\sin\underline{\phi} \odot W_1(\underline{m}), \tag{6.26}$$

$$\nabla R_{\text{TV}}(\underline{\theta}) = [\mathbf{Q_x}^T \text{sign}(\mathbf{Q_x}\underline{f}) + \mathbf{Q_y}^T \text{sign}(\mathbf{Q_y}\underline{f})] \odot (\underline{1} + \cos\underline{\phi}) \odot W_2(\underline{m}), \tag{6.27}$$

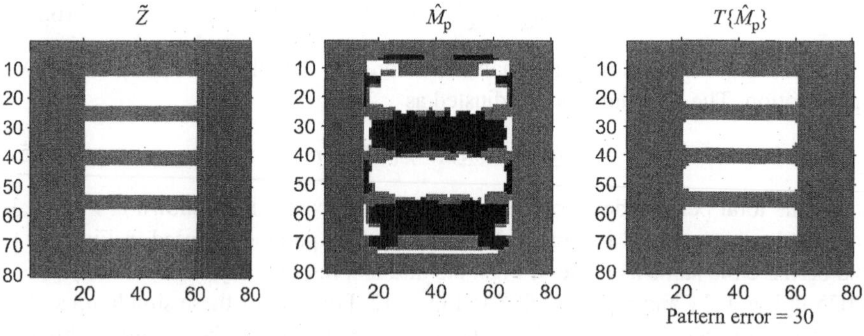

Figure 6.12 Two-phase PSM optimization using discretization penalty and total variation penalty. *Left to right:* Desired pattern, optimized binary mask, and binary output patterns. $\gamma_D = 0.0025$ and $\gamma_{\text{TV}} = 0.008$. Black, gray, and white represent -1, 0, and 1, respectively.

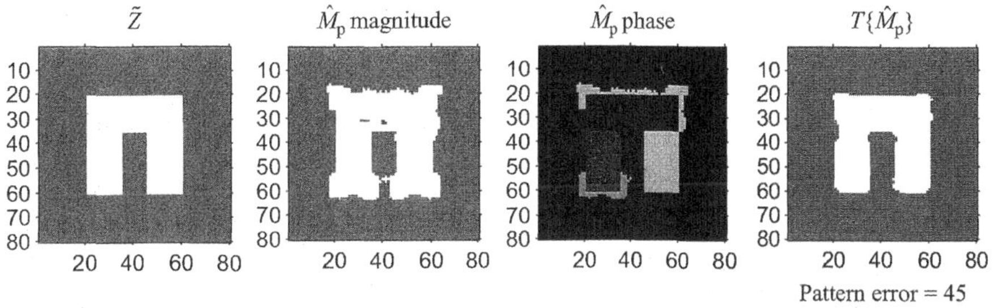

Figure 6.13 Generalized PSM optimization using discretization penalty and total variation penalty. *Left to right:* Desired pattern, pole-level optimized mask magnitude, and phase and binary output patterns. $\gamma_A = 0.01$ and $\gamma_P = 0.001$. Total variation regularization uses $\gamma_{TV,\phi} = 0.1$ for the update of ϕ, and $\gamma_{TV,\theta} = 0.001$ for the update of θ.

where $\underline{f}$ is the activation pattern and $\underline{f}_k = |\underline{m}_k - \underline{\tilde{z}}_k|$ for $k = 1, \ldots, N^2$. $W_1(\underline{m}) = \mathrm{Re}[(\underline{m} - \underline{\tilde{z}}) \odot e^{-j\underline{\theta}}] \odot \frac{1}{2\underline{f}}$. $W_2(\underline{m}) = \mathrm{Re}[(\underline{m} - \underline{\tilde{z}}) \odot e^{-j\underline{\theta}} \odot (-j)] \odot \frac{1}{2\underline{f}}$. Using the total variation penalty in the simulation of Fig. 6.9, the results are presented in Fig. 6.13. The total variation regularization uses $\gamma_{TV,\phi} = 0.1$ for the update of $\underline{\phi}$, and $\gamma_{TV,\theta} = 0.001$ for the update of $\underline{\theta}$. It is observed that the optimized four-phase PSM includes several singular pixels and small features. To overcome the limitation of the total variation penalty, a more effective detail reduction approach referred to as "wavelet penalty" regularization can be used.

Since the typical mask patterns encountered in circuitry are piecewise smooth images, the Haar wavelet is used as the building block. Consider a $N \times N$ (assume N is even) image $M_{N\times N}$, where m_{ij} represents the (i, j) matrix element. The 1-depth Haar wavelet transform of the image above, ignoring the scale parameters, leads to the level-1 approximation coefficient block and three detail coefficient blocks, each block of size $\frac{N}{2} \times \frac{N}{2}$. Figure. 6.14 illustrates the 1-depth Haar wavelet transform

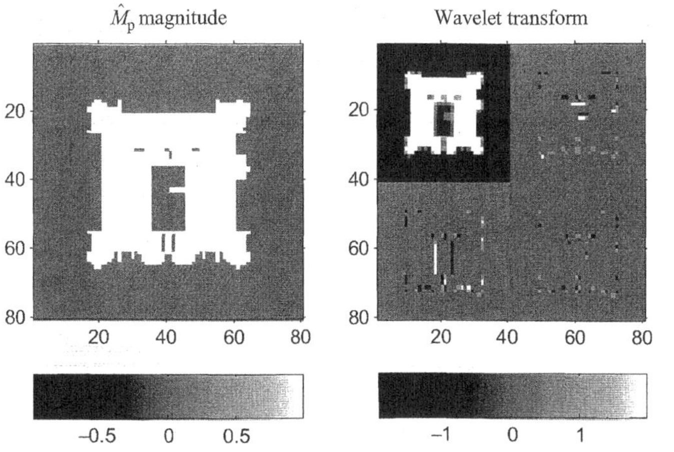

Figure 6.14 The 1-depth Haar wavelet transform of the optimized mask magnitude in Fig. 5.13.

of the magnitude of the optimized pole-level mask in Fig. 5.13. Specifically, the approximation coefficient block is $A_{\frac{N}{2}\times\frac{N}{2}}$, where

$$
\begin{aligned}
a_{ij} = {} & m_{(2(i-1)+1)(2(j-1)+1)} + m_{(2(i-1)+1)(2(j-1)+2)} + m_{(2(i-1)+2)(2(j-1)+1)} \\
& + m_{(2(i-1)+2)(2(j-1)+2)},
\end{aligned} \tag{6.28}
$$

for $i, j = 1, \ldots, \frac{N}{2}$. The elements of the horizontal, vertical, and diagonal detail coefficient blocks are respectively,

$$
\begin{aligned}
h_{ij} = {} & m_{(2(i-1)+1)(2(j-1)+1)} - m_{(2(i-1)+1)(2(j-1)+2)} + m_{(2(i-1)+2)(2(j-1)+1)} \\
& - m_{(2(i-1)+2)(2(j-1)+2)},
\end{aligned} \tag{6.29}
$$

$$
\begin{aligned}
v_{ij} = {} & m_{(2(i-1)+1)(2(j-1)+1)} + m_{(2(i-1)+1)(2(j-1)+2)} - m_{(2(i-1)+2)(2(j-1)+1)} \\
& - m_{(2(i-1)+2)(2(j-1)+2)},
\end{aligned} \tag{6.30}
$$

$$
\begin{aligned}
d_{ij} = {} & m_{(2(i-1)+1)(2(j-1)+1)} - m_{(2(i-1)+1)(2(j-1)+2)} - m_{(2(i-1)+2)(2(j-1)+1)} \\
& + m_{(2(i-1)+2)(2(j-1)+2)},
\end{aligned} \tag{6.31}
$$

for $i, j = 1, \ldots, \frac{N}{2}$. The approximation coefficient block represents the low-frequency component of the image and the other three detail coefficient blocks represent the high-frequency components or the details of the image. Further, using Eq. (6.29) in Eq. (6.31), the total energy in the detail components is

$$
\begin{aligned}
E_{\text{detail}} = {} & h_{11}h_{11}^{*} + h_{12}h_{12}^{*} \cdots + h_{\left(\frac{N}{2}\right)\left(\frac{N}{2}\right)} h_{\left(\frac{N}{2}\right)\left(\frac{N}{2}\right)}^{*} + v_{11}v_{11}^{*} + v_{12}v_{12}^{*} \cdots \\
& + v_{\left(\frac{N}{2}\right)\left(\frac{N}{2}\right)} v_{\left(\frac{N}{2}\right)\left(\frac{N}{2}\right)}^{*} + d_{11}d_{11}^{*} + d_{12}d_{12}^{*} \cdots + d_{\left(\frac{N}{2}\right)\left(\frac{N}{2}\right)} d_{\left(\frac{N}{2}\right)\left(\frac{N}{2}\right)}^{*}.
\end{aligned} \tag{6.32}
$$

To remove details in the mask, the energy of the detail components should be reduced during the optimization process. Although E_{detail} contains many terms, there are just three terms related to a specific mask element m_{ij}. This property is convenient for calculating the energy differential of the detail components with respect to each pixel value m_{ij}. We refer to this property as the "localization property." As shown in Appendix C, the partial derivatives of E_{detail} with respect to $\underline{\phi}$ and $\underline{\theta}$ are as follows:

$$
\begin{aligned}
\frac{\partial E_{\text{detail}}}{\partial \underline{\phi}_{(2(i-1)+p)(2(j-1)+q)}} = {} & -\sin\underline{\phi}_{(2(i-1)+p)(2(j-1)+q)} \\
& \times \operatorname{Re}\Big[e^{-j\underline{\theta}_{(2(i-1)+p)(2(j-1)+q)}} \\
& \times \big(3\underline{m}_{(2(i-1)+p)(2(j-1)+q)} - \underline{m}_{(2(i-1)+p_1)(2(j-1)+q)} \\
& - \underline{m}_{(2(i-1)+p)(2(j-1)+q_1)} - \underline{m}_{(2(i-1)+p_1)(2(j-1)+q_1)} \big) \Big],
\end{aligned} \tag{6.33}
$$

$$\frac{\partial E_{\text{detail}}}{\partial \underline{\theta}_{(2(i-1)+p)(2(j-1)+q)}} = (1+\cos\underline{\phi}_{(2(i-1)+p)(2(j-1)+q)})$$
$$\times \text{Re}\left[(-j)e^{-j\underline{\theta}_{(2(i-1)+p)(2(j-1)+q)}}\right.$$
$$\times \left(3\underline{m}_{(2(i-1)+p)(2(j-1)+q)} - \underline{m}_{(2(i-1)+p_1)(2(j-1)+q)}\right.$$
$$\left.\left. - \underline{m}_{(2(i-1)+p)(2(j-1)+q_1)} - \underline{m}_{(2(i-1)+p_1)(2(j-1)+q_1)}\right)\right], \tag{6.34}$$

where $i, j = 1, \ldots, \frac{N}{2}$; $p, q = 1$ or 2; $p_1 = (p+1) \bmod 2$, and $q_1 = (q+1) \bmod 2$. From Eqs. (6.33) and (6.34), the gradient of E_{detail} can be calculated and the cost function can be adjusted as

$$J(\underline{m}) = F(\underline{m}) + \gamma_{\text{A}} R_{\text{A}}(\underline{\theta}) + \gamma_{\text{P}} R_{\text{P}}(\underline{\phi}) + \gamma_{\text{WA}} E_{\text{detail}}(\underline{m}). \tag{6.35}$$

The experiment of Fig. 6.9 is then repeated using the wavelet penalty. The results are illustrated in Fig. 6.15, where $\gamma_{\text{A}} = 0.01$ and $\gamma_{\text{P}} = 0.001$. The weights of wavelet penalty assigned to $\underline{\phi}$ and $\underline{\theta}$ are $\gamma_{\text{WA},\phi} = 0.2$ and $\gamma_{\text{WA},\theta} = 0.001$, respectively. Other parameters are the same as those in Fig. 6.9. It can be seen that the wavelet penalty removes many small peaks at the bottom of the two parallel bars in Fig. 6.9. These results illustrate the efficiency of the wavelet penalty. However, regularization invariably will have a trade-off reducing the pattern details while often increasing the pattern errors. Since the wavelet penalty removes small assisting blocks, the distortions of the output images are increased.

Comparing the results in Figs. 6.13 and 6.15, the following observations can be made. First, the Haar wavelet is suitable for the piecewise smooth image. Second, given a set error range in the attained mask patterns in all simulations tested for the generalized PSM optimization, the Haar wavelet penalty removes more details than the total variation penalty. Finally, the Haar wavelet penalty makes the shape of the blocks constructing the mask more regular and closer to the Haar basis functions. The

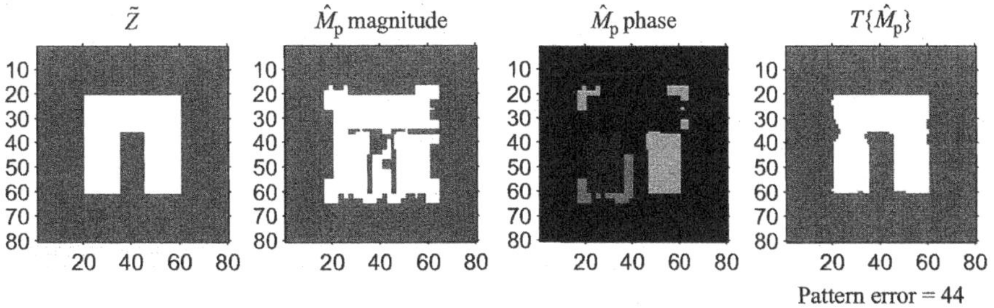

Figure 6.15 Generalized PSM optimization using discretization penalty and wavelet penalty. *Left to right:* Desired pattern, pole-constrained optimized mask magnitude, and phase and binary output patterns. $\gamma_{\text{A}} = 0.01$ and $\gamma_{\text{P}} = 0.001$. Wavelet regularization uses $\gamma_{\text{WA},\phi} = 0.2$ and $\gamma_{\text{WA},\theta} = 0.001$.

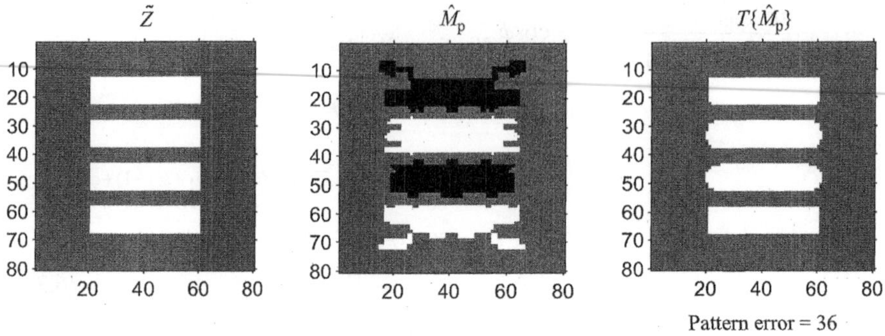

Figure 6.16 Generalized PSM optimization using discretization penalty and wavelet penalty, where two phase levels are used. *Left to right:* Desired pattern, pole-constrained optimized mask magnitude, and phase and binary output patterns. $\gamma_A = 0.001$ and $\gamma_P = 0.0001$. Wavelet regularization uses $\gamma_{WA,\phi} = 0.03$ and $\gamma_{WA,\theta} = 0.001$.

Haar wavelet penalty thus leads to a set of advantages that can be attributed to the rectangular shape of the Haar basis waveforms. It should be emphasized that the Haar wavelet is one choice for the wavelet penalty approaches. Different basis sets could be selected so as to synthesize a desired set of features. Furthermore, while a 1-depth wavelet transform was used in the experiments above to measure the energy of the details, a deeper wavelet transform can be investigated and used to attain simpler mask patterns.

The wavelet penalty is also valid when the generalized PSM optimization algorithm is used to design two-phase masks. The simulation in Fig. 6.10 is repeated in Fig. 6.16, where $\gamma_A = 0.001$ and $\gamma_P = 0.0001$. The weights of wavelet penalty assigned to $\underline{\phi}$ and $\underline{\theta}$ are $\gamma_{WA,\phi} = 0.03$ and $\gamma_{WA,\theta} = 0.001$, respectively. Other parameters are the same as those in Fig. 6.10. It can be seen that the complexity of the mask pattern is effectively reduced.

6.2.3 Localized Wavelet Penalty

In the experiments above, the wavelet penalty is applied to the entire mask pattern without any local discrimination. Thus, an equal penalty was assigned to details in all regions. In practice, details may be intolerable in some special mask regions, while some details may be permissible in other regions. Because of the "localization property" of the Haar wavelet penalty, we can assign regional weights to the penalty term. Thus, a regional weighted wavelet penalty is effective for achieving local discrimination. To this end, the cost function is adjusted as

$$J(\underline{m}_i) = F(\underline{m}_i) + \gamma_A r_A(\underline{\theta}_i) + \gamma_P r_P(\underline{\phi}_i) + \omega(i)\gamma_{WA} E_{\text{detail}}(\underline{m}_i), \qquad (6.36)$$

where $\omega(i)$, $i = 1, \ldots, N^2$, are the weight coefficients and may be changed in different spatial regions. The experimental result using regional weighted wavelet penalty is presented in Fig. 6.17. The experiment shown in Fig. 6.17 has the same parameters

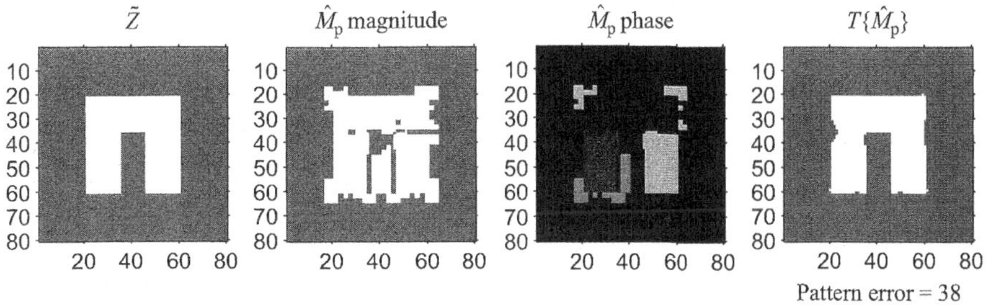

Figure 6.17 Generalized PSM optimization using discretization penalty and localized wavelet penalty. *Left to right:* Desired pattern, pole-level optimized mask magnitude, and phase and binary output patterns. $\gamma_A = 0.01$ and $\gamma_P = 0.001$. Wavelet regularization uses $\gamma_{WA,\phi} = 0.2$ and $\gamma_{WA,\theta} = 0.001$. The gap between the vertical bars has regional weight of 1.6. The other regions have regional weight of 0.7.

as the one shown in Fig. 6.15, except for the regional weights. We placed a higher cost, $\omega(i) = 1.6$, to the areas in the gap between the vertical bars, and we assigned lower cost, $\omega(i) = 0.7$, to other regions. Comparing the results in Figs. 6.15 and 6.17, it is observed that the localized wavelet penalty removes more details in the gap, but tolerates slightly more details in other regions.

The simulation of the parallel bar pattern in Fig. 6.16 is repeated in Fig. 6.18 using the localized wavelet penalty. The experiment shown in Fig. 6.18 has the same parameters as the one shown in Fig. 6.16, except for the regional weights. We placed a higher cost, $\omega(i) = 1.1$, to the regions for the second and the third bars. $\omega(i) = 0.8$ was assigned to the regions for the first and the fourth bars, and we assigned $\omega(i) = 0.7$ to other regions. Comparing the results in Figs. 6.16 and 6.18, it is observed that the localized wavelet penalty removes more details around the second and the third horizontal bars.

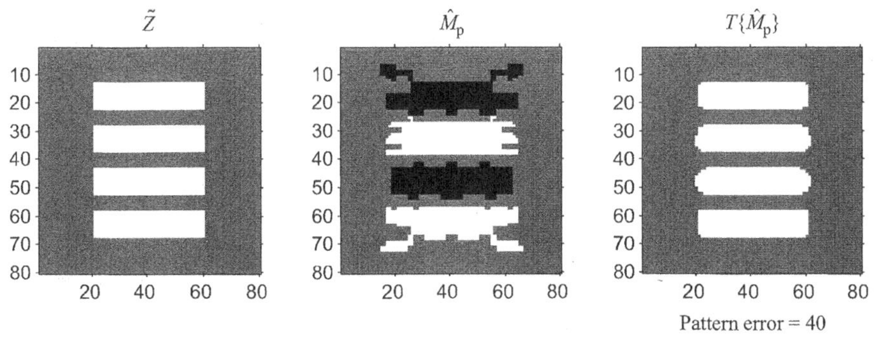

Figure 6.18 Generalized PSM optimization using discretization penalty and localized wavelet penalty, where two phase levels are used. *Left to right:* Desired pattern, pole-level optimized mask magnitude, and phase and binary output patterns. $\gamma_A = 0.001$ and $\gamma_P = 0.0001$. Wavelet regularization uses $\gamma_{WA,\phi} = 0.03$ and $\gamma_{WA,\theta} = 0.001$. The regions for the first and the fourth bars have regional weight of 0.8. The regions for the second and the third bars have regional weight of 1.1. The other regions have regional weight of 0.7.

6.3 SUMMARY

This chapter presented several regularization frameworks to reduce the output pattern error and complexity of the optimized masks. The discretization penalty was extended to the PSM optimization. In addition, a wavelet penalty was described to reduce the mask complexity, which can be more effective than the total variation penalty in the generalized PSM optimization.

7

Computational Lithography with Partially Coherent Illumination

In Chapter 5, a set of gradient-based OPC and PSM optimization methods have been developed to solve the inverse lithography problem under coherent illumination. Most practical illumination sources, however, have a nonzero line width and their radiation is more generally described as partially coherent [75]. While the inverse lithography methods derived in Chapter 5 are effective in coherent illumination, these algorithms fail to account for the nonlinearities of partially coherent illumination (PCI), and produce inadequate results when applied to a partially coherent illumination system. PCI is desired, since it can improve the theoretical resolution limit. PCI is thus introduced in practice through modified illumination sources having large coherent factors or through off-axis illuminations. In partially coherent imaging, the mask is illuminated by light traveling in various directions. The source points giving rise to these incident rays are incoherent with one another, such that there is no interference that could lead to nonuniform light intensity impinging on the mask [92, 93]. According to the Hopkins diffraction model, the light intensity distribution exposed on the wafer in PCI is bilinear and described by [74]

$$I(\mathbf{r}) = \iint M(\mathbf{r_1})M^*(\mathbf{r_2})\gamma(\mathbf{r_1} - \mathbf{r_2})h(\mathbf{r} - \mathbf{r_1})h^*(\mathbf{r} - \mathbf{r_2})d\mathbf{r_1}d\mathbf{r_2}, \tag{7.1}$$

where $\mathbf{r} = (x, y)$, $\mathbf{r_1} = (x_1, y_1)$, and $\mathbf{r_2} = (x_2, y_2)$. $M(\mathbf{r})$ is the mask pattern, $\gamma(\mathbf{r_1} - \mathbf{r_2})$ is the complex degree of coherence, and $h(\mathbf{r})$ represents the amplitude impulse response of the optical system. The complex degree of coherence $\gamma(\mathbf{r_1} - \mathbf{r_2})$ is generally a complex number, whose magnitude represents the extent of optical interaction between two spatial locations $\mathbf{r_1} = (x_1, y_1)$ and $\mathbf{r_2} = (x_2, y_2)$ of the light source [92]. The complex degree of coherence in the spatial domain is the inverse 2D Fourier transform of the illumination shape. In general, this equation is tedious to compute, both analytically and numerically [75].

This chapter focuses on the development of computationally effective inverse optimization algorithms for the design of OPC and PSM for lithography under partially coherent illumination. Three nonlinear models are used in the optimization. The first

Computational Lithography By Xu Ma and Gonzalo R. Arce

relies on a Fourier representation, which approximates the partially coherent system as a sum of coherent systems. The second model is based on an average coherent approximation, which is computationally faster. The third is a singular value decomposition (SVD) model, which is tailored to PSM optimization.

7.1 OPC OPTIMIZATION

7.1.1 OPC Design Algorithm Using the Fourier Series Expansion Model

Let $M(x, y)$ be the input binary mask to an optical lithography system $T\{\cdot\}$, with a partially coherent illumination. The PCI optical system is approximated by a Hopkins diffraction model described in Eq. (7.1). The effect of the photoresist is modeled by a hard threshold operation and approximated by a sigmoid function. The output pattern is denoted as $Z(x, y) = T\{M(x, y)\}$. Given a $N \times N$ desired output pattern $\tilde{Z}(x, y)$, the OPC optimization problem under PCI can be formulated as the search of $\hat{M}(x, y)$ over the $N \times N$ real space $\Re^{N\times N}$ such that

$$\hat{M}(x, y) = \arg \min_{M(x,y)\in\Re^{N\times N}} d\left(T\{M(x, y)\}, \tilde{Z}(x, y)\right), \tag{7.2}$$

where $d(\cdot, \cdot)$ is the square of the l^2 norm criterion.

The forward imaging process based on the Fourier series expansion model is illustrated in Fig. 7.1, where the Fourier series expansion model is used in the image formation stage. The details of the Fourier series expansion model are discussed in Section 2.2.1. Light propagating through the mask pattern is affected by diffraction and mutual interference—a phenomenon described by the Hopkins diffraction model [7, 14, 42, 43]. Light that is transmitted through the mask reaches a light-sensitive photoresist, which is subsequently developed through the use of solvents. Assume that the positive photoresist is used in the partially coherent illumination system, which is represented by a sigmoid function in Fig. 7.1. $H^{\mathbf{m}}$ is $N^2 \times N^2$ convolution matrix with an equivalent two-dimensional filter $h^{\mathbf{m}}$, which is described in Eq. (2.19). Note that the output of the convolution and the absolute-square operation is the intensity distribution of the aerial image. On the other hand, in Fig. 5.1, the output of the absolute operation is the amplitude of the electric field.

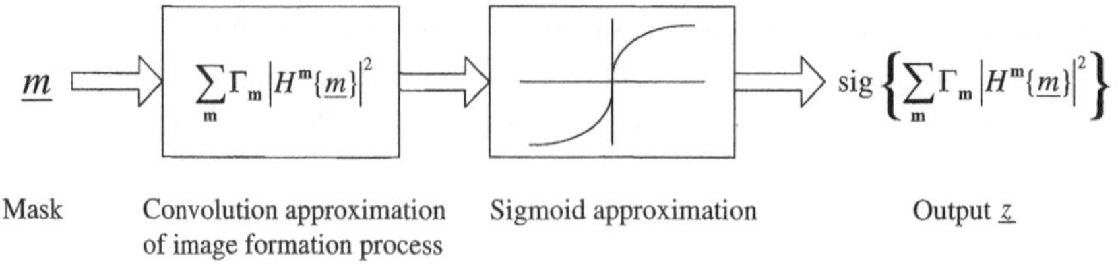

Figure 7.1 Approximated forward process model based on the Fourier series expansion model.

Following the definitions above, the output of the sigmoid function is the $N \times N$ image denoted as

$$Z = \text{sig}\left\{\sum_{\mathbf{m}} \Gamma_{\mathbf{m}} |H^{\mathbf{m}}\{\underline{m}\}|^2\right\}. \tag{7.3}$$

In Eq. (7.3), $|H^{\mathbf{m}}\{\underline{m}\}|^2$ is the output aerial image of the mth coherent component of the entire partially imaging system. $\Gamma_{\mathbf{m}}$ is the coefficient of the 2D Fourier series expansion of the complex degree of coherence $\gamma(\mathbf{r})$. The sigmoid function is defined as

$$\text{sig}(x) = \frac{1}{1 + \exp[-a(x - t_r)]}, \tag{7.4}$$

where t_r is the process threshold, and a dictates the steepness of the sigmoid function. The equivalent vector of Z is denoted as $\underline{z}$. The ith entry of vector $\underline{z}$ can be represented as

$$\underline{z}_i = \frac{1}{1 + \exp\left[-a \sum_{\mathbf{m}} \Gamma_{\mathbf{m}} \left|\sum_{j=1}^{N^2} h_{ij}^{\mathbf{m}} \underline{m}_j\right|^2 + at_r\right]}, \quad i = 1, \ldots N^2, \tag{7.5}$$

where $h_{ij}^{\mathbf{m}}$ is the i, jth entry of the filter $h^{\mathbf{m}}$. The binary output pattern Z_{b} is evaluated as

$$Z_{\text{b}} = \Lambda\left\{\sum_{\mathbf{m}} \Gamma_{\mathbf{m}} |H^{\mathbf{m}}\{\underline{m}\}|^2\right\}, \tag{7.6}$$

where $\Lambda(\cdot)$ is the hard threshold function described in Eq. (5.4). The equivalent vector is denoted as $\underline{z}_{\text{b}}$. In the optimization process, $\hat{\underline{m}}$ is searched to minimize the square of the l^2 norm of the difference between $\underline{z}$ and $\tilde{\underline{z}}$. Therefore,

$$\hat{\underline{m}} = \arg\min_{\hat{\underline{m}}}\{F(\underline{m})\}, \tag{7.7}$$

where the cost function $F(\cdot)$ is defined as

$$F(\underline{m}) = \|\tilde{\underline{z}} - \underline{z}\|_2^2 = \sum_{i=1}^{N^2} (\tilde{\underline{z}}_i - \underline{z}_i)^2, \tag{7.8}$$

where $\underline{z}_i$ is already represented in Eq. (7.5). To reduce the above bound-constrained optimization problem to an unconstrained optimization problem, we adopt the parameter transformation in Ref. [68]. Let

$$\underline{m}_k = \frac{1 + \cos(\theta_k)}{2}, \quad k = 1, \ldots, N^2, \tag{7.9}$$

where $\underline{\theta}_k \in (-\infty, \infty)$ and $\underline{m}_k \in [0, 1]$. Assuming the vector $\underline{\theta} = [\theta_1, \ldots, \theta_{N^2}]^T$, the optimization problem is formulated as

$$(\hat{\underline{\theta}}) = \arg\min_{\underline{\theta}}\{F(\underline{\theta})\}$$

$$= \arg\min_{\underline{\theta}} \left\{ \sum_{i=1}^{N^2} \left(\tilde{z}_i - \frac{1}{1+\exp\left[-a\sum_{\mathbf{m}}\Gamma_{\mathbf{m}}|\sum_{j=1}^{N^2} h_{ij}^{\mathbf{m}}\frac{1+\cos\underline{\theta}_j}{2}|^2 + at_r\right]} \right)^2 \right\}. \tag{7.10}$$

The steepest descent method is used to optimize the above problem. The gradient $\nabla F(\underline{\theta})_{\underline{\theta}}$ derived in Appendix D can be calculated as follows:

$$\begin{aligned} \nabla F(\underline{\theta}) &= \underline{d}_{\underline{\theta}} = a\left\{\sum_{\mathbf{m}}\Gamma_{\mathbf{m}}(H^{\mathbf{m}})^{*T}\left[(\underline{\tilde{z}} - \underline{z}) \odot \underline{z} \odot (\underline{1} - \underline{z})\right.\right. \\ &\quad \left.\left.\odot (H^{\mathbf{m}})^{*}(\underline{m})\right]\right\} \odot \sin\underline{\theta} + a\left\{\sum_{\mathbf{m}}\Gamma_{\mathbf{m}}(H^{\mathbf{m}})^{T}\left[(\underline{\tilde{z}} - \underline{z})\right.\right. \\ &\quad \left.\left.\odot \underline{z} \odot (\underline{1} - \underline{z}) \odot (H^{\mathbf{m}})(\underline{m})\right]\right\} \odot \sin\underline{\theta}, \end{aligned} \tag{7.11}$$

where $\nabla F(\underline{\theta}) \in \Re^{N^2\times 1}$, $*$ is the conjugate operation, and T is the conjugate transposition. $\underline{1} = [1, \ldots, 1]^T \in \Re^{N^2\times 1}$. The element-by-element multiplication operator $\odot$ between two $N \times 1$ vectors $\underline{x}$ and $\underline{y}$ is defined as

$$\underline{x} \odot \underline{y} = \left[\underline{x}_1\underline{y}_1, \underline{x}_2\underline{y}_2, \ldots, \underline{x}_N\underline{y}_N\right]^T. \tag{7.12}$$

Assuming $\underline{\theta}^k$ is the kth iteration result, at the $k+1$th iteration,

$$\underline{\theta}^{k+1} = \underline{\theta}^k - s_{\underline{\theta}}\underline{d}_{\underline{\theta}}^k, \tag{7.13}$$

where $s_{\underline{\theta}}$ is the step size.

The iterative optimization above, in general, leads to gray masks with pixel values between 0 and 1. Therefore, a post-processing step described in Eq. (5.18) is needed to obtain the binary optimized mask, $\hat{\underline{m}}_{bk} = U(\hat{\underline{m}}_k - t_m)$, $k = 1, \ldots, N^2$, where t_m is a global threshold. We define the pattern error E as the distance between the desired output image $\tilde{Z}$ and the actual binary output pattern Z_b, that is,

$$\begin{aligned} E &= \sum_{i=1}^{N^2}|\underline{\tilde{z}}_i - \underline{z}_{bi}| = \sum_{i=1}^{N^2}\left|\underline{\tilde{z}}_i - U_i\left(\sum_{\mathbf{m}}\Gamma_{\mathbf{m}}|H^{\mathbf{m}}\underline{m}_b|^2 - t_r\right)\right| \\ &= \sum_{i=1}^{N^2}\left|\underline{\tilde{z}}_i - \Lambda_i\left(\sum_{\mathbf{m}}\Gamma_{\mathbf{m}}|H^{\mathbf{m}}\underline{m}_b|^2\right)\right|. \end{aligned} \tag{7.14}$$

When the pattern error is reduced to a tolerable level, the steepest descent iteration is stopped. Note that if the complex degree of coherence approaches the value of 1, $\gamma \rightarrow 1$, the system becomes completely coherent and $H^{\mathbf{m}} \rightarrow H$ in Eq. (7.11). The gradient in Eq. (7.11) then reduces to

$$\nabla F(\underline{\theta}) = \underline{d}_{\theta} = a\{H^T[(\underline{z}^* - \underline{z}) \odot \underline{z} \odot (\underline{1} - \underline{z}) \odot H(\underline{m})]\} \odot \sin \underline{\theta}, \quad (7.15)$$

which is the result obtained in Ref. [68] for binary mask optimization in the completely coherent case.

In the prior simulation settings, the post-processing (binarization) of the gray optimized mask pattern is suboptimal with no guarantee that the pattern error is under the goal. Furthermore, the optimized mask patterns contain numerous details, which may bring difficulties to mask fabrication. To obtain near-binary and low-complexity mask patterns, the discretization penalty and wavelet penalty are applied in the OPC optimization algorithm. The details of the discretization penalty and wavelet penalty are discussed in Sections 6.1.1 and 6.2.2, respectively.

7.1.2 Simulations Using the Fourier Series Expansion Model

To demonstrate the validity of the OPC optimization algorithm based on the Fourier series expansion model, consider the desired pattern shown in Fig. 7.2a, having dimension of 1035 nm × 1035 nm. The desired pattern consists of 45 nm features. The pitch $p = 90$ nm is indicated by dashed lines. To prove the universality of the algorithm for different sizes of illuminations, the simulations are repeated based on annular illuminations with large, medium, and small partial coherence factors. The values of the inner and outer partial coherence factors of the illuminations are the same as those in Fig. 2.7. In Fig. 7.3, the top row illustrates the simulation results applying the large annular illumination with $\sigma_{\text{inner}} = 0.8$ and $\sigma_{\text{outer}} = 0.975$: (left)

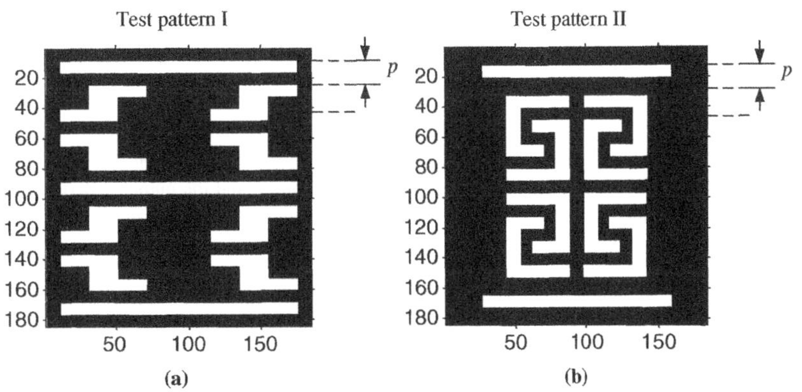

Figure 7.2 Test patterns for OPC optimization in partially coherent imaging systems. (a) Test pattern I. (b) Test pattern II. Both patterns contain 45 nm features with pitch $p = 90$ nm as indicated by the dashed lines.

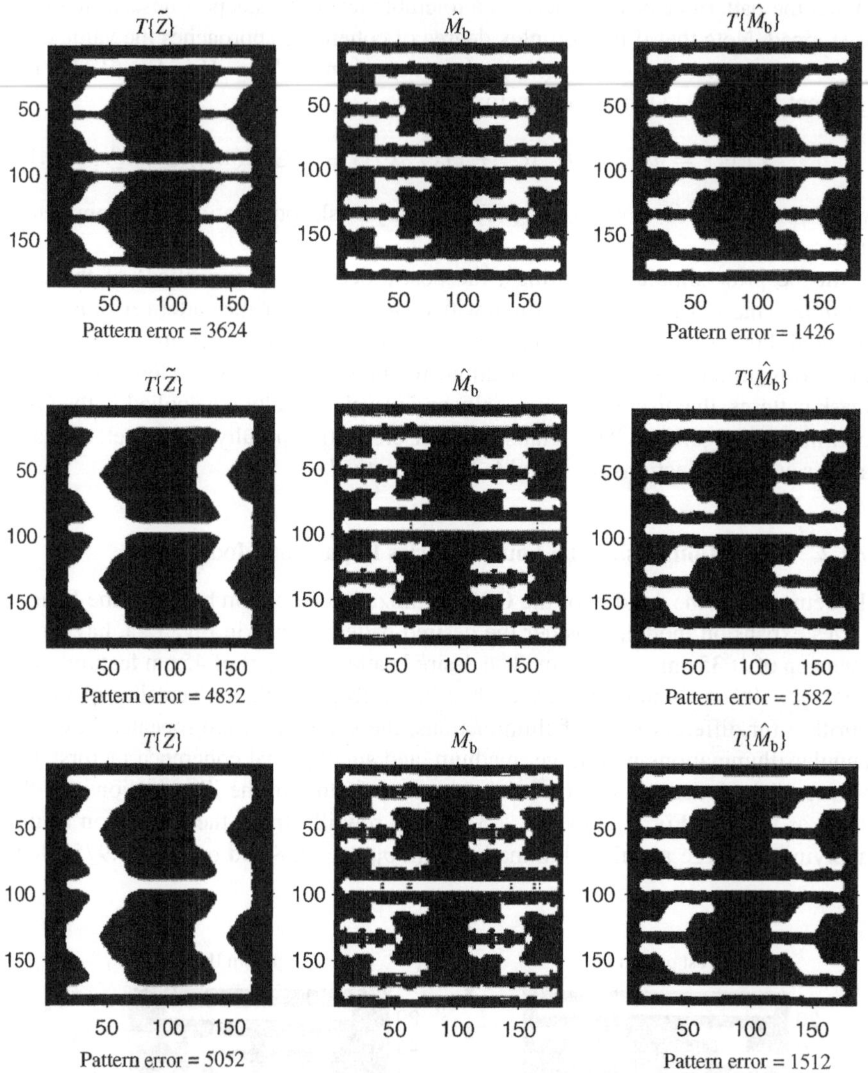

Figure 7.3 OPC optimization using the Fourier series expansion model. *Left to right*: The output pattern when the target pattern is used as input, the binary optimized mask, and the output pattern of binary optimized mask. *Top row*: Illustrates the simulations using the annular illumination with large partial coherence factor ($\sigma_{\text{inner}} = 0.8$ and $\sigma_{\text{outer}} = 0.975$). *Middle row*: Illustrates the simulations using the annular illumination with medium partial coherence factor ($\sigma_{\text{inner}} = 0.5$ and $\sigma_{\text{outer}} = 0.6$). *Bottom row*: Illustrates the simulations using the annular illumination with small partial coherence factor ($\sigma_{\text{inner}} = 0.3$ and $\sigma_{\text{outer}} = 0.4$).

the output pattern when the desired pattern is inputted $\left(T\{\tilde{Z}\}\right)$, (center) the binary optimized mask ($\hat{M}_b$) using the Fourier series expansion model, and (right) the output pattern of binary optimized mask $\left(T\{\hat{M}_b\}\right)$. The middle row shows the simulation results applying the medium annular illumination with $\sigma_{\text{inner}} = 0.5$ and $\sigma_{\text{outer}} = 0.6$.

The bottom row shows the simulation results applying the small annular illumination with $\sigma_{\text{inner}} = 0.3$ and $\sigma_{\text{outer}} = 0.4$. The Fourier series expansion coefficients of the annular illumination is

$$\Gamma_{\mathbf{m}} = \begin{cases} \frac{4D_{\text{cu}}^2 D_{\text{cl}}^2}{\pi D^2 \left(D_{\text{cl}}^2 - D_{\text{cu}}^2\right)}, & \text{for } D/2D_{\text{cl}} \leq |\mathbf{m}| \leq D/2D_{\text{cu}}, \\ 0, & \text{elsewhere,} \end{cases} \tag{7.16}$$

where NA $= 1.25$ and $\lambda = 193$ nm. D_{cl} and D_{cu} are the coherent lengths of the inner and outer circles, respectively. $\sigma_{\text{inner}} = \frac{\lambda}{2D_{\text{cl}}\text{NA}}$ and $\sigma_{\text{outer}} = \frac{\lambda}{2D_{\text{cu}}\text{NA}}$ are the corresponding inner and outer partial coherence factors. The convolution kernel is

$$h(\mathbf{r}) = \frac{J_1(2\pi r \text{NA}/\lambda)}{2\pi r \text{NA}/\lambda}. \tag{7.17}$$

In the simulations, we assume $h(\mathbf{r})$ vanishes outside the area A_h defined by $x, y \in [-56.25\text{ nm}, 56.25\text{ nm}]$. In the sigmoid function, we assign parameters $a = 25$ and $t_r = 0.19$. The global threshold is $t_m = 0.5$, the constant $k = 0.29$, and the pixel size is $5.625\text{ nm} \times 5.625\text{ nm}$. Step length and the regularization weights are $s_{\underline{\theta}} = 2$, $\gamma_D = 0.025$, and $\gamma_{WA} = 0.025$. The shape of the initial mask pattern is the same as that of the desired binary output pattern $\tilde{Z}$. For $\underline{\theta}$, we assign the value $\frac{\pi}{5}$ corresponding to the areas having a magnitude of 1 and $\frac{4\pi}{5}$ for the areas having magnitude of 0. The number of Fourier series terms to represent the SOCS model is 52, 12, and 12 for the large, medium, and small annular illuminations, respectively.

To show the stability of the described OPC optimization algorithm, consider another desired pattern shown in Fig. 7.2b, having dimension of $1035\text{ nm} \times 1035\text{ nm}$. The desired pattern consists of 45 nm features. The pitch $p = 90$ nm is indicated by dashed lines. The simulations are repeated in Fig. 7.4. All the parameters are the same as those in Fig. 7.3. As shown in Figs. 7.3 and 7.4, this approach has proved efficient in our extensive simulation analysis. Figures 7.5 and 7.6 illustrate the convergence of the OPC optimization algorithm based on Fourier series expansion model. In Figs. 7.5 and 7.6, the pattern errors are evaluated by the Fourier series expansion model. Solid, dashed, and dotted lines represent the convergence curves corresponding to large, medium, and small annular illumination, respectively.

7.1.3 OPC Design Algorithm Using the Average Coherent Approximation Model

The optimization process based on the Fourier series expansion model suffers from expensive computational cost. For a mask pattern with dimension $N \times N$, the number of multiplication operations in each iteration is $(2T^2 + 4T)N^4 + (10T + 2)N^2 + 2$, where T is the number of Fourier series terms used to represent the PCI. According to Eq. (2.23), $T \sim CN^2$. The computational cost is thus in the order of $O(2C^2N^8) \sim O(C'N^8)$, where $C' = 2C^2$. In general, N is large. Therefore, the development of a fast algorithm is desired. Based on the average coherent approximation model, a fast algorithm, referred to as the average coherence approximation algorithm (ACAA), is presented in this section to reduce the computation complexity.

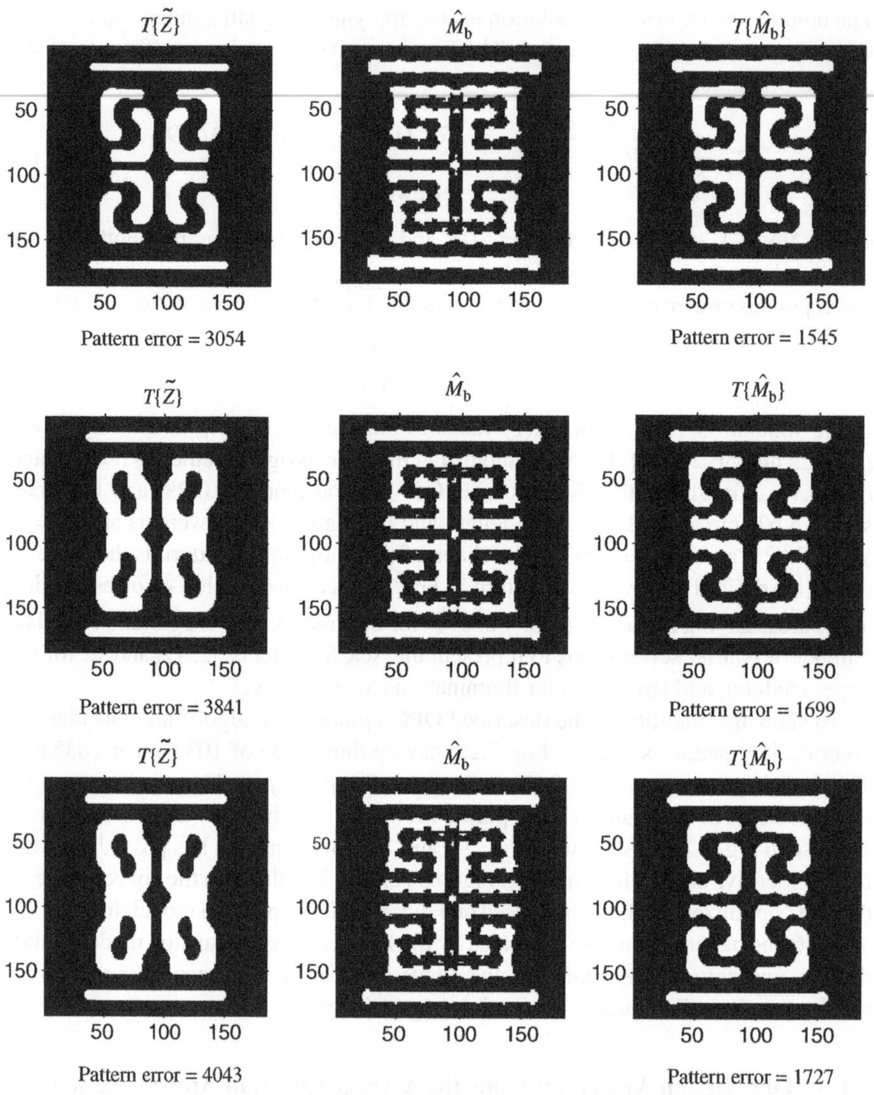

Figure 7.4 OPC optimization using the Fourier series expansion model. *Left to right*: The output pattern when the target pattern is used as input, the binary optimized mask, and the output pattern of binary optimized mask. *Top row*: Illustrates the simulations using the annular illumination with large partial coherence factor ($\sigma_{\text{inner}} = 0.8$ and $\sigma_{\text{outer}} = 0.975$). *Middle row*: Illustrates the simulations using the annular illumination with medium partial coherence factor ($\sigma_{\text{inner}} = 0.5$ and $\sigma_{\text{outer}} = 0.6$). *Bottom row*: Illustrates the simulations using the annular illumination with small partial coherence factor ($\sigma_{\text{inner}} = 0.3$ and $\sigma_{\text{outer}} = 0.4$).

According to the average coherent approximation model described in Section 2.2.3, the forward imaging process based on the average coherent approximation model is illustrated in Fig. 7.7, where the image formation stage is decomposed into the superposition of a coherent and an incoherent illumination component.

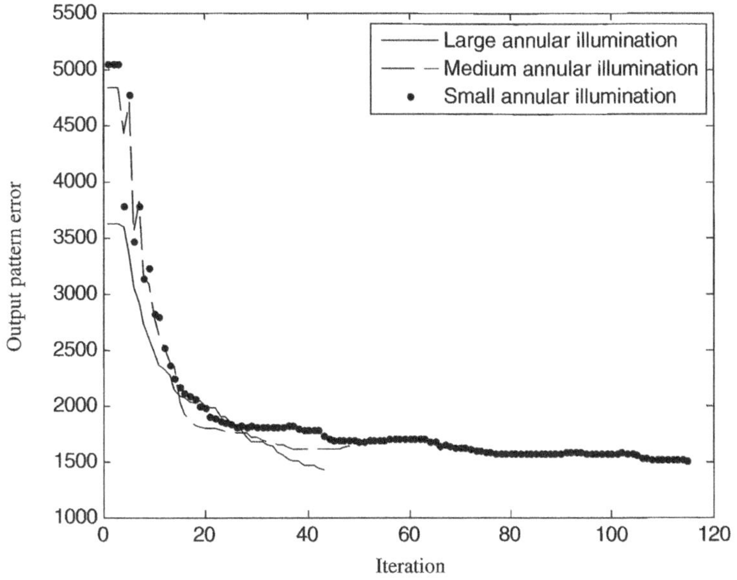

Figure 7.5 Convergence of the OPC optimization algorithm versus steepest descent iterations for Fig. 7.3.

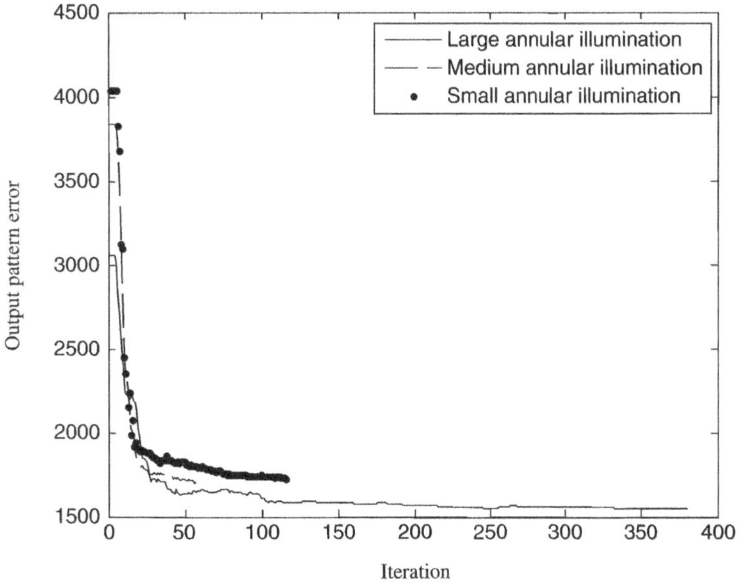

Figure 7.6 Convergence of the OPC optimization algorithm versus steepest descent iterations for Fig. 7.4.

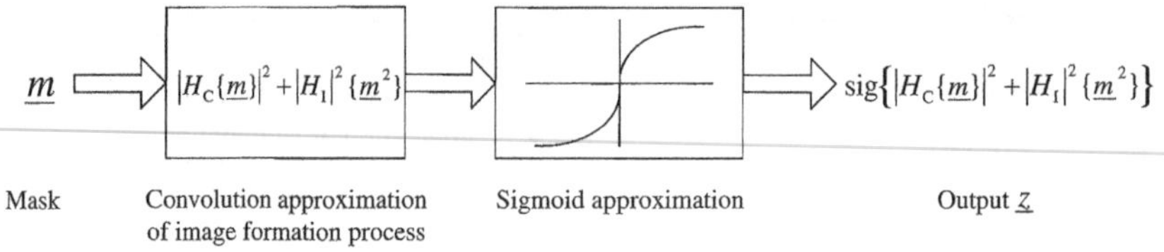

Figure 7.7 Approximated forward process model based on the average coherent approximation model.

Equations (7.5), (7.8), and (7.11) are thus modified to account for the use of average coherence approximation, leading to

$$\underline{z}_i = \frac{1}{1+\exp\left[-a\left(\left|\sum_{j=1}^{N^2} h_{Cij}\underline{m}_j\right|^2 + \sum_{j=1}^{N^2} |h_{Iij}|^2 |\underline{m}_j|^2\right) + at_r'\right]}, \qquad i = 1, \ldots, N^2, \tag{7.18}$$

where h_{Cij} is the i, jth entry of the equivalent amplitude impulse response of the coherent component in the entire partially coherent imaging system. h_{Iij} is the i, jth entry of the equivalent amplitude impulse response of the incoherent component. The cost function is then formulated as

$$\begin{aligned} F(\underline{\theta}) &= \|\underline{\tilde{z}} - \underline{z}\|_2^2 = \sum_{i=1}^{N^2} (\tilde{z}_i - z_i)^2 \\ &= \sum_{i=1}^{N^2} \left(\tilde{z}_i - \frac{1}{1+\exp\left[-a\left(\left|\sum_{j=1}^{N^2} h_{Cij}\underline{m}_j\right|^2 + \sum_{j=1}^{N^2} |h_{Iij}|^2 |\underline{m}_j|^2\right) + at_r'\right]} \right)^2. \end{aligned} \tag{7.19}$$

The gradient of the cost function is

$$\begin{aligned} \nabla F(\underline{\theta}) &= \underline{d}_\theta = 2a\left\{\left(H_C^T\left[(\underline{\tilde{z}} - \underline{z}) \odot \underline{z} \odot (\underline{1} - \underline{z}) \odot (H_C(\underline{m})\right]\right\} \odot \sin\underline{\theta} \\ &\quad + 2a\left\{\left(H_I^{2T}\left[(\underline{\tilde{z}} - \underline{z}) \odot \underline{z} \odot (\underline{1} - \underline{z}) \odot (\underline{m})\right]\right)\right\} \odot \sin\underline{\theta}, \end{aligned} \tag{7.20}$$

where t_r' is the process threshold for the average coherent approximation model. The derivation of Eq. (7.20) is shown in Appendix D. In general $t_r' \neq t_r$ and it must be estimated *a priori* such that it leads to similar binary output pattern as that of the Fourier series expansion model. The aerial imaging synthesis can be implemented based on the two models. Thus, given the t_r for the Fourier series expansion model,

t_r' can be found using a line search process. The two-dimensional filters h_c and h_i are described in Eqs.(2.35) and (2.36). The convolution matrices H_C and H_I are each of size $N^2 \times N^2$ with equivalent two-dimensional filters h_c and h_i, respectively. For a mask pattern with dimension $N \times N$, the number of multiplication operations in each iteration is equal to $8N^4 + 12N^2 + 4$. The computational cost of this second approach is in the order of $O(8N^4)$. Compared with the computational complexity of the algorithm based on Fourier series expansion model, and ignoring the lower order of large number, the reduction of the computational complexity is in the order of $\frac{1}{C''N^4}$, where $C'' = \frac{C'}{8}$. When N is much larger than 1, the fast algorithm is significantly more efficient. The drawback of ACAA is that the error of the corresponding optimized output pattern can be higher, because of the inaccuracy of the average coherent approximation model. Nevertheless, the simulations in Section 7.1.4 will show that ACAA is effective for the inverse lithography problem.

7.1.4 Simulations Using the Average Coherent Approximation Model

Although the accuracy of the approximation model depends on various parameters, it is shown in the following simulations that it is in general effective for the inverse lithography problem. In Fig. 7.8, the simulations shown in Fig. 7.3 are repeated using ACAA, with $s_{\underline{\theta}} = 0.5$. The weight of the discretization penalty $\gamma_D = 0.025$, and the weight of the wavelet penalty $\gamma_{WA} = 0.025$. $t_r' = 0.09$, 0.095, and 0.17 for the large, medium, and small annular illuminations, respectively. The simulation time is effectively reduced by an order of magnitude, indicating that the ACAA is indeed computationally efficient. However, it is found that the optimized output pattern errors somewhat increase, due to the reduced accuracy of the average coherent approximation model. Nevertheless, this approach can also effectively add subresolution blocks to the optimized mask patterns and compensate for the distortion in the optical system.

To show the stability of the described OPC optimization algorithm, the simulations in Fig. 7.4 are repeated in Fig. 7.9. In Fig. 7.9, all the parameters are the same as those in Fig. 7.8, except for that $t_r' = 0.19$ for the small annular illumination. As shown in Figs. 7.8 and 7.9, ACAA has proved efficient in the extensive simulation analysis. Figures 7.10 and 7.11 illustrate the convergence of the OPC optimization algorithm based on the average coherent approximation model. In Figs. 7.10 and 7.11, the pattern errors are evaluated by the average coherent approximation model. Solid, dashed, and dotted lines represent the convergence curves corresponding to large, medium, and small annular illuminations, respectively.

7.1.5 Discussion and Comparison

As mentioned in Section 7.1.3, the computational complexity reduction of the algorithm based on the average coherent approximation model is in the order of $\frac{1}{C''N^4}$, compared with the algorithm based on the Fourier series expansion model. To

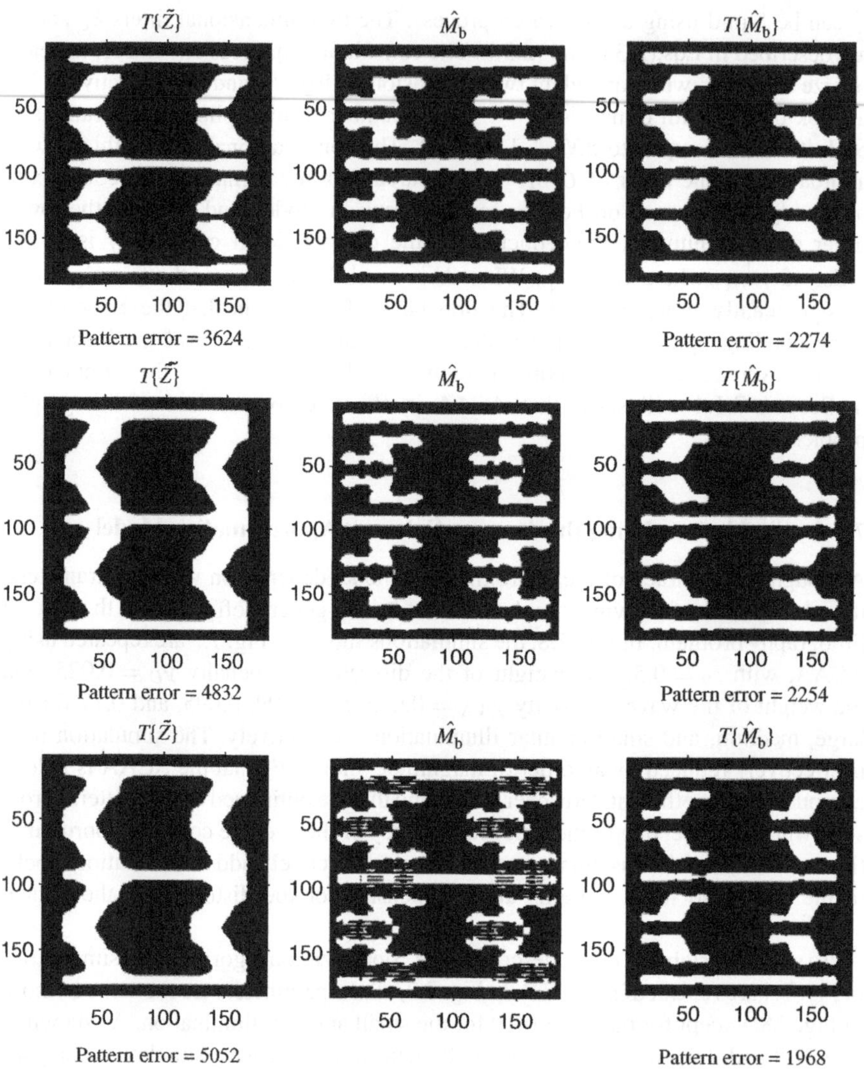

Figure 7.8 OPC optimization using the average coherent approximation model. *Left to right*: The output pattern when the target pattern is used as input, the binary optimized mask, and the output pattern of binary optimized mask. *Top row*: Illustrates the simulations using the annular illumination with large partial coherence factor ($\sigma_{\text{inner}} = 0.8$ and $\sigma_{\text{outer}} = 0.975$). *Middle row*: Illustrates the simulations using the annular illumination with medium partial coherence factor ($\sigma_{\text{inner}} = 0.5$ and $\sigma_{\text{outer}} = 0.6$). *Bottom row*: Illustrates the simulations using the annular illumination with small partial coherence factor ($\sigma_{\text{inner}} = 0.3$ and $\sigma_{\text{outer}} = 0.4$).

illustrate the computational efficiency of both algorithms, the computational running time is illustrated here. The constructions of the inverse masks shown in Figs. 7.3 and 7.8 with medium partial coherence factor took 18 min for the first approach and 55 s for the second approach. The constructions of the inverse masks shown in Figs. 7.4

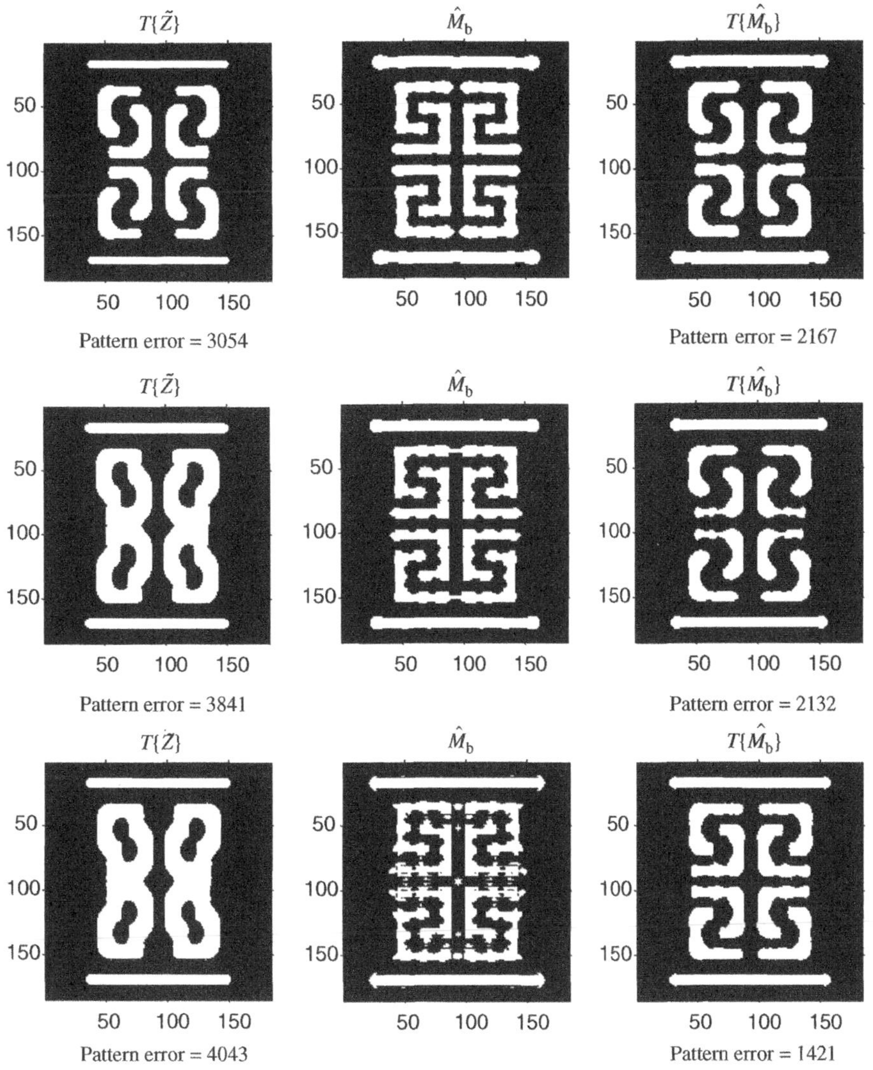

Figure 7.9 OPC optimization using the average coherent approximation model. *Left to right*: The output pattern when the target pattern is used as input, the binary optimized mask, and the output pattern of binary optimized mask. *Top row*: Illustrates the simulations using the annular illumination with large partial coherence factor ($\sigma_{\text{inner}} = 0.8$ and $\sigma_{\text{outer}} = 0.975$). *Middle row*: Illustrates the simulations using the annular illumination with medium partial coherence factor ($\sigma_{\text{inner}} = 0.5$ and $\sigma_{\text{outer}} = 0.6$). *Bottom row*: Illustrates the simulations using the annular illumination with small partial coherence factor ($\sigma_{\text{inner}} = 0.3$ and $\sigma_{\text{outer}} = 0.4$).

and 7.9 with medium partial coherence factor took 5.5 min for the first approach and 62 s for the second approach. The computation was done on an Intel Pentium4 CPU 3.40 GHz, 1.00 GB of RAM. This difference would be scaled with the dimension of the mask being constructed.

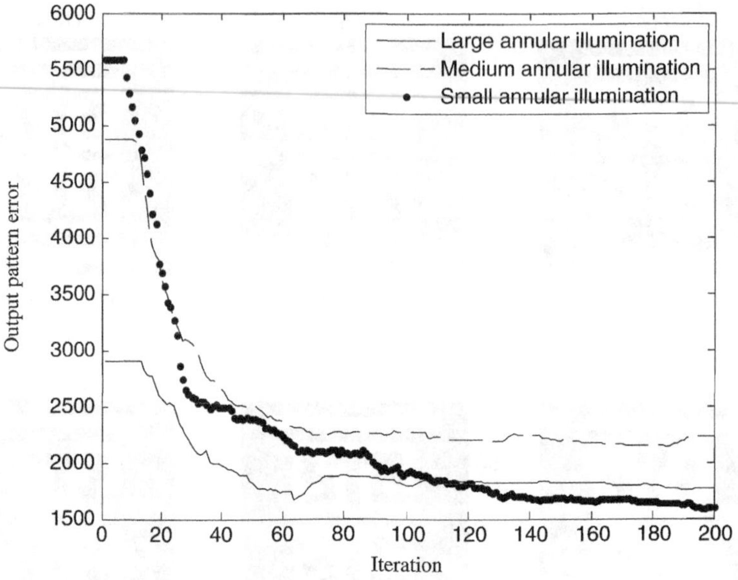

Figure 7.10 Convergence of the OPC optimization algorithm versus steepest descent iterations for Fig. 7.8.

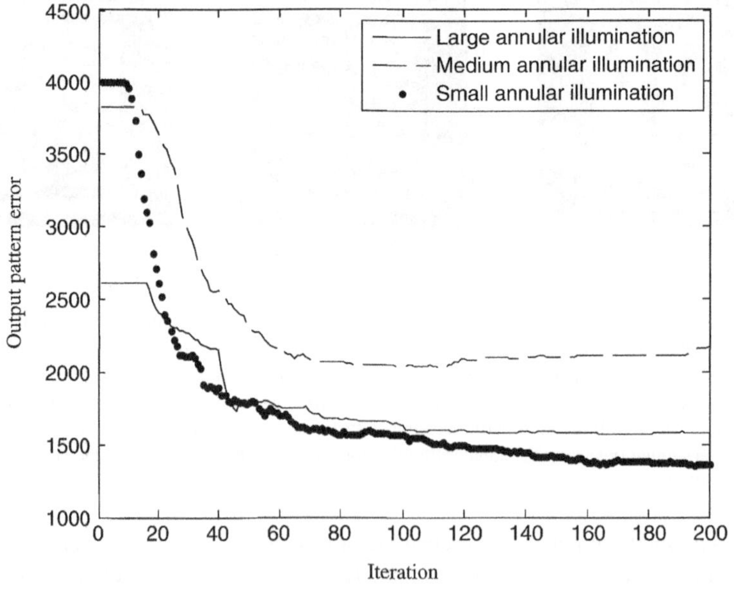

Figure 7.11 Convergence of the OPC optimization algorithm versus steepest descent iterations for Fig. 7.9.

On the other hand, based on the simulations in Figs. 7.3, 7.4, 7.8, and 7.9, the error reduction of the first approach is about 60% on average, while 48% on average for the second approach. The error increase of the second approach comes from the reduced accuracy of average coherent approximation model. In conclusion, the Fourier series expansion model gives an accurate representation of PCI, while the average coherent approximation model supplies approaches of faster imaging synthesis and analysis. Therefore, the second algorithm outperforms the first one in the computational complexity comparison. However, the first algorithm outperforms the second one in the performance comparison. It should be noted that other factors such as the treatment of boundary regions and hierarchy management can affect the overall run time and should be taken into account for the large- scale mask designs.

7.2 PSM OPTIMIZATION

In optical lithography systems, phase-shifting masks provide no advantages in the completely incoherent case, while they make their most significant contributions to the output intensity in the completely coherent case [92]. Common partially coherent illumination modes lie between these two limits, and include dipole, quadrupole, and annular shapes, which provide small to large partial coherence factors. Illumination with large partial coherence factors is closer to the completely incoherent illumination case, while small partial coherence factors approach completely coherent illumination. There are some trade-offs in the extent that partial coherence is used. Large partial coherence factors, such as $\sigma = 0.9$, lead to improvements on resolution and contrast. On the other hand, small partial coherence factors, such as $\sigma = 0.3$, have the advantage to form sparse patterns, which can be exploited effectively by phase-shifting masks. Medium partial coherence factors such as $\sigma = 0.5$ and $\sigma = 0.6$ are preferred for mask pattern containing both sparse and dense patterns. The smallest usable partial coherence factor is approximately $\sigma = 0.3$ [92].

While gradient-based inverse lithography optimization methods have been studied extensively in the past two decades for the case of coherent illumination, equivalent methods for inverse lithography under partially coherent illumination have not been addressed until recently [17]. In Section 7.1, the Fourier series expansion model and the average coherent approximation model are used to develop gradient-based binary mask design algorithms for inverse lithography under partially coherent illuminations. The goal of this section is to extend the concepts in Section 7.1 to focus on the development of gradient-based inverse optimization algorithms for the design of PSM under partially coherent illumination. The following PSM optimization algorithm just takes the alternating PSMs into account, where the optimized two-phase PSM consists of pixel values of -1, 0, or 1. Thus, this PSM optimization algorithm is the extension of the two-phase PSM optimization algorithm discussed in Section 5.3. As it will be described later, the described methods are most effective with small to medium partial coherence factors [44, 47]. Since the imaging synthesis and analysis of partially coherent systems are much more complex than the coherent case, SVD is applied to expand the partially coherent imaging equation by eigenfunctions into a sum of coherent systems [13, 44, 47, 58]. An iterative optimization framework for the PSM design is formulated when the partially coherent imaging system is approximated by the first-order coherent approximation corresponding to the largest eigenvalue. The first-order coherent approximation removes the influence among different coherent components during the inverse optimization process and reduces the computational complexity of the algorithms.

7.2.1 PSM Design Algorithm Using the Singular Value Decomposition Model

According to Section 2.1.2, the Hopkins diffraction model in frequency domain is formulated as

$$I(x, y) = \iiiint_{-\infty}^{+\infty} \mathrm{TCC}(f_1, g_1; f_2, g_2)\tilde{M}(f_1, g_1)\tilde{M}^*(f_2, g_2) \times \exp\{-i2\pi[(f_1 - f_2)x + (g_1 - g_2)y]\}\mathrm{d}f_1\mathrm{d}g_1\mathrm{d}f_2\mathrm{d}g_2, \quad (7.21)$$

where $\tilde{M}(f_1, g_1)$ and $\tilde{M}(f_2, g_2)$ are the Fourier transforms of $M(x_1, x_2)$ and $M(x_2, y_2)$, respectively. $\mathrm{TCC}(f_1, g_1; f_2, g_2)$ is the transmission cross-coefficient, indicating the interaction between $\tilde{M}(f_1, g_1)$ and $\tilde{M}(f_2, g_2)$. Specifically,

$$\mathrm{TCC}(f_1, g_1; f_2, g_2) = \iint_{-\infty}^{+\infty} \tilde{\gamma}(f, g)\tilde{h}(f + f_1, g + g_1)\tilde{h}^*(f + f_2, g + g_2)\mathrm{d}f\mathrm{d}g, \quad (7.22)$$

where $\tilde{\gamma}(f, g)$, referred to as the effective source, is the Fourier transform of $\gamma(x, y)$. $\tilde{h}(f, g)$ is the Fourier transform of $h(x, y)$. Let $M(x, y)$ be the input phase-shifting mask to an optical lithography system $T\{\cdot\}$, with a partially coherent illumination. The PCI optical system is represented by the first-order coherent approximation corresponding to the largest eigenvalue in the SVD model. The details of the SVD model are discussed in Section 2.2.2. Given a $N \times N$ desired output pattern $\tilde{Z}(x, y)$, the PSM inverse lithography optimization problem can thus be formulated as the search of $\hat{M}(x, y)$ over the $N \times N$ real space $\Re^{N\times N}$ such that

$$\hat{M}(x, y) = \arg \min_{M(x,y)\in\Re^{N\times N}} d\left(T\{M(x, y)\}, \tilde{Z}(x, y)\right), \quad (7.23)$$

where $d(\cdot, \cdot)$ is the square of the l^2 norm criterion. The pixel values of $M(x, y)$ lie in the range of $[-1, 1]$.

Figure 7.12 depicts the approximated forward process model based on the SVD model [44, 47]. In this optimization approach, the convolution matrix H_1 is a $N^2 \times N^2$ matrix. Its equivalent two-dimensional filter is the first equivalent kernel $h_1(x, y)$ corresponding to the first-order coherent approximation in the SVD model. The output of the sigmoid function is the $N \times N$ image denoted as

$$Z = \mathrm{sig}\{|H_1\{\underline{m}\}|^2\}. \quad (7.24)$$

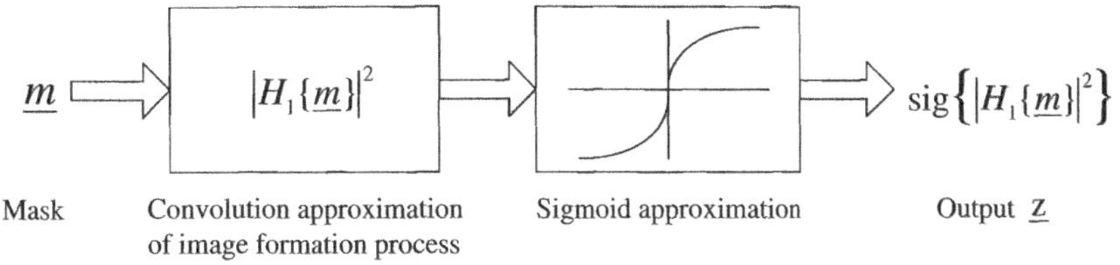

Figure 7.12 Approximated forward process model based on the SVD model.

The equivalent vector is denoted as $\underline{z}$. According to Eq. (7.24), the ith entry of vector $\underline{z}$ can be represented as

$$\underline{z}_i = \frac{1}{1+\exp\left[-a\left|\sum_{j=1}^{N^2} h_{1,ij}\underline{m}_j\right|^2 + at_r\right]}, \quad i = 1, \ldots, N^2, \tag{7.25}$$

where $h_{1,ij}$ is the i, jth entry of the first equivalent kernel $h_1(x, y)$. The optimized $N \times N$ real-valued mask denoted as $\hat{M}$ minimizes the distance between Z and $\tilde{Z}$. Its equivalent vector is denoted as $\underline{\hat{m}} \in [-1, 1]$. The optimized two-phase mask $\hat{M}_\mathrm{p}$ is the quantization of $\hat{M}$. Its equivalent vector is denoted as $\underline{\hat{m}}_\mathrm{p}$, with all entries constrained to -1, 0, or 1. Thus, the two-phase PSM optimization can be formulated as

$$\underline{\hat{m}} = \arg\min_{\underline{\hat{m}}}\{F(\underline{m})\}, \tag{7.26}$$

where the cost function $F(\cdot)$ is defined as

$$F(\underline{m}) = \|\underline{\tilde{z}} - \underline{z}\|_2^2 = \sum_{i=1}^{N^2}(\underline{\tilde{z}}_i - \underline{z}_i)^2, \tag{7.27}$$

where $\underline{z}_i$ is already represented in Eq. (7.25). To reduce the above bound-constrained optimization problem to an unconstrained optimization problem, we adopt the following parameter transformation: $\underline{m}_k = \cos(\underline{\theta}_k)$, $k = 1, \ldots, N^2$, where $\underline{\theta}_k \in (-\infty, \infty)$ and $\underline{m}_k \in [-1, 1]$. Defining the vector $\underline{\theta} = [\theta_1, \ldots, \theta_{N^2}]^T$, Eq. (7.26) is adjusted as

$$(\underline{\hat{\theta}}) = \arg\min_{\underline{\theta}}\{F(\underline{\theta})\}$$

$$= \arg\min_{\underline{\theta}}\left\{\sum_{i=1}^{N^2}\left(\tilde{z}_i - \frac{1}{1+\exp\left[-a\left|\sum_{j=1}^{N^2} h_{1,ij}\cos\underline{\theta}_j\right|^2 + at_r\right]}\right)^2\right\}. \tag{7.28}$$

The steepest descent method is used to optimize the above problem. The gradient $\nabla F(\theta)_{\theta}$ can be calculated as follows:

$$\nabla F(\underline{\theta}) = 2a\left\{\left(H_1^{*H}\left[(\underline{\tilde{z}} - \underline{z}) \odot \underline{z} \odot (\underline{1} - \underline{z}) \odot (H_1^{*}\underline{m})\right]\right)\right\} \odot \sin \underline{\theta}$$
$$+ 2a\left\{\left(H_1^{H}\left[(\underline{\tilde{z}} - \underline{z}) \odot \underline{z} \odot (\underline{1} - \underline{z}) \odot (H_1\underline{m})\right]\right)\right\} \odot \sin \underline{\theta}, \tag{7.29}$$

where $\nabla F(\underline{\theta}) \in \Re^{N^2 \times 1}$, $\odot$ is the element-by-element multiplication operator, and $\underline{1} = [1, \ldots, 1]^T \in \Re^{N^2 \times 1}$.

The iterative optimization above, in general, leads to a real-valued mask with pixel values between -1 and 1. Therefore, a post-processing step described in Eq. (5.27) is used to obtain the optimized pole-level PSM, that is, $\hat{\underline{m}}_{pk} = \mathrm{sgn}(\hat{\underline{m}}_k)U(|\hat{\underline{m}}|_k - t_m)$, $k = 1, \ldots, N^2$, where t_m is the global threshold. We define the pattern error E as the distance between the desired output image $\tilde{Z}$ and the actual binary output pattern Z_b evaluated by Eq. (7.21) and a hard threshold operator,

$$E = \sum_{i=1}^{N^2} |\tilde{\underline{z}}_i - \underline{z}_{bi}| = \sum_{i=1}^{N^2} \left|\tilde{\underline{z}}_i - U_i(|H_1\underline{m}_b|^2 - t_r)\right|$$
$$= \sum_{i=1}^{N^2} \left|\tilde{\underline{z}}_i - \Lambda_i(|H_1\underline{m}_b|^2)\right|. \tag{7.30}$$

7.2.2 Discretization Regularization for PSM Design Algorithm

In the prior simulation settings, the fact that the estimated mask pattern should be trinary is not considered. To attain a near-trinary optimized mask pattern through the optimization process, we adopt the discretization penalty in Ref. [67]. This discretization penalty is used to reduce the error increase incurred by the post-processing step, which is different from the discretization penalty discussed in Section 6.1.2. The formulation of the discretization penalty is summarized as following. For each pixel value, the discretization penalty term is

$$r_D(\underline{m}_k) = -4.5\underline{m}_k^4 + \underline{m}_k^2 + 3.5, \quad k = 1, \ldots, N^2. \tag{7.31}$$

Thus, the cost function in Eq. (7.27) is adjusted as $J(\underline{m}) = F(\underline{m}) + \gamma_D R_D(\underline{m})$. In our simulations for $\sigma = 0.3$, discretization regularization attains near-trinary optimized mask and reduces 30% output pattern error. For $\sigma = 0.6$, discretization regularization reduces 32% of the output pattern error.

7.2.3 Simulations

To demonstrate the validity of the optimization algorithms, consider the desired pattern shown in Fig. 7.13 having dimension of 561 nm × 561 nm. The matrices representing all the patterns have dimension of $N \times N$, where $N = 51$. The pixel size

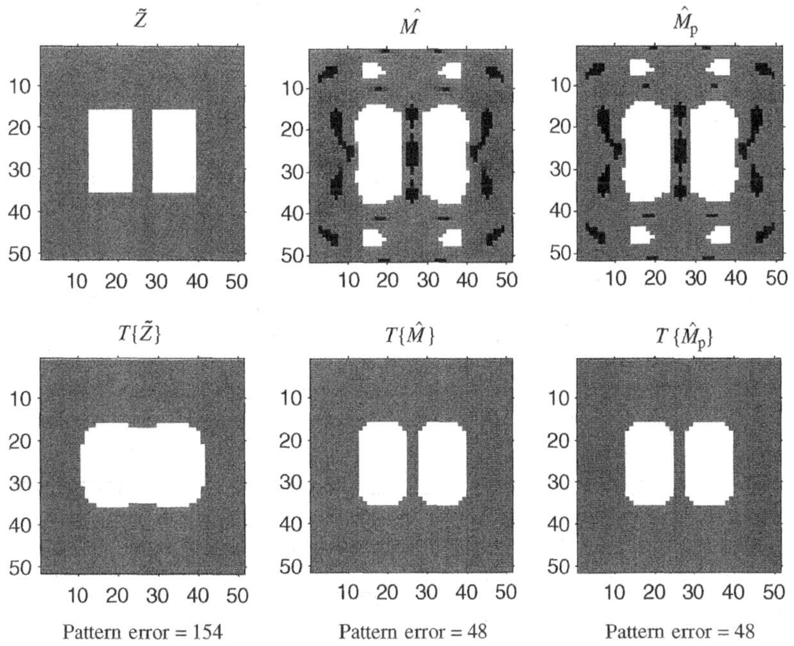

Figure 7.13 Two-phase PSM optimization using SVD model. *Top row* (input masks) (left to right): Desired pattern, optimized real-valued mask, and optimized trinary mask. *Bottom row*: Illustrates the corresponding binary output patterns. White, gray, and black represent 1, 0, and −1, respectively. $\sigma = 0.3$.

is 11 nm × 11 nm. The partially coherent illumination is a circular illumination with small partial coherence factor $\sigma = 0.3$.

In Fig. 7.13, the top row illustrates the input masks of (left) the desired pattern $\tilde{Z}$, (center) the optimized real-valued mask $\hat{M}$, and (right) the optimized trinary mask $\hat{M}_\mathrm{p}$. The optimized trinary mask, referred to as the alternating phase-shifting mask, includes clear areas and shifting areas, which introduce 180° phase difference with each other. The binary output patterns are shown in the bottom row. White, gray, and black represent 1, 0, and −1, respectively. The effective source is

$$\tilde{\gamma}(f, g) = \frac{\lambda^2}{\pi(\sigma \mathrm{NA})^2}\mathrm{circ}\left(\frac{\lambda\sqrt{f^2+g^2}}{\sigma \mathrm{NA}}\right)$$
$$= \begin{cases} \frac{\lambda^2}{\pi(\sigma \mathrm{NA})^2}, & \text{for } \sqrt{f^2+g^2} \le \frac{\sigma \mathrm{NA}}{\lambda} \\ 0, & \text{elsewhere,} \end{cases} \quad (7.32)$$

where NA = 1.35 and $\lambda = 193$ nm. The amplitude impulse response is

$$h(\mathbf{r}) = h(x, y) = \frac{J_1(2\pi r \mathrm{NA}/\lambda)}{2\pi r \mathrm{NA}/\lambda}. \quad (7.33)$$

The Fourier transform of $h(x, y)$ is

$$\tilde{h}(f, g) = \frac{\lambda^2}{\pi(\mathrm{NA})^2}\mathrm{circ}\left(\frac{\lambda\sqrt{f^2+g^2}}{\mathrm{NA}}\right)$$
$$= \begin{cases} \frac{\lambda^2}{\pi(\mathrm{NA})^2}, & \text{for } \sqrt{f^2+g^2} \leq \frac{\mathrm{NA}}{\lambda}, \\ 0, & \text{elsewhere.} \end{cases} \quad (7.34)$$

In the sigmoid function, we assign parameters $a = 200$ and $t_r = 0.003$. The binary output patterns in the bottom row are evaluated by Eq. (7.21) followed by a hard threshold operator with threshold $\bar{t}_r = tr \times \sum_{k=1}^{N^2} \alpha_k$. The global threshold is $t_m = 0.33$. The step length and the regularization weights are $s_{\underline{\theta}} = 0.2$ and $\gamma_D = 0.1$. The shape of the initial mask pattern is the same as that of the desired binary output pattern $\tilde{Z}$. For $\underline{\theta}$, we assign $\frac{\pi}{5}$ corresponding to the areas having a magnitude of 1 and $\frac{\pi}{2}$ for the areas having a magnitude of 0. As shown in Fig. 7.13, this approach is effective for small partial coherence factors. Figure 7.14 illustrates the convergence of the two-phase PSM optimization algorithm, where the pattern error is evaluated by the first-order coherent approximation of the SVD model. These results are consistent with those obtained in other numerous simulations we have ran with small partial coherence factors.

These simulations are then repeated in Fig. 7.15 with the same parameters, except for the value of the partial coherence factor which in this case is raised to $\sigma = 0.6$. Figure 7.16 illustrates the convergence of the two-phase PSM optimization algorithm,

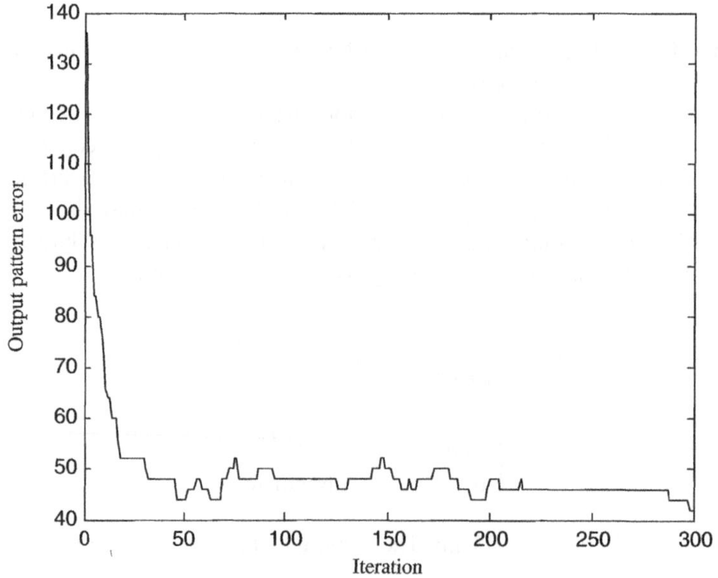

Figure 7.14 Convergence of the two-phase PSM optimization algorithm versus steepest descent iterations for Fig. 7.13.

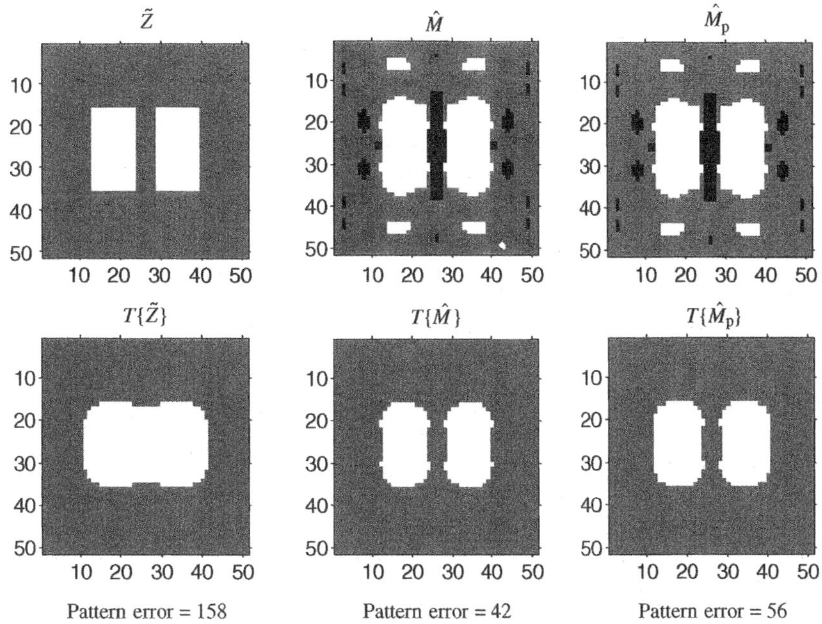

Figure 7.15 Two-phase PSM optimization using SVD model. *Top row* (input masks) (left to right): Desired pattern, optimized real-valued mask, and optimized trinary mask. *Bottom row*: Illustrates the corresponding binary output patterns. White, gray, and black represent 1, 0, and −1, respectively. $\sigma = 0.6$.

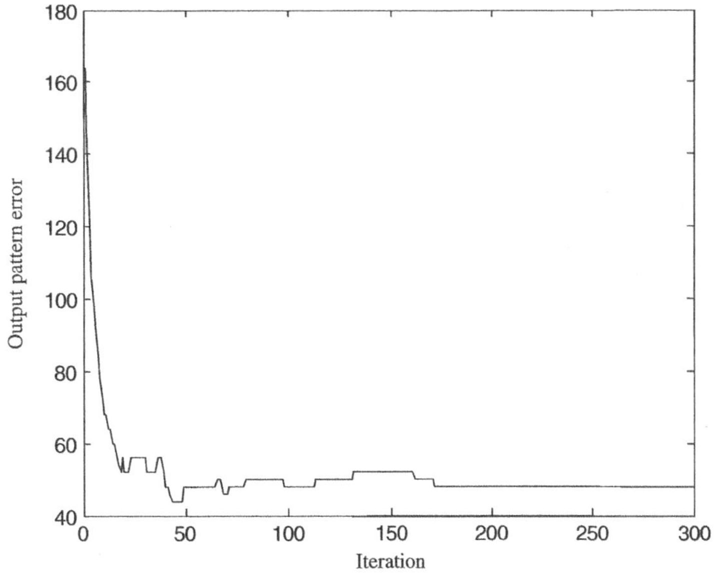

Figure 7.16 Convergence of the two-phase PSM optimization algorithm versus steepest descent iterations for Fig. 7.15.

where the pattern error is evaluated by the first-order coherent approximation of the SVD model. In Fig. 7.15, the output pattern error corresponding to the optimized trinary mask in this case is increased compared with that of Fig. 7.13. This degradation results from the less accurate first-order coherent approximation to the partially coherent system when medium or large partial coherence factors are used. In fact, the SVD approach taken here gradually degrades as the partial coherent factor increases from small to large values. Nevertheless, the optimized PSM attains a 65% reduction of the output pattern error even with the medium partial coherent factor values.

7.3 SUMMARY

This chapter developed a set of RET optimization algorithms in the partially coherent imaging systems. Based on the Fourier series expansion model and the average coherent approximation model, two kinds of OPC optimization algorithms were described in the partially coherent imaging systems. Based on the SVD model, a PSM optimization algorithm was developed, which is most effective with small to medium partial coherence factors.

8

Other RET Optimization Techniques

As a supplemental to the OPC and PSM optimization algorithms discussed in Chapters 5 and 7, a variety of techniques to improve the performance of OPC and PSM optimizations are described in this chapter. First, a double patterning method is developed based on the generalized PSM optimization framework described in Section 5.4. Second, a post-processing based on 2D discrete cosine transform (DCT) is presented to simultaneously reduce the complexity of the mask pattern and the output pattern error. Finally, a photoresist tone reversing method is described to print extremely sparse patterns on the wafer.

8.1 DOUBLE-PATTERNING METHOD

In Section 5.4, a generalized PSM optimization algorithm is developed to design multiphase PSMs capable of generating mask patterns with arbitrary Manhattan geometries. Although this approach is effective in eliminating the phase conflict, it usually results in a phase-shifting mask, which is difficult to fabricate [40, 41]. In the event that multiphase masks are difficult to fabricate, but the goal is to still synthesize masks with arbitrary geometry, a double patterning PSM optimization method is developed in this section. At each stage the PSM masks are constrained to have two phases. The two-stage patterning method can lead to high-fidelity output pattern reproduction at the expense of a more complex exposure process. Thus, the double patterning optimization method is an alternative to the multiphase PSM to avoid the phase conflict. Although the double patterning method is developed under coherent illumination, this approach can be extended to the case of partially coherent imaging systems. This extension is beyond the scope of this book. Poonawala et al. have developed a double exposure method under coherent illumination [65, 66] aimed at avoiding the limitations of the two-phase PSM approach described in Ref. [67]. Their approach used one etching process after the two exposures, whereas the double patterning method presented in this section uses an etching processes after each exposure.

The steps involved in a typical double patterning lithography with positive photoresist process is shown in Fig. 8.1. In this method, the photoresist layer is exposed

Computational Lithography By Xu Ma and Gonzalo R. Arce

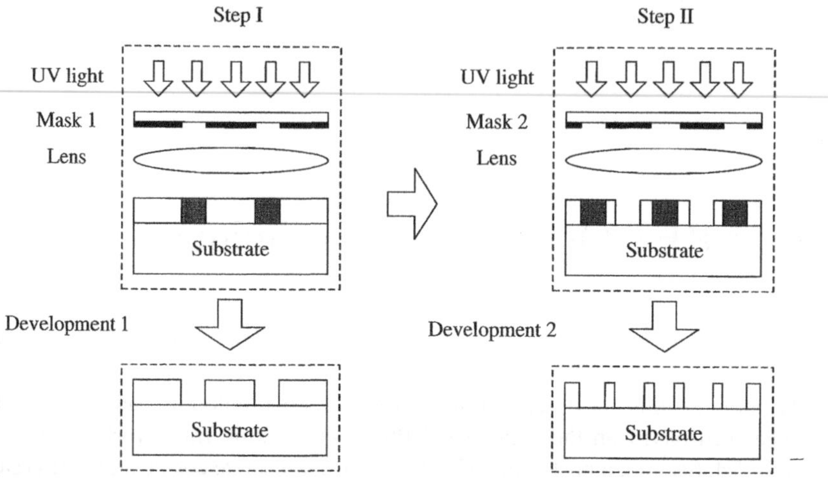

Figure 8.1 The steps involved in a typical double patterning lithography with positive photoresist process.

twice, each with a two-phase mask. After each exposure, an etching process is applied to remove the exposed photoresist. The approximated forward process model of the double patterning lithography system is shown in Fig. 8.2. Assume that the two masks are M_1 and M_2. Their equivalent vectors are $\underline{m}_1$ and $\underline{m}_2$. The optical lithography system is represented by $T\{\cdot\}$, which is illustrated in Fig. 8.2. Considering the coherent illumination case, the image formation process is approximated by a Hopkins diffraction model described in Eq. (2.14). The effect of the photoresist is modeled by a hard threshold operation and approximated by a sigmoid function. The output patterns of the first and second patternings are denoted as $Z_1(x, y)$ and $Z_2(x, y)$, with the equivalent vectors $\underline{z}_1$ and $\underline{z}_2$, respectively. The overall output pattern after two patterning steps is denoted as $Z(x, y) = T\{M_1(x, y), M_2(x, y)\} = U(Z_1(x, y) + Z_2(x, y))$. Given a $N \times N$ desired output pattern $\tilde{Z}(x, y)$, the double patterning optimization problem under coherent illumination can be formulated as the search of $\hat{M}_1(x, y)$ and $\hat{M}_2(x, y)$ over the $N \times N$ complex space $C^{N\times N}$ such that

$$\begin{aligned}&\left(\hat{M}_1(x, y), \hat{M}_2(x, y)\right)\\&= \arg \min_{M_1(x,y), M_2(x,y) \in C^{N\times N}} d\left(T\{M_1(x, y), M_2(x, y)\}, \tilde{Z}(x, y)\right),\end{aligned} \tag{8.1}$$

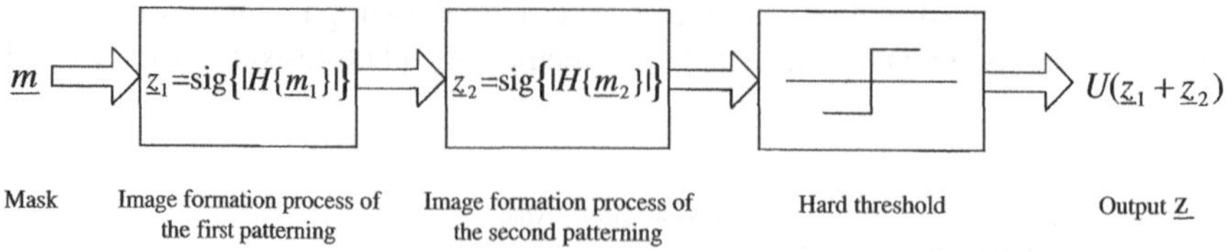

Figure 8.2 Approximated forward process model of the double patterning lithography system.

where $d(\cdot,\cdot)$ is the square of the l^2 norm criterion. The iterative optimization above leads to complex-valued masks that are not constrained to a discrete number of magnitude and phase levels. Additional post-processing steps described in Eqs. (5.40) and (5.42) are used to obtain the pole-level optimized PSMs, referred to as $\hat{M}_{\text{p1}}$ and $\hat{M}_{\text{p2}}$.

Using the parameter transformations as described in Eqs. (5.31) and (5.32),

$$\underline{m}_{1k} = \underline{r}_{1k} e^{j\underline{\theta}_{1k}} = \frac{1+\cos\underline{\phi}_{1k}}{2} e^{j\underline{\theta}_{1k}}, \quad k = 1,\ldots,N^2, \tag{8.2}$$

$$\underline{m}_{2k} = \underline{r}_{2k} e^{j\underline{\theta}_{2k}} = \frac{1+\cos\underline{\phi}_{2k}}{2} e^{j\underline{\theta}_{2k}}, \quad k = 1,\ldots,N^2. \tag{8.3}$$

The corresponding output images are

$$\underline{z}_{1k} = \frac{1}{1+\exp\left[-a\left|\sum_{i=1}^{N^2} h_{ki}\frac{1+\cos\underline{\phi}_{1i}}{2} e^{j\underline{\theta}_{1i}}\right| + at_r\right]}, \quad k = 1,\ldots,N^2, \tag{8.4}$$

$$\underline{z}_{2k} = \frac{1}{1+\exp\left[-a\left|\sum_{i=1}^{N^2} h_{ki}\frac{1+\cos\underline{\phi}_{2i}}{2} e^{j\underline{\theta}_{2i}}\right| + at_r\right]}, \quad k = 1,\ldots,N^2. \tag{8.5}$$

The superposition of $\underline{z}_1$ and $\underline{z}_2$ is the final output pattern $\underline{z}$,

$$\underline{z}_k = U(\underline{z}_{1k} + \underline{z}_{2k} - 1), \quad k = 1,\ldots,N^2, \tag{8.6}$$

where $U(\cdot)$ is a unit step function. Since the derivative of the step function will introduce a Dirac impulse term that is inconvenient for further analysis, a simple approximation is given by the hyperbolic tangent function,

$$\underline{z}_k = U(\underline{z}_{1k} + \underline{z}_{2k} - 1) \approx \frac{1}{2}[\tanh(\underline{z}_{1k} + \underline{z}_{2k} - 1) + 1], \quad k = 1,\ldots,N^2. \tag{8.7}$$

The curves of unit step function and its approximation using hyperbolic tangent function are shown in Fig. 8.3, where the solid and dashed lines represent the unit step function and its approximation, respectively. The cost function is then calculated as

$$\begin{aligned} F &= F(\underline{\theta}_1, \underline{\phi}_1, \underline{\theta}_2, \underline{\phi}_2) = \|\underline{\tilde{z}} - \underline{z}\|_2^2 \\ &= \sum_{i=1}^{N^2}\left\{\underline{\tilde{z}}_i - \frac{1}{2}[\tanh(\underline{z}_{1i} + \underline{z}_{2i} - 1) + 1]\right\}^2. \end{aligned} \tag{8.8}$$

Therefore, the gradients $\nabla F_{\underline{\theta}_1}$, $\nabla F_{\underline{\phi}_1}$, $\nabla F_{\underline{\theta}_2}$, and $\nabla F_{\underline{\phi}_2}$ can be calculated as

$$\begin{aligned} \nabla F_{\underline{\theta}_\text{p}} &= a \times \frac{1+\cos\underline{\phi}_p}{2} \odot \sin\underline{\theta}_p \odot \{H^T[(\underline{\tilde{z}} - \underline{z}) \odot \text{sech}^2(\underline{z}_1 + \underline{z}_2 - \underline{1}) \\ &\odot \underline{z}_\text{p} \odot (\underline{1} - \underline{z}_\text{p}) \odot H(\underline{m}_\text{pR}) \odot T(\underline{m}, p)]\} \\ &- a \times \frac{1+\cos\underline{\phi}_p}{2} \odot \cos\underline{\theta}_p \odot \{H^T[(\underline{\tilde{z}} - \underline{z}) \odot \text{sech}^2(\underline{z}_1 + \underline{z}_2 - \underline{1}) \\ &\odot \underline{z}_\text{p} \odot (\underline{1} - \underline{z}_\text{p}) \odot H(\underline{m}_\text{pI}) \odot T(\underline{m}, p)]\}, \end{aligned} \tag{8.9}$$

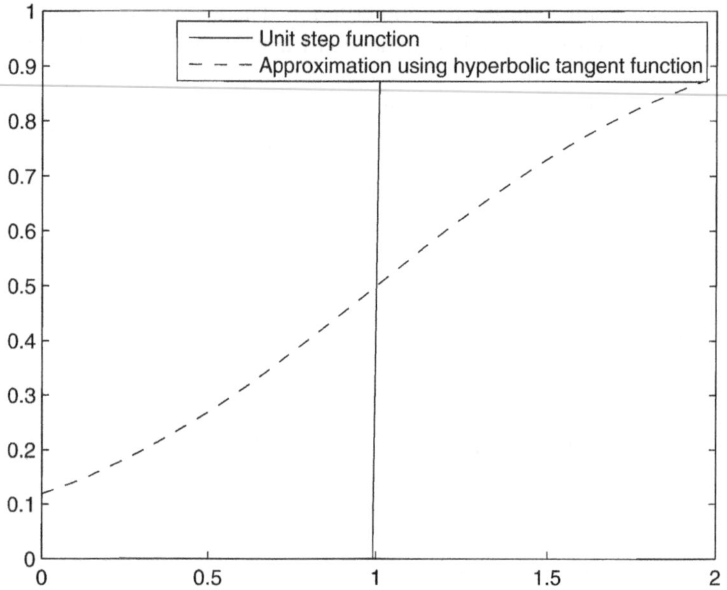

Figure 8.3 Unit step function and its approximation using hyperbolic tangent function.

$$\begin{aligned}\nabla F_{\underline{\phi}_p} &= \frac{a}{2} \times \sin\underline{\phi}_p \odot \cos\underline{\theta}_p \odot \{H^T[(\tilde{\underline{z}} - \underline{z}) \odot \text{sech}^2(\underline{z}_1 + \underline{z}_2 - \underline{1}) \\ &\quad \odot \underline{z}_{\text{p}} \odot (\underline{1} - \underline{z}_{\text{p}}) \odot H(\underline{m}_{\text{pR}}) \odot T(\underline{m}, p)]\} \\ &\quad + \frac{a}{2} \times \sin\underline{\phi}_p \odot \sin\underline{\theta}_p \odot \{H^T[(\tilde{\underline{z}} - \underline{z}) \odot \text{sech}^2(\underline{z}_1 + \underline{z}_2 - \underline{1}) \\ &\quad \odot \underline{z}_{\text{p}} \odot (\underline{1} - \underline{z}_{\text{p}}) \odot H(\underline{m}_{\text{pI}}) \odot T(\underline{m}, p)]\},\end{aligned} \tag{8.10}$$

where $p = 1$ or 2 and $T(\underline{m}, p) = [H(\underline{m}_{\text{pR}})^2 + H(\underline{m}_{\text{pI}})^2]^{-\frac{1}{2}}$. The derivations of Eqs. (8.9) and (8.10) are shown in Appendix E. Both discretization and wavelet penalties can be applied to the double patterning optimization method. The details of the discretization penalty and wavelet penalty are described in Sections 6.1.3 and 6.2.2, respectively. Considering a desired pattern of U-junction in Fig. 5.13, the experiment is repeated in Fig. 8.4, where a double patterning method is used.

Figure 8.4, from left to right, shows the optimized pole-level mask for the first patterning $\hat{M}_{\text{p1}}$, for the second patterning $\hat{M}_{\text{p2}}$, and the binary output pattern $Z_{\text{b}} = T\{\hat{M}_{\text{p1}}, \hat{M}_{\text{p2}}\}$. The aerial image formation process is approximated by a 11×11 Gaussian low-pass filter with $k = 14$, and in the sigmoid function, $a = 80$ and $t_r = 0.5$. The global threshold is $t_m = 0.5$. The step sizes and penalty weights for the first patterning are $s_{\phi_1} = 4$, $s_{\theta_1} = 0.01$, $\gamma_{\text{A1}} = 0.015$, $\gamma_{\text{P1}} = 0.001$, $\gamma_{W\text{A},\phi_1} = 0.5$, and $\gamma_{W\text{A},\theta_1} = 0.003$. The step sizes and penalty weights for the second patterning are $s_{\phi_2} = 4$, $s_{\theta_2} = 0.01$, $\gamma_{\text{A2}} = 0.015$, $\gamma_{\text{P2}} = 0.001$, $\gamma_{W\text{A},\phi_2} = 0.5$, and $\gamma_{W\text{A},\theta_2} = 0.003$. In Fig. 8.4, gray, white, and black represent 0, 1, and -1, respectively.

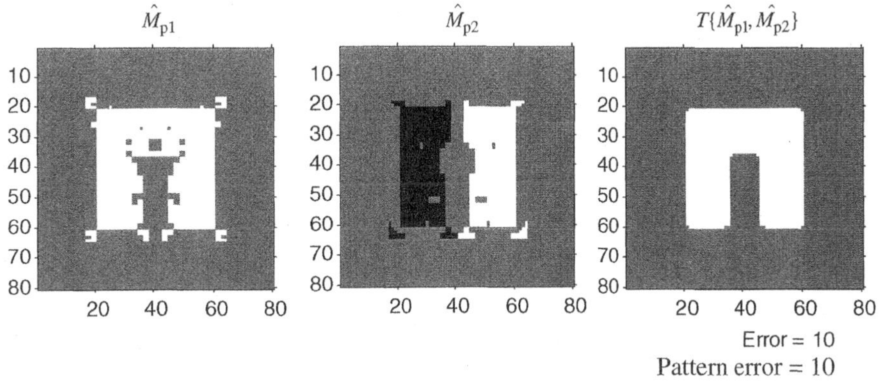

Figure 8.4 *Left to right*: Mask for first patterning, mask for second patterning, and the output pattern. $s_{\phi_1} = 4, s_{\theta_1} = 0.01, s_{\phi_2} = 4, s_{\theta_2} = 0.01, \gamma_{A1} = 0.015, \gamma_{P1} = 0.001, \gamma_{WA,\phi_1} = 0.5, \gamma_{WA,\theta_1} = 0.003$, $\gamma_{A2} = 0.015, \gamma_{P2} = 0.001, \gamma_{WA,\phi_2} = 0.5$, and $\gamma_{WA,\theta_2} = 0.003$. Gray, white, and black represent 0, 1, and -1, respectively.

The initial pattern for the first mask optimization is shown in Fig. 8.5a, where white and gray represent the transmission and opaque areas, respectively. $\underline{\phi}_{1k}$ is assigned to be $\frac{\pi}{5}$ corresponding to the areas having a magnitude of 1 and $\frac{4\pi}{5}$ for the areas having magnitude of 0. $\underline{\theta}_{1k}$ is assigned to be $\frac{\pi}{5}$ for the entire pattern. The initial pattern for the second mask optimization is shown in Fig. 8.5b, where white and black represent transmission areas with phases of 0 and π, respectively. Gray represents the opaque areas. $\underline{\phi}_{2k}$ is assigned to be $\frac{\pi}{5}$ corresponding to the areas having a magnitude of 1 and $\frac{4\pi}{5}$ for the areas having the magnitude of 0. $\underline{\theta}_{2k}$ is assigned to be $\frac{6\pi}{5}$ for the left half of the entire pattern and $\frac{\pi}{5}$ for the right half of the entire pattern. Comparing Figs. 5.13

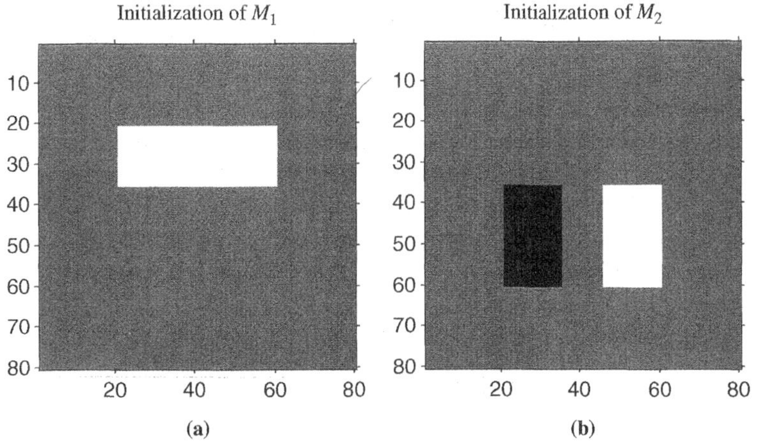

Figure 8.5 (a) Initial pattern for the first mask optimization. (b) Initial pattern for the second mask optimization. Gray, white, and black represent 0, 1, and -1, respectively.

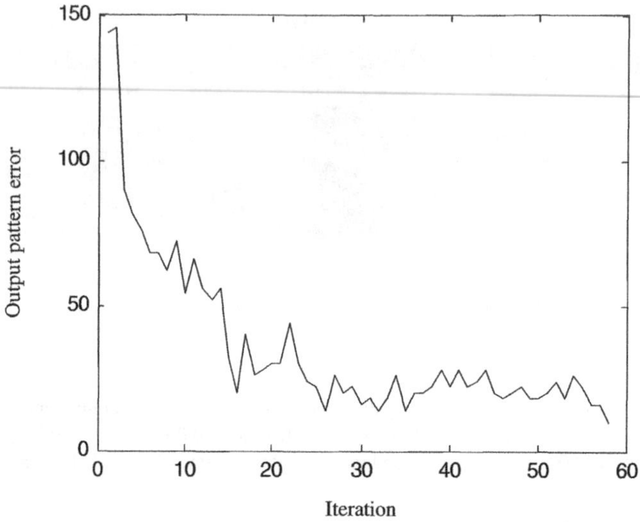

Figure 8.6 Convergence of the double patterning optimization algorithm versus steepest descent iterations for Fig. 8.4.

and 8.4, if a single four-phase mask is used, the output pattern error is 18. The double patterning method replaces one complex four-phase mask with two simple masks, and reduces the output pattern error to 10. Figure 8.6 shows the successful convergence of the double patterning optimization algorithm versus steepest descent iterations for Fig. 8.4. The double patterning optimization method is indeed effective, but requires more complicated processing and longer fabrication time.

8.2 POST-PROCESSING BASED ON 2D DCT

Sections 6.1 and 6.2 presented a set of regularization frameworks, such as discretization penalty, total variation penalty, and wavelet penalty, to reduce the complexity of mask patterns. These approaches extend the cost function by penalty terms, thus automatically influencing the solution patterns during the iterative optimization process. Although effective, regularization invariably will have a trade-off reducing the pattern details while often increasing the pattern errors. The balance between the complexity of masks and output pattern errors is controlled by the weights of the penalties. In general, larger penalty weights will lead to simpler mask patterns, however, to larger output pattern errors. To overcome this limitation, a post-processing based on 2D DCT is developed to simultaneously reduce the complexity of the mask pattern and the output pattern error.

Recently, Zhang et al. proposed an efficient mask design for inverse lithography based on 2D DCT [101]. The solution space is greatly reduced by cutting off the high-frequency components of the desired pattern in discrete cosine spectrum.

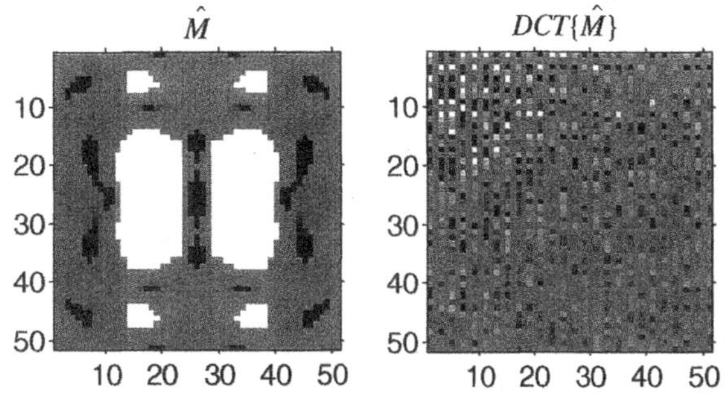

Figure 8.7 *Left*: Optimized real-valued mask in Fig. 7.13. *Right*: 2D DCT of the optimized real-valued mask.

Low-frequency components of the mask pattern were proven to have more influence on the fidelity of the output pattern than the high-frequency components. From this point of view, a post-processing of the mask pattern based on 2D DCT is introduced in this section. Figure 8.7 illustrates the optimized real-valued mask shown in Fig. 7.13 and its 2D DCT. It is observed that most of the energy of the mask pattern concentrates on the low-frequency components. To reduce the complexity of the mask pattern, a subset of high-frequency components of the optimized real-valued or complex-valued mask are cut off. The post-processed real-valued or complex-valued mask $\hat{M}'$ is the inverse 2D DCT of the maintained low-frequency components. The post-processed pole-level mask $\hat{M}'_{\mathrm{p}}$ is the discretization of $\hat{M}'$. The following simulations of the post-processing is based on the two-phase PSM optimization under partially coherent illumination (PCI), which is described in Section 7.2. It is straightforward to extend this post-processing step to other optimization algorithms.

Figure 8.8 illustrates the relationship between the number of maintained DCT low-frequency components and the output pattern errors with partial coherence factors $\sigma = 0.3$ (solid line) and $\sigma = 0.6$ (dotted line). Since the inverse lithography is an ill-posed problem, numerous input patterns can lead to the same binary output pattern. Thus, the post-processing based on 2D DCT can even simultaneously reduce the complexity of the masks and the output pattern errors. It is shown in Fig. 8.8 that the fidelity of the output patterns is improved by maintaining just 136 low-frequency components with $\sigma = 0.3$ and 666 low-frequency components with $\sigma = 0.6$. Figure 8.9 illustrates the simulations of the PSM optimization using the DCT post-processing under PCI with $\sigma = 0.3$. All the parameters are the same as those of the simulations shown in Fig. 7.13. In Fig. 8.9, the left figure shows the output pattern of the desired pattern. The middle figure shows the post-processed two-phase PSM using DCT post-processing maintaining 136 low-frequency components. The right figure shows the binary output pattern of the post-processed optimized PSM. White, gray, and black represent 1, 0, and -1, respectively. It is obvious that the post-processed two-phase PSM in Fig. 8.9 is much simpler than the optimized PSM in Fig. 7.13. In addition, the output pattern error is reduced from 48 to 26.

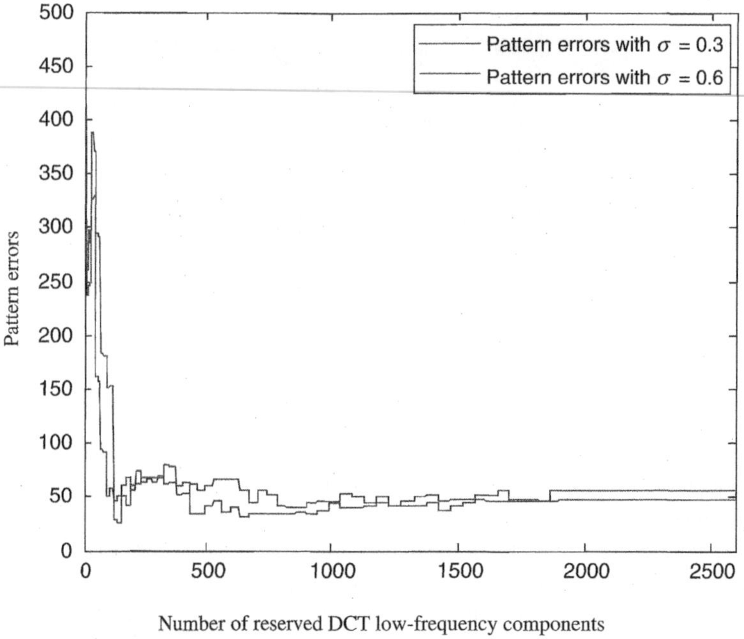

Figure 8.8 The relationship between the number of maintained DCT low-frequency components and the output pattern errors.

Figure 8.10 repeats the simulation in Fig. 7.15 using the DCT post-processing, where $\sigma = 0.6$. All the parameters are the same as those of the simulations shown in Fig. 7.15. The left figure shows the output pattern of the desired pattern. The middle figure shows the post-processed two-phase PSM using DCT post-processing maintaining 666 low-frequency components. The right figure shows the binary output pattern of the post-processed optimized PSM. White, gray, and black represent 1, 0,

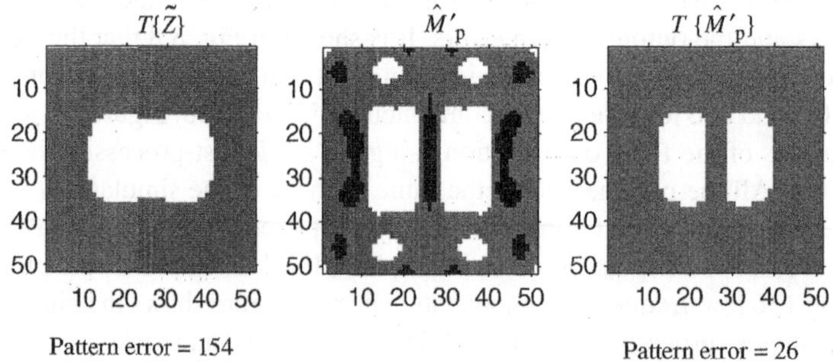

Figure 8.9 *Left to right*: Output pattern when the desired pattern is inputted, post-processed two-phase PSM with the DCT post-processing maintaining 136 low-frequency components, and the binary output pattern of the post-processed two-phase PSM. White, gray, and black represent 1, 0, and -1, respectively.

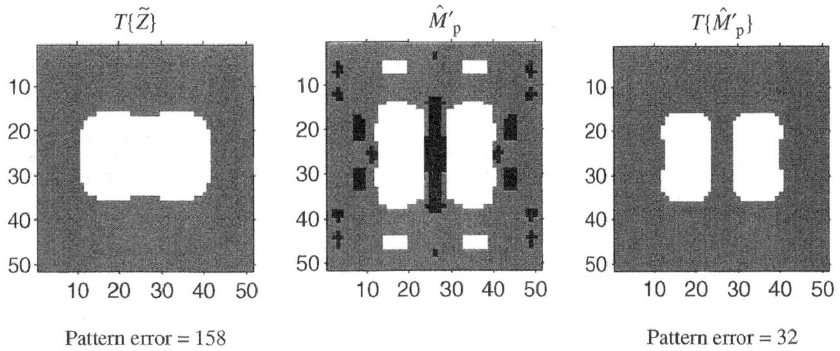

Figure 8.10 *Left to right*: Output pattern when the desired pattern is inputted, post-processed two-phase PSM with the DCT post-processing maintaining 666 low-frequency components, and the binary output pattern of the post-processed two-phase PSM. White, gray, and black represent 1, 0, and −1, respectively.

and −1, respectively. It is obvious that the post-processed optimized PSM in Fig. 8.10 is simpler than the optimized PSM in Fig. 7.15. In addition, the output pattern error is reduced from 56 to 32.

8.3 PHOTORESIST TONE REVERSING METHOD

In this section, we focus on the inverse lithography optimization where photoresist tone reversing is used. Photoresist can be divided by its polarity. In a positive photoresist process, more photoresist material remains in the low-exposure area on the wafer and less in the high-exposure area. Negative photoresist responds in the opposite manner. Photoresist tone reversing method exploits both kinds of photoresist materials on the wafer. Reversion of the photoresist tone is addressed to improve the lithography performance of subresolution features [92]. In this section, the photoresist tone reversing is used to image a desired resolution and contrast for the sparse pattern, whose resolution limit is much higher than the traditional case without application of the photoresist tone reversing. The following simulations of photoresist tone reversing method are based on the two-phase PSM optimization under partially coherent illumination, which is described in Section 7.2. It is straightforward to extend this approach to other optimization algorithms.

Consider an optical lithography system applying monotonous positive photoresist with the parameters $k = 0.29$, $\lambda = 193$ nm, and NA $= 1.35$. The resolution limit is

$$R = k\frac{\lambda}{\text{NA}} = 41.5\,\text{nm}. \tag{8.11}$$

PSM optimization without use of photoresist tone reversing fails to print an aerial image containing features with dimensions smaller than the limit in Eq. (8.11). An example is illustrated in Fig. 8.11, where the desired image of dimension 561 nm × 561 nm contains two pairs of vertical bars, each with critical dimension of 22 nm $< R$.

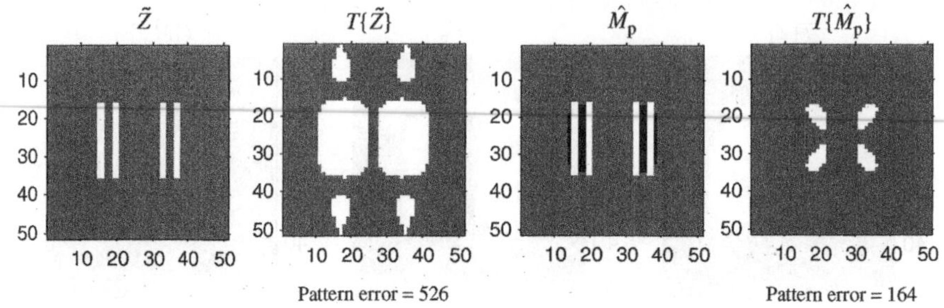

Figure 8.11 *Left to right*: Desired pattern, output pattern when the desired pattern is inputted, optimized two-phase PSM, and the output pattern of the optimized PSM. White, gray, and black represent 1, 0, and -1, respectively. $\sigma = 0.3$.

In this simulation, all the parameters are the same as those of the simulations shown in Fig. 7.13, except for $t_r = 0.001$. The illumination is a circular illumination with $\sigma = 0.3$. In Fig. 8.11, from left to right, the first figure shows the desired pattern. The second figure shows the binary output pattern of the desired pattern. The third figure shows the optimized two-phase PSM. The fourth figure shows the binary output pattern of the optimized PSM. White, gray, and black represent 1, 0, and -1, respectively. Note that the binary output pattern of the optimized PSM is totally different from the desired pattern, indicating that the PSM optimization approach cannot attain the desired output pattern on the wafer. The reason is that the dimension of the features in the desired pattern is smaller than the resolution limit without application of the photoresist tone reversing.

To overcome the limit, photoresist tone reversing is exploited to find an adequate distribution of positive and negative photoresists on the wafer. One possibility of the distribution is shown in the left figure in Fig. 8.12. White and black represent positive and negative photoresists, respectively. Assign negative photoresist in the gaps in

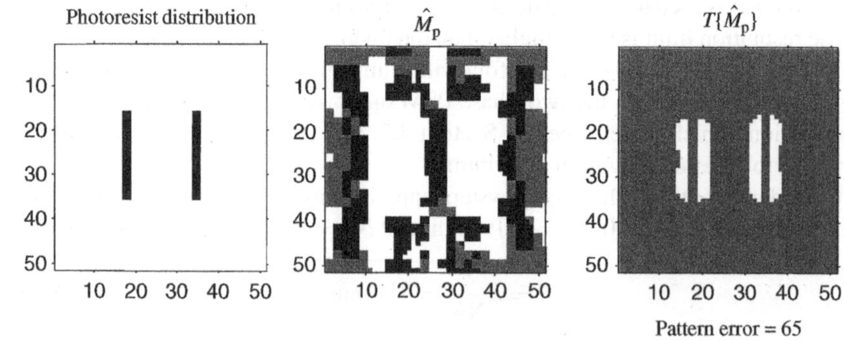

Figure 8.12 *Left to right*: Photoresist distribution, optimized two-phase PSM using the photoresist tone reversing method, and the output pattern of the optimized PSM. White and black represent positive and negative photoresists, respectively, in the left figure. White, gray, and black represent 1, 0, and -1, respectively, in the middle and right figures. $\sigma = 0.3$.

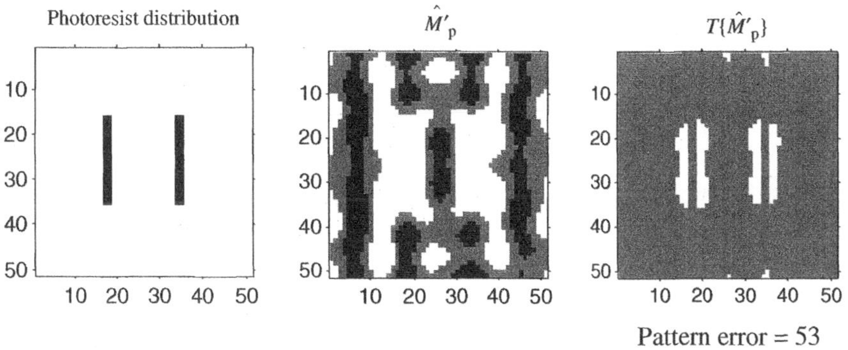

Figure 8.13 *Left to right*: Photoresist distribution, post-processed trinary mask using the DCT post-processing maintaining 91 low-frequency components, and the binary output pattern of the post-processed trinary mask. White and black represent positive and negative photoresists, respectively, in the first figure. White, gray, and black represent 1, 0, and -1, respectively, in the second and the third figures. $\sigma = 0.3$.

each pair of the bars and positive photoresist in other areas. If the optimized mask is able to expose a rectangular aerial image in the area over each pair of bars, the negative photoresist will prevent the exposure in the gaps. Thus, the binary output pattern is the same as the desired pattern. Therefore, the PSM optimization approach is used to search the binary output pattern of two rectangles on the wafer. In Fig. 8.12, $s_{\underline{\theta}} = 2$, $t_r = 0.01$, $\gamma_D = 0.1$, and $\gamma_{WA} = 0$. Other parameters are the same as those in Fig. 8.11. The optimized two-phase PSM is shown in the middle figure. The binary output pattern of the optimized PSM is shown in the right figure. White, gray, and black represent 1, 0, and -1, respectively. It is obvious that photoresist tone reversing method is effective to expose a sparse feature, whose resolution limit is much higher than the traditional case without application of photoresist tone reversing.

The DCT post-processing of the mask developed in Section 8.2 can be simultaneously exploited with the photoresist tone reversing method. The simulation is illustrated in Fig. 8.13, where all the parameters are the same as those in Fig. 8.12. The left figure shows the photoresist distribution. The middle figure shows the post-processed optimized PSM using the DCT post-processing maintaining 91 low-frequency components. The right figure shows the binary output pattern of the post-processed optimized PSM. Note that the DCT post-processing successfully obtains a simpler mask pattern and reduces the error of the binary output pattern from 65 to 53.

To show the stability of the photoresist tone reversing method, another example is illustrated in Fig. 8.14, with dimension of 561 nm $\times$ 561 nm. The desired pattern is a hole contact having a width of 22 nm $< R$. In this simulation, all the parameters are the same as the simulations shown in Fig. 7.13, except for $t_r = 0.0003$. The wavelet penalty is not used in this simulation. The illumination is a circular illumination with $\sigma = 0.3$. Note that, in Fig. 8.14, the binary output pattern of the optimized PSM is a platform with amplitude of 0, indicating that the PSM optimization approach cannot attain the desired output pattern on the wafer.

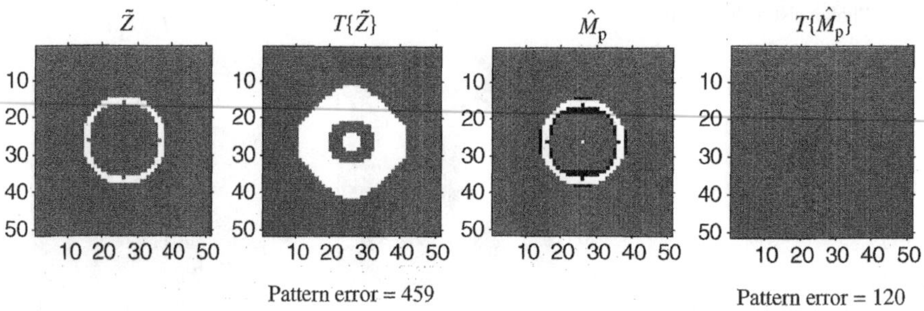

Figure 8.14 *Left to right*: Desired pattern, output pattern when the desired pattern is inputted, optimized two-phase PSM, and the output pattern of the optimized PSM. White, gray, and black represent 1, 0, and −1, respectively. $\sigma = 0.3$.

Using the photoresist tone reversing method, the distribution of photoresist is shown in the left figure in Fig. 8.15, where all the parameters are the same as those in Fig. 8.14, except for $s_\theta = 0.2$, $t_r = 0.01$, $\gamma_D = 0.1$, and $\gamma_{WA} = 0$. Assign negative photoresist to the inscribed circle of the hole contact and positive photoresist to the other areas. If the optimized mask is able to expose a circular aerial image, the same as the circum circle of the hole contact, the negative photoresist will prevent the exposure in the inscribed circle. Thus, the binary output pattern is the same as the desired pattern. Therefore, the PSM optimization approach is used to search the same binary output pattern as the circum circle of the hole contact on the wafer. In Fig. 8.15, the optimized mask is a binary mask shown in the middle figure. The binary output pattern of the optimized mask is shown in the right figure. White and gray represent 1 and 0, respectively. It is obvious that photoresist tone reversing method is effective to expose the hole contact, whose spatial frequency is higher than the resolution limit.

The simulation using DCT post-processing is illustrated in Fig. 8.16, where all the parameters are the same as those in Fig. 8.15. The left figure shows the photoresist

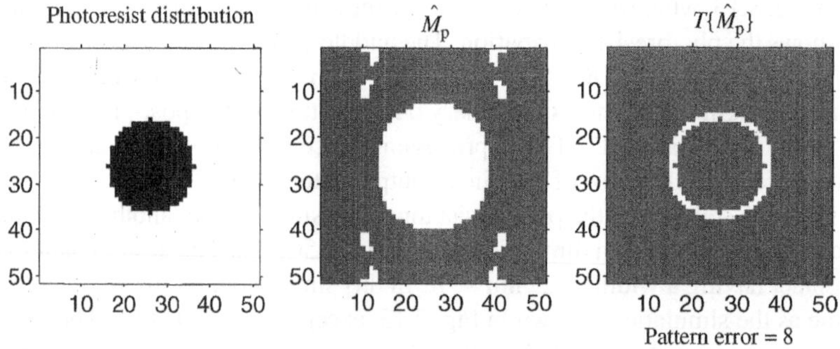

Figure 8.15 *Left to right*: Photoresist distribution, optimized mask using the photoresist tone reversing method, and the output pattern of the optimized mask. White and black represent positive and negative photoresists, respectively, in the first figure. White and gray represent 1 and 0, respectively, in the second and the third figures. $\sigma = 0.3$.

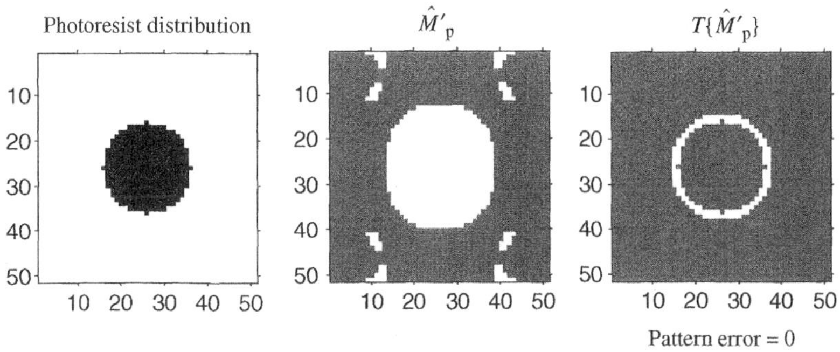

Figure 8.16 *Left to right*: Photoresist distribution, post-processed optimized mask using the DCT post-processing maintaining 1128 low-frequency components, and the binary output pattern of the post-processed optimized mask. White and black represent positive and negative photoresists, respectively, in the first figure. White and gray represent 1 and 0, respectively, in the second and the third figures. $\sigma = 0.3$.

distribution. The middle figure shows the post-processed optimized mask using the DCT post-processing maintaining 1128 low-frequency components. The right figure shows the binary output pattern of the post-processed optimized mask. Note that the DCT post-processing successfully reduces the error of the binary output pattern from 8 to 0.

The heuristic photoresist distribution design approach described above is suitable for simple target patterns. The joint optimization of the photoresist distribution and mask pattern is desirable and is of interest for future research work.

8.4 SUMMARY

This chapter described three techniques to improve the performance of OPC and PSM optimization algorithms, such as double patterning optimization method, post-processing based on 2D DCT, and photoresist tone reversing technique. Simulations were presented to prove the validity of these approaches.

9

Source and Mask Optimization

In Chapters 5–8, a set of computationally efficient pixel-based OPC and PSM optimization algorithms based on gradient-based searches have been introduced for inverse lithography. These optimization methods, as well as other traditional RETs, fix the source during the optimization and limit the degrees of freedom that can be optimized in the mask patterns. OPC design, for instance, is usually limited by the competing requirements of lithography optimization and has to strike a balance between image contrast and pattern length when printing dense patterns [92]. To overcome these limitations, a set of simultaneous source and mask optimization (SMO) methods have been developed recently, where the synergy is exploited in the joint optimization of the source and mask patterns. The optimized source and mask patterns of SMO algorithms fall well outside the realm of known design forms and lead to solutions closer to global minimums.

Several source and mask optimization algorithms have been proposed in the literature. Burkhardt et al. introduced an algorithm to analytically predict the pupil pattern for an arbitrary periodic mask feature, where the optimized illumination depends only on stepper parameters and mask geometry [10]. Gau et al. proposed an algorithm to optimize the source for features at many pitches [24]. Recently, Rosenbluth et al. introduced the idea of simultaneous optimization of the source and mask [73]. Progler et al. presented an automated cooptimization algorithm for the embedded phase-shift mask transmission factor and illumination source [70]. Robert et al. proposed SMO algorithms to improve the process window by optimizing the mask in the frequency domain [72]. All the methods mentioned above, however, are not pixel based and the searching process for a suitable solution is computationally expensive.

This chapter describes computationally efficient pixel-based algorithms for jointly optimizing the source and mask patterns in partially coherent imaging systems [46]. Algorithms for both binary and phase-shifting mask designs are discussed. This is accomplished as follows. The Fourier series expansion model is first applied to decompose the partially coherent imaging systems as the sum of coherent systems [58, 74]. Then, the simultaneous source and mask design is formulated as an optimization problem, where the cost function is the square of the l^2 norm of the difference between the desired output pattern and the output intensity. The output intensity is

Computational Lithography By Xu Ma and Gonzalo R. Arce

usually referred to as the aerial image printed on the wafer. Cost sensitivity is calculated and applied to drive the cost function in the descent direction during the optimization process. Subsequently, the computationally efficient pixel-based SMO algorithms are described. To control the complexity of the source and mask patterns, topological constraints are added to the optimization framework. It is noted that the SMO algorithms are capable of designing both binary and alternating PSMs.

9.1 LITHOGRAPHY PRELIMINARIES

According to the Hopkins diffraction model, the light intensity distribution exposed on the wafer in partially coherent illumination (PCI) is described as

$$I(\mathbf{r}) = \iint M(\mathbf{r_1})M^*(\mathbf{r_2})\gamma(\mathbf{r_1} - \mathbf{r_2})h(\mathbf{r} - \mathbf{r_1})h^*(\mathbf{r} - \mathbf{r_2})\mathrm{d}\mathbf{r_1}\mathrm{d}\mathbf{r_2}, \qquad (9.1)$$

where $M(\mathbf{r})$ is the mask pattern, $\gamma(\mathbf{r_1} - \mathbf{r_2})$ is the complex degree of coherence, and $h(\mathbf{r})$ is the amplitude impulse response of the optical system. The complex degree of coherence in the spatial domain is the inverse 2D Fourier transform of the image of the illumination shape $\Gamma(m_x, m_y)$ in the lens pupil. Applying the Fourier series expansion model described in Section 2.2.1, $\gamma(\mathbf{r})$ can be rewritten as

$$\gamma(\mathbf{r}) = \sum_{\mathbf{m}} \Gamma_{\mathbf{m}} \exp(j\omega_0 \mathbf{m} \cdot \mathbf{r}) \qquad (9.2)$$

and

$$\Gamma_{\mathbf{m}} = \frac{1}{D^2} \int_{A_\gamma} \gamma(\mathbf{r}) \exp(j\omega_0 \mathbf{m} \cdot \mathbf{r}) \mathrm{d}\mathbf{r}, \qquad (9.3)$$

where $\omega_0 = \pi/D$, $\mathbf{m} = (m_x, m_y)$, m_x and m_y are integers, and $\cdot$ is the inner-product operation. Thus, the light intensity on the wafer is decomposed as

$$I(\mathbf{r}) = \sum_{\mathbf{m}} \Gamma_{\mathbf{m}} |M(\mathbf{r}) \otimes h^{\mathbf{m}}(\mathbf{r})|^2, \qquad (9.4)$$

where

$$h^{\mathbf{m}}(\mathbf{r}) = h(\mathbf{r}) \exp(j\omega_0 \mathbf{m} \cdot \mathbf{r}). \qquad (9.5)$$

Two examples of the Fourier series expansion models corresponding to annular and circular illuminations are used in the following simulations. For the annular illumination, the complex degree of coherence is

$$\gamma(\mathbf{r}) = \frac{J_1(2\pi r/2D_{\mathrm{cu}})}{2\pi r/2D_{\mathrm{cu}}} - \frac{D_{\mathrm{cu}}^2}{D_{\mathrm{cl}}^2} \frac{J_1(2\pi r/2D_{\mathrm{cl}})}{2\pi r/2D_{\mathrm{cl}}}, \qquad (9.6)$$

where $r = \sqrt{x^2 + y^2}$. The corresponding Fourier series coefficients are

$$\Gamma_{\mathbf{m}} = \begin{cases} \frac{4D_{\mathrm{cu}}^2 D_{\mathrm{cl}}^2}{\pi D^2 (D_{\mathrm{cl}}^2 - D_{\mathrm{cu}}^2)}, & \text{for } D/2D_{\mathrm{cl}} \le |\mathbf{m}| \le D/2D_{\mathrm{cu}}, \\ 0, & \text{elsewhere}, \end{cases} \qquad (9.7)$$

where D_{cl} and D_{cu} are the coherent lengths of the inner and outer circles, respectively. $\sigma_{inner} = \frac{\lambda}{2D_{cl}NA}$ and $\sigma_{outer} = \frac{\lambda}{2D_{cu}NA}$ are the corresponding inner and outer partial coherence factors.

For the circular illumination, the complex degree of coherence is

$$\gamma(\mathbf{r}) = \frac{J_1(2\pi r/2D_c)}{2\pi r/2D_c}. \tag{9.8}$$

The corresponding Fourier series coefficients are

$$\Gamma_{\mathbf{m}} = \begin{cases} \frac{4D_c^2}{\pi D^2}, & |\mathbf{m}| \leq D/2D_c, \\ 0, & \text{elsewhere}, \end{cases} \tag{9.9}$$

where D_c is the coherent length of the circle. $\sigma = \frac{\lambda}{2D_c NA}$ is the corresponding partial coherence factor. The convolution kernel $h(\mathbf{r})$ is

$$h(\mathbf{r}) = \frac{J_1(2\pi r NA/\lambda)}{2\pi r NA/\lambda}. \tag{9.10}$$

Assuming that the illumination source $\Gamma(m_x, m_y)$ is binary, the Fourier series representation in Eq. (9.3) restricts its coefficient values such that $\Gamma_{\mathbf{m}} = 0$ or 1. Let $M(x, y) = 0$ or 1 be the input binary mask, and $M(x, y) = -1, 0,$ or 1 be the input two-phase PSM. $T\{\cdot, \cdot\}$ denotes an optical lithography system, which is formulated in Eq. (9.4). The output intensity is denoted as $I(x, y) = T\{\Gamma(m_x, m_y), M(x, y)\}$. $\tilde{I}(x, y) \in \Re^{N\times N}$ is the desired output pattern, whose pixel values are constrained to 0 or 1. The goal of the simultaneous source–mask optimization is to find the optimized $\Gamma(m_x, m_y)$ and $M(x, y)$ called $\hat{\Gamma}(m_x, m_y)$ and $\hat{M}(x, y)$ such that

$$D = d\left(I(x, y), \tilde{I}(x, y)\right) = d\left(T\{\Gamma(m_x, m_y), M(x, y)\}, \tilde{I}(x, y)\right) \tag{9.11}$$

is minimized, where $d(\cdot, \cdot)$ is the square of the l^2 norm criterion. The simultaneous source and mask optimization problem can thus be formulated as

$$\begin{aligned} &\left(\hat{\Gamma}(m_x, m_y), \hat{M}(x, y)\right) \\ &= \arg \min_{\Gamma(m_x, m_y), M(x,y)} d\left(T\{\Gamma(m_x, m_y), M(x, y)\}, \tilde{I}(x, y)\right). \end{aligned} \tag{9.12}$$

Following the definitions above, the following notations are used:

(1) The $M_{N\times N}$ matrix represents the mask pattern, with entry values equal to 0 or 1 for the binary mask and $-1, 0,$ or 1 for the phase-shifting mask. The $N^2 \times 1$ equivalent raster-scanned vector representation is denoted as $\underline{m}$.

(2) A convolution matrix $H^{\mathbf{m}} \in C^{N^2\times N^2}$ represents the point spread function (PSF) of the mth coherent component of the partially coherent imaging system, where C is the complex domain. The equivalent two-dimensional filter of $H^{\mathbf{m}}$ is $h^{\mathbf{m}}$.

(3) The desired $N \times N$ binary output pattern is denoted as $\tilde{I}$. It is the desired intensity distribution sought on the wafer. Its raster-scanned vector representation is denoted as $\tilde{\underline{i}}$.

(4) The initial source and mask patterns of the optimization are $\tilde{\Gamma}$ and $|\tilde{M}| = \tilde{I}$, respectively.

(5) The virtual output intensity is the $N \times N$ image denoted as

$$I = \sum_{\mathbf{m}} \Gamma_{\mathbf{m}} |H^{\mathbf{m}} \underline{m}|^2. \tag{9.13}$$

The equivalent raster-scanned vector is denoted as $\underline{i}$.

(6) The optimized source and mask denoted as $\hat{\Gamma}$ and $\hat{M}$ minimize the cost function, that is,

$$(\hat{\Gamma}, \hat{M}) = \arg\min_{\Gamma, M} d\left(\sum_{\mathbf{m}} \Gamma_{\mathbf{m}} |H^{\mathbf{m}} \underline{m}|^2, \tilde{I}\right). \tag{9.14}$$

Given the output intensity $\underline{i} = \sum_{\mathbf{m}} \Gamma_{\mathbf{m}} |H^{\mathbf{m}} \underline{m}|^2$, the ith entry in this vector can be represented as

$$\underline{i}_p = \sum_{\mathbf{m}} \Gamma_{\mathbf{m}} \left| \sum_{q=1}^{N^2} h_{pq}^{\mathbf{m}} \underline{m}_q \right|^2, \quad p = 1, \ldots, N^2, \tag{9.15}$$

where $h_{pq}^{\mathbf{m}}$ is the p, qth entry of the filter $h^{\mathbf{m}}$. The cost function is the square of the l^2 norm of the difference between $\underline{i}$ and $\tilde{\underline{i}}$. Therefore,

$$d = F(\underline{m}) = \|\tilde{\underline{i}} - \underline{i}\|_2^2 = \sum_{p=1}^{N^2} (\tilde{\underline{i}}_p - \underline{i}_p)^2, \tag{9.16}$$

where $\underline{i}_p$ is already represented in Eq. (9.15).

In the following, the sensitivity of the cost function F with respect to source and mask changes will be used to guide the optimization process. The change of F with respect to the change of the source is $\frac{\partial F}{\partial \Gamma}$. According to Appendix F, $\frac{\partial F}{\partial \Gamma}$ can be calculated as

$$\frac{\partial F}{\partial \Gamma_{\mathbf{m}}} = -2(\tilde{\underline{i}} - \underline{i})^T |H^{\mathbf{m}}(\underline{m})|^2, \tag{9.17}$$

where T is the conjugate transposition. The change of F with respect to the change of the mask is $\frac{\partial F}{\partial M}$. According to Appendix F, $\frac{\partial F}{\partial M}$ can be calculated as

$$\frac{\partial F}{\partial M} = -4\mathrm{Re}\left\{\sum_{\mathbf{m}} \Gamma_{\mathbf{m}} (H^{\mathbf{m}})^T [(\tilde{\underline{i}} - \underline{i}) \odot H^{\mathbf{m}}(\underline{m})]\right\}, \tag{9.18}$$

where Re{·} denotes the real part of the argument, and $\odot$ means the element-by-element multiplication. The combined cost sensitivity is denoted as

$$\nabla F = \left(\frac{\partial F}{\partial \Gamma}^T, \frac{\partial F}{\partial M}^T \right)^T . \tag{9.19}$$

9.2 TOPOLOGICAL CONSTRAINT

To attain the desired manufacturability properties of the optimized source and mask patterns, some topological constraints are imposed on the optimization process [98]. Yu et al. constrained the optimized binary masks to be topological invariant by the relationships between the neighbor pixels. Some of these operations and constraints have been defined with the goal of maintaining shape topologies [33]. To reduce the computational complexity of the source and mask optimization algorithms, a simplified version of the topological constraints proposed by Yu et al. is introduced in this section. In the following, some modified definitions of the shape topologies are listed.

Definition 9.1 (Flipping-off and flipping-on operations). A pixel p can have value 0 or 1. Turning from pixel value 1 to 0 and from 0 to 1 are respectively called flipping-off and flipping-on of that pixel. In general, if a pixel p can have value -1, 0, or 1, decreasing and increasing the pixel value are respectively called flipping-off and flipping-on operations.

Definition 9.2 (Neighbor pixels). As shown in Fig. 9.1, the pixels x_2, x_4, x_5, x_7 are the 4-neighbors of the pixel p. The pixels x_1, $x_2, \ldots, x_8$ are the 8-neighbors of p.

Definition 9.3 (Boundary pixels). A 4- (or 8-) boundary pixel is a pixel with at least one 4- (or 8-) neighbor pixel having a different value.

Definition 9.4 (Singular pixels). A singular pixel is a pixel whose value is different from that of all of its 4-neighbors.

X_1	X_2	X_3
X_4	p	X_5
X_6	X_7	X_8

Figure 9.1 4 and 8-Neighbors of pixel p.

Definition 9.5 (Changeable pixels). A changeable pixel is a 4-boundary pixel, flipping of which does not introduce singular pixels. The set of all changeable pixels is denoted as $\mathcal{S}$.

In the pixel-based simultaneous source and mask optimization approach, only the changeable pixels in the source pattern Γ and mask pattern M are considered to be flipped-on or flipped-off. This topological constraint guarantees lower complexity of the optimized source and mask patterns.

9.3 SOURCE–MASK OPTIMIZATION ALGORITHM

In this section, computationally efficient and effective pixel-based simultaneous source–mask optimization algorithms are described for both binary mask and PSM designs. In these optimization algorithms, the amplitudes of the mask patterns are initialized as the desired patterns, and the source patterns are initialized as the traditional partially coherent illuminations (annular or circular illuminations). Subsequently, changeable pixels on the mask and source patterns are searched and cost sensitivity is calculated to drive the cost function in the descent direction during the optimization process. The changeable pixels are flipped only when the cost function is reduced and the topological constraints are satisfied. Algorithms are terminated when no changeable pixel exists. The pixel-based simultaneous source and binary mask optimization algorithm is shown in Table 9.1 [46].

The pixel-based simultaneous source and PSM optimization algorithm is similar to the above algorithm. However, the pixel values in the mask pattern can be $-1, 0$, or 1. Therefore, flipping-on or flipping-off operation means to increase or decrease the pixel value by 1, respectively. In step 6, $p(x_{\max}, y_{\max})$ is allowed to be $-1, 0$, or 1.

9.4 SIMULATIONS

The simulations of gradient-based simultaneous source and binary mask optimizations are shown in Fig. 9.2. In these simulations, the initial mask pattern $\tilde{M}$ has dimension of $1035\,\text{nm} \times 1035\,\text{nm}$ and is the same as the target output pattern. The critical dimension of the initial mask pattern is 45 nm. The pixel size is assigned based on the critical dimension. Since singular pixels will increase the complexity of the optimized masks and are difficult to fabricate, the pixel size should be large enough. In addition, the high-frequency components of the mask will be removed by the low-pass filtering effect of the lens. Therefore, the small singular pixel does not contribute to the output aerial image on the wafer. Based on the above analysis, the pixel size is assigned to be $15\,\text{nm} \times 15\,\text{nm}$ in our simulations. The initial source is an annular illumination. The corresponding complex degree of coherence and Fourier series coefficients are shown in Eqs. (9.6) and (9.7). The convolution kernel is shown in Eq. (9.10) with $\text{NA} = 1.25$ and $\lambda = 193\,\text{nm}$, and it is assumed to vanish outside the area A_h defined by $x, y \in [-150\,\text{nm}, 150\,\text{nm}]$. In Fig. 9.2, the top row (from left to

Table 9.1 The Simultaneous Source and Binary Mask Optimization Algorithm

Step 1	Initialization of source and mask pattern: $\lvert M\rvert = \lvert\tilde{M}\rvert = \tilde{I}$, and $\Gamma = \tilde{\Gamma}$.
Step 2	Find the changeable pixels in the source and mask patterns (see *Definition 9.5*).
Step 3	Calculate the combinational cost sensitivity ∇F of the changeable pixels (see Eqs. (9.17)–(9.19)).
Step 4	Find the pixel $p(x_{\max}, y_{\max}) \in S$ in the source or mask pattern having the maximum absolute cost sensitivity: $\max(\lvert\nabla F\rvert) = \lvert\nabla F(p(x_{\max}, y_{\max}))\rvert$.
Step 5	Update the value of $p(x_{\max}, y_{\max})$ according to the sign of $\nabla F(p(x_{\max}, y_{\max}))$: $p(x_{\max}, y_{\max}) = p(x_{\max}, y_{\max}) - \text{sgn}(\nabla F(p(x_{\max}, y_{\max})))$.
Step 6	If($p(x_{\max}, y_{\max}) \neq 0$ or 1) or (cost function F is increased) $p(x_{\max}, y_{\max}) = p(x_{\max}, y_{\max}). + \text{sgn}(\nabla F(p(x_{\max}, y_{\max})))$.
Step 7	Clear the cost sensitivity of $p(x_{\max}, y_{\max})$: $\nabla F(p(x_{\max}, y_{\max})) = 0$.
Step 8	If $\nabla F \neq \mathbf{0}$ Go to step 4. Otherwise If no pixel is flipped in the current iteration End. Otherwise Go to step 2.

right) shows the initial source pattern ($\sigma_{\text{inner}} = 0.4$ and $\sigma_{\text{outer}} = 0.5$), the initial binary mask pattern, and the corresponding output aerial image intensity. The middle row (from left to right) shows the initial source pattern ($\sigma_{\text{inner}} = 0.4$ and $\sigma_{\text{outer}} = 0.5$), the optimized binary mask pattern without simultaneous optimization of source pattern, and the corresponding output aerial image intensity, where only the mask pattern is optimized using the algorithm in Table 9.1. The bottom row (from left to right) shows the optimized source pattern, the optimized binary mask pattern, and the corresponding output intensity, where mask and source patterns are simultaneously optimized. In the source and mask patterns, black and white represent 0 and 1, respectively. It is shown that optimization of the mask pattern alone reduces the output pattern error by 24%. The optimized mask contains more small assisting features, and the output pattern has gaps on the middle horizontal bar. On the other hand, the SMO algorithm reduces the output pattern error by 28%. In addition, the SMO algorithm leads to simpler optimized masks and better fidelity of the output pattern. Note the significant improvement on the separation of the horizontal bars and the objects between the bars.

The simulations of pixel-based simultaneous source and PSM optimization are shown in Fig. 9.3. The initial source pattern is a circular illumination with $\sigma = 0.4$. The corresponding complex degree of coherence and Fourier series coefficients are

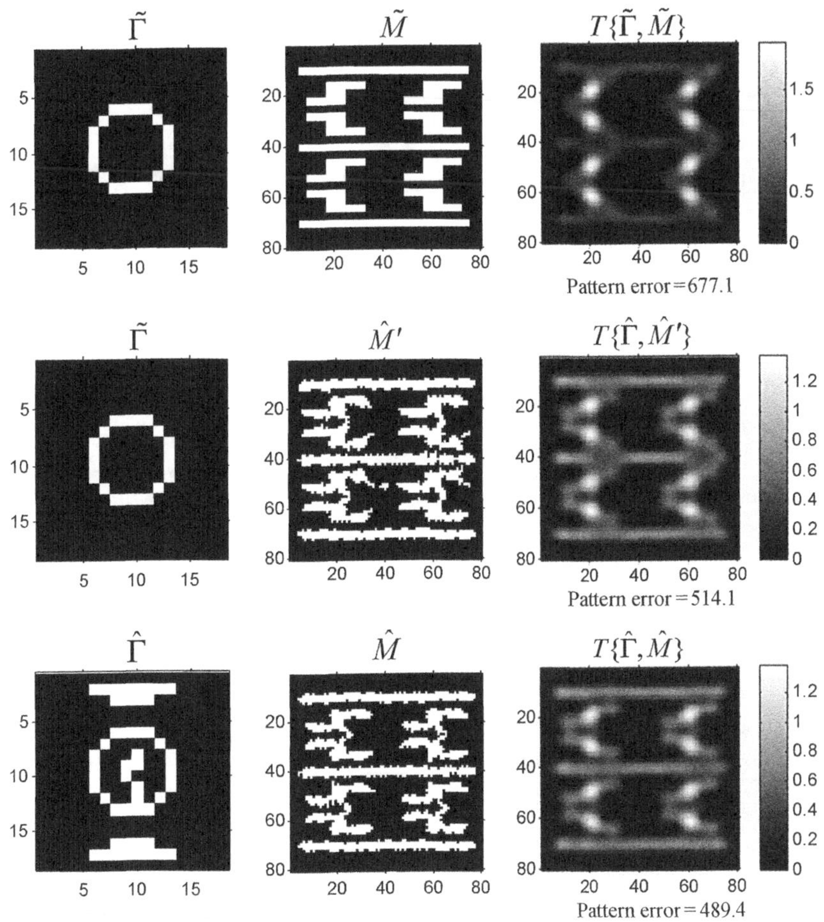

Figure 9.2 Pixel-based simultaneous source and binary mask optimization. *Top row* (from left to right): The initial source pattern ($\sigma_{\text{inner}} = 0.4$ and $\sigma_{\text{outer}} = 0.5$), the initial binary mask pattern (critical dimension = 45 nm), and the corresponding output intensity. *Middle row* (from left to right): The initial source pattern ($\sigma_{\text{inner}} = 0.4$ and $\sigma_{\text{outer}} = 0.5$), the optimized binary mask pattern without simultaneous optimization of source pattern, and the corresponding output intensity. *Bottom row* (from left to right): The optimized source pattern, the optimized binary mask pattern, and the corresponding output intensity. In the source and mask patterns, black and white represent 0 and 1, respectively.

shown in Eqs. (9.8) and (9.9). Other parameters are the same as the simulations shown in Fig. 9.2. In the source and mask patterns, black, gray, and white represent −1, 0, and 1, respectively. It is shown that optimization of the mask pattern alone reduces the output pattern error by 65%. On the other hand, the SMO algorithm reduces the output pattern error by 71%. In addition, the SMO algorithm leads to simpler optimized masks and better fidelity of the output pattern. The performance differences between the optimization of only the mask patterns and the joint optimization of source and mask patterns clearly show the advantages of the SMO algorithms. As shown in

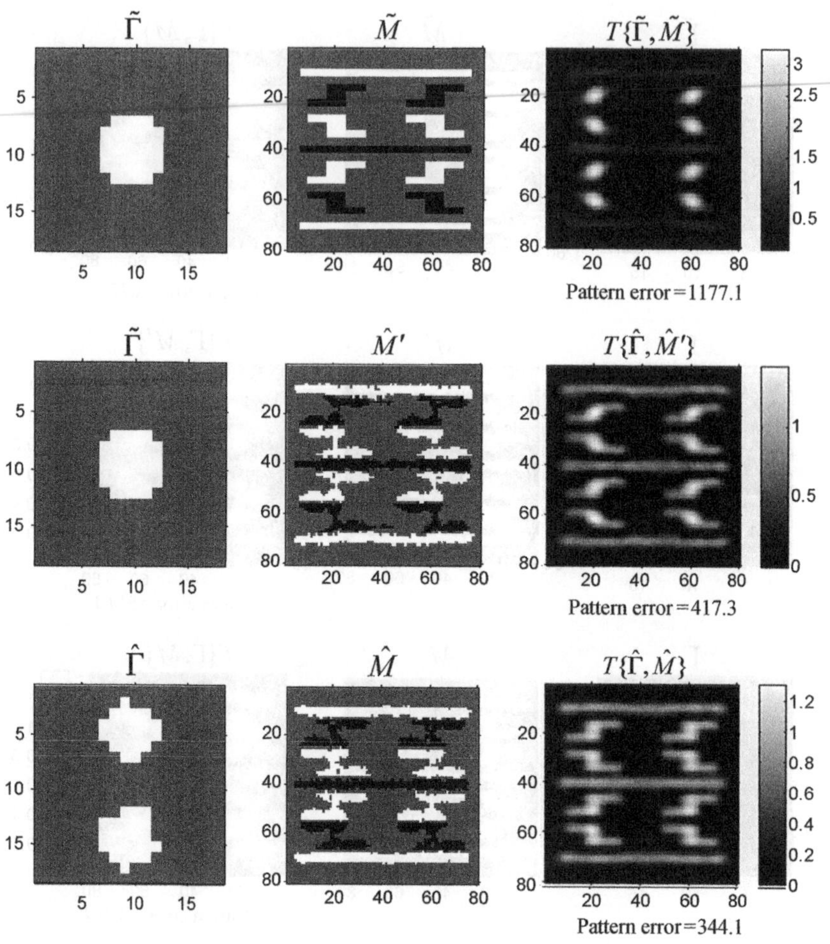

Figure 9.3 Pixel-based simultaneous source and phase-shifting mask optimization. *Top row* (from left to right): The initial source pattern ($\sigma = 0.4$), the initial phase-shifting mask pattern (critical dimension $= 45$ nm), and the corresponding output intensity. *Middle row* (from left to right): The initial source pattern ($\sigma = 0.4$), the optimized PSM without simultaneous optimization of source pattern, and the corresponding output intensity. *Bottom row* (from left to right): The optimized source pattern, the optimized phase-shifting mask pattern, and the corresponding output intensity. In the source and mask patterns, black, gray, and white represent -1, 0, and 1, respectively.

Figs. 9.2 and 9.3, the described SMO algorithms have been proven effective for both binary mask and PSM. It is noted that the described algorithms result in asymmetric structures in the optimized source and mask shapes. These asymmetric structures introduce higher degree of freedom in the optimization process and lead to small output pattern errors. To obtain symmetric structures, the described algorithms can be easily modified to optimize just the top half part of the source and mask patterns with respect to the midline. The symmetric pixels in the bottom half part are flipped in the same way.

9.5 SUMMARY

This chapter developed computationally efficient and effective simultaneous source–mask optimization algorithms, where both the OPC and PSM optimization frameworks were considered. Cost sensitivity was used to drive the cost function in the descent direction and topological constraints were applied to reduce the mask complexity.

10

Coherent Thick-Mask Optimization

In the previous chapters, a variety of OPC and PSM optimization algorithms are developed in coherent and partially coherent imaging systems. In addition, these algorithms are extended to the realm of joint optimization of source and mask. All these algorithms, however, have been developed under the thin-mask assumption, where Kirchhoff's boundary condition is directly applied to the mask topology and consequently the mask is treated as a 2D object [4, 81, 82]. As the critical dimension (CD) printed on the wafer shrinks into the subwavelength regime, the thick-mask effects become very pronounced such that these effects should be taken into account in the mask optimization. Thick-mask effects include polarization dependence due to the different boundary conditions for the electric and magnetic fields, transmission error in small openings, diffraction edge effects or electromagnetic coupling, and so on [4]. The thick-mask effects can be rigorously represented by the near-field pattern of the mask, which is different from the Kirchhoff approximation of the mask topography. Two decades ago, Wong and Neureuther discovered the intensity imbalance of alternating PSM, and applied the finite-difference time-domain (FDTD) method to study the mask topography effects in the projection printing of PSM [89, 90]. This phenomenon was proven by experimental results later [62]. Yuan exploited the waveguide (WG) method to model the light diffraction of 2D phase-shifting masks [99], which was subsequently generalized by Lucas et al. to the 3D topography [39]. Erdmann et al. evaluated and compared the FDTD and WG methods for the simulation of typical hyper NA (NA $>$ 1) imaging problem [20]. Adam and Neureuther introduced domain decomposition methods for the simulation of photomask scattering [3]. Nevertheless, these approaches are too complex to be applied in gradient-based OPC and PSM designs [45, 48].

Recently, Azpiroz et al. introduced a novel boundary layer (BL) model for the fast evaluation of the near field of a thick mask [4, 81, 82]. Different from other computationally complex and resource-consuming rigorous mask models, the BL model treats the near field of the mask as the superposition of the interior transmission areas and the boundary layers, which have fixed dimensions and determined locations. The BL model effectively compensates for the inaccuracy of Kirchhoff's approximation, which is attributed to the thick-mask effects, different polarizations, and phase errors.

Computational Lithography By Xu Ma and Gonzalo R. Arce

The simplicity and accuracy of the BL model enable the formulation of model-based optimization algorithms for binary and phase-shifting masks. This chapter thus focuses on the formulation of gradient-based OPC and PSM optimization algorithms based on the BL model to take into account the thick-mask effects under coherent illumination [45, 48]. These are accomplished as follows: First, the optical lithography process under coherent illumination is formulated as the combination of the BL and the Hopkins diffraction models. The cost functions of the OPC and PSM optimization problems are formulated as the square of the l^2 norm of the difference between the virtual aerial image and the desired pattern on the wafer. Then the gradient of the cost function, referred to as the cost sensitivity function, is developed and used to drive the cost function in the descent direction during the optimization process. Topological constraints of the mask pattern are introduced and used to limit the minimum opening size of the optimized mask pattern.

10.1 KIRCHHOFF BOUNDARY CONDITIONS

The Kirchhoff boundary condition has been extensively used in the development of OPC and PSM optimization methods, where the mask thickness is assumed to be infinitesimal and the mask is considered as a 2D object. Direct application of the Kirchhoff boundary condition leads to a thin-mask approximation of the field on the exit surface of the mask [4, 7]. The thin-mask model ignores diffraction and polarization effects due to the 3D topography of the mask. The exiting field of the mask is approximated as the multiplication between the incident field and an ideal transmission function of the mask pattern. Thus, the exiting field is the same as the 2D topography of the mask. The thin-mask model provides accurate results when the mask features are much larger than the wavelength, and the field is evaluated several wavelengths away from the mask apertures. However, the thick-mask effect becomes an increasing source of simulation errors when the size of mask features reduces into the order of the wavelength [61, 96]. In addition, alternating PSM also employs etching profiles with abrupt discontinuities and trench depths in the order of the wavelength [4]. This 3D topography of PSM aggravates the influence of the thick-mask effects. In this case, rigorous resource-consuming 3D simulations will be needed to evaluate the virtual electromagnetic field exiting from the mask surface. The inaccuracy of the Kirchhoff boundary condition motivates the development of rigorous mask models. The boundary layer model, with its advantages of simplicity and accuracy, is described in Section 10.2.

10.2 BOUNDARY LAYER MODEL

10.2.1 Boundary Layer Model in Coherent Imaging Systems

As the CD printed on the wafer shrinks into the subwavelength regime, the thick-mask effects have become significant and thus these need to be taken into account in the design of OPC and PSM optimization methods. Although numerous rigorous mask models simulating the 3D electromagnetic field of the mask were developed, these models are resource consuming and too complex to be applied in the gradient-based binary and phase-shifting mask designs for inverse lithography.

Recently, Azpiroz et al. introduced a novel boundary layer model for the fast evaluation of the near field of the thick mask in coherent and partially coherent imaging systems [4, 81, 82]. The near field is modeled as the superposition of the interior transmission areas and the boundary layers with fixed dimensions and determined locations. The concepts of the BL model under coherent illumination are illustrated in Fig. 10.1, where the polarization of the impinging electric field $\mathbf{E}$ is assigned to be in the horizontal direction. Figure 10.1 shows a typical rectangular opening of the mask with width equal to a and height equal to b. The harmonic mean of the area's width a and height b is $d = \frac{2ab}{a+b}$. The near field of the opening is divided into five areas: A, B, C, D, and E. A is the interior transmission area with transmission coefficient $\eta_{\mathrm{I}} = 1$ for the binary mask, and $\eta_{\mathrm{I}} = 1$ or -1 for the phase-shifting mask. The transmission coefficients of the boundary layers depend on the polarization of the electric field of the impinging light. Since the polarization of the electric field is assigned to be in the horizontal direction, B and D are the tangential boundary areas with width w and transmission coefficient η_{T}. C and E are the normal boundary areas with width w and transmission coefficient η_{N}. In the BL model, the relative error of amplitude of the electric field on the wafer produced by the thin-mask approximation is measured by the deviation of its real component from the rigorously FDTD-calculated EM field value. Experimental results show that the relative error of amplitude is in proportion to the width of the boundary layer w and inversely proportional to the harmonic mean

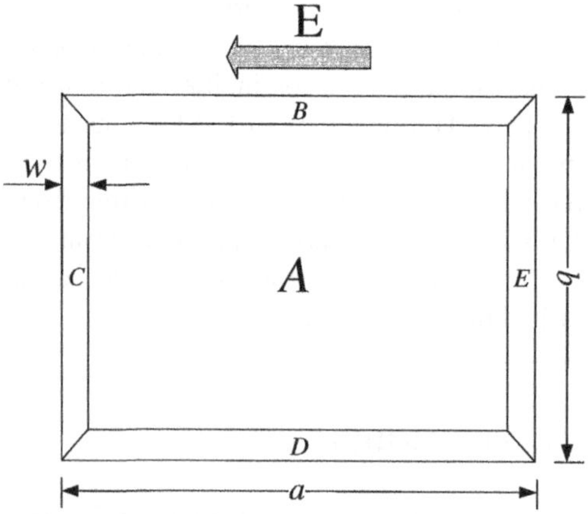

Figure 10.1 BL model under coherent illumination, where the polarization of the electric field is assigned to be in the horizontal direction. w is the width of the boundary areas. a and b are the width and height of the entire opening area, respectively.

d, represented as

$$\mathrm{Re}\left\{\frac{\Delta\mathbf{E}}{\mathbf{E}}\right\} = \frac{4w}{d} = \frac{(2a+2b)w}{ab} = \frac{\text{Boundary Layer Area (real part)}}{\text{Total Area}}, \tag{10.1}$$

where Re{·} denotes the real part of the argument. $\mathbf{E}$ is the total electric field from the rigorously FDTD-calculated EM field value. $\Delta\mathbf{E}$ is the electric field error from the thin-mask assumption, and $\mathrm{Re}\{\frac{\Delta\mathbf{E}}{\mathbf{E}}\}$ is the relative error of amplitude. Given the a and b, w can be calculated from Eq. (10.1). The deviation of the real component is compensated by the opaque boundary layers surrounding all openings on the mask whose transmission coefficients are zero. This real part of the boundary layer model is independent of the opening sizes and regardless of polarization. Given the value of w, it is shown experimentally that the relative error of phase is in proportion to w and transmission coefficient magnitude $|\eta_T|$, and is inversely proportional to the height of the opening b (dimension normal to polarization), represented as

$$\begin{aligned}\mathrm{Im}\left\{\frac{\Delta\mathbf{E}}{\mathbf{E}}\right\} &= |\eta_T|\frac{2w}{b} = |\eta_T|\frac{2aw}{ab} \\ &= |\eta_T|\frac{\text{Boundary Layer Area (imaginary part)}}{\text{Total Area}},\end{aligned} \tag{10.2}$$

where Im{·} denotes the imaginary part of the argument. Subsequently, the relative error of phase is compensated by the boundary layers with complex transmission coefficient η_T and width of w on the opening edges tangent to the electric field of impinging light. The transmission coefficient of the tangential boundary areas, η_T, is calculated from the slope of the linear relation described in Eq. (10.2). For arbitrary mask opening geometries, where the included angle between the edge and electric field polarization direction is α, $|\eta_T|$ is proportional to $|\cos(\alpha)|$ [4]. Thus, $|\eta_T|$ reaches its maximum value when the edge is tangent to the electric field polarization direction, while reduces to zero when they are normal to each other. This rule accounts for the polarization dependence of the boundary condition at the metal edges. Vanishing of the tangential components of electric field on metal surfaces mainly contributes to the relative error of phase. Although lithographic masks cannot be considered as perfect conductors, a similar concept of the surface effect can be applied to account for the relative error of phases. On the other hand, normal components of the electric field exhibit a discontinuity due to the accumulation of charges on the chrome surface, but this effect is mostly filtered out by the optical lens [4, 86]. As shown in Fig. 10.2, the inaccuracy of the thin-mask approximation is effectively offset by the superposition of complex-valued boundary layers. The real values of the boundary layers (Fig. 10.2a) are zero (opaque) around the area A. The complex values (Fig. 10.2b) are η_T on the tangential direction and zero on the normal direction. Figure 10.2c shows the final boundary layer model. For rectangular mask openings, these values for η_T and η_N have been shown in Refs. [4, 81, 82] to effectively compensate for the thin-mask distortion. However, the relationships described in Eqs. (10.1) and (10.2) are not accurate when the opening size reduces

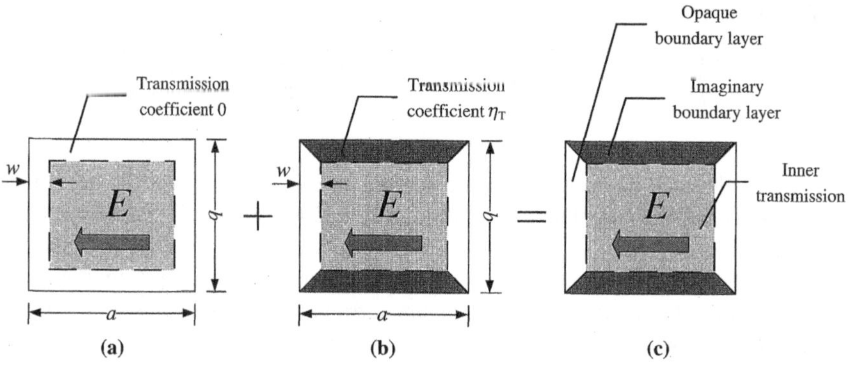

Figure 10.2 (a) Real component of the boundary layer model. (b) Imaginary component of the boundary layer model. (c) Final boundary layer model as the superposition of both real and imaginary components.

below the wavelength. For the validity of the BL model, the minimum size of the opening is constrained to be larger than the wavelength. The simplicity and accuracy of the BL model are suitable for the gradient-based OPC and PSM optimization algorithms.

Azpiroz et al. studied two types of optical lithography systems. The first one is a 4× projection system with NA = 0.68 and $\lambda = 248$ nm, while the second one is with NA = 0.85 and $\lambda = 193$ nm. For each type of optical lithography system, three kinds of openings on the mask were studied, as illustrated in Fig. 10.3. Figure 10.3, from left to right, shows the cross sections of (a) clear opening, (b) 180° phase-shifting opening, and (c) 180° phase-shifting opening with 35 nm undercut. For all these types of optical lithography systems and openings, the boundary widths, transmission coefficients, and corresponding minimum opening sizes of the BL model are summarized in Table 10.1.

In the following, we will use the two types of lithography systems described by Azpiroz et al. to develop OPC and PSM optimization algorithms. Binary mask consists of the clear openings, while phase-shifting mask consists of the clear openings and the 180° phase-shifting openings. The utilization of the 180° phase-shifting openings with 35 nm undercut is beyond the scope of this book and hence not considered here.

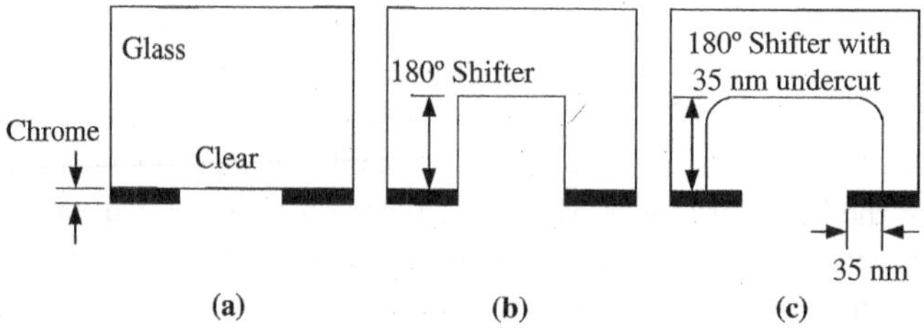

Figure 10.3 Cross sections of three types of openings on the mask. (a) Clear opening. (b) 180° Phase-shifting opening. (c) 180° Phase-shifting opening with 35 nm undercut.

Table 10.1 The Boundary Widths, Transmission Coefficients, and Corresponding Minimum Opening Sizes of the BL Model

Opening Type	Boundary Width (nm)		Tangential Boundary Coefficient	
	I	II	I	II
Clear	24.8	14.5	0.0j	0.8j
180° Shifter	55.8	53.0	−0.52j	−0.30j
180° Shifter with 35 nm undercut	37.2	33.7	−0.66j	−0.635j

Opening Type	Normal Boundary Coefficient	Interior Coefficient	Minimum Opening Size (nm)	
			I	II
Clear	0	1	248	200
180° Shifter	0	−1	300	250
180° Shifter with 35 nm undercut	0	−1	350	200

10.2.2 Boundary Layer Model in Partially Coherent Imaging Systems

Most practical illumination sources have a nonzero line width and their radiation is more generally described as partially coherent. Compared with the coherent illumination having a deterministic polarization, the partially coherent illumination consists of an unpolarized source. For the unpolarized source, the field polarization varies randomly, and the field components generated by different source points are not correlated and are added incoherently [4]. Assume that each source point of the partially coherent illumination generates a plane wave impinging on the mask plane with incident azimuth angle ϕ and elevation angle θ, which is shown in Fig. 10.4. It is shown in Fig. 10.5a that the unpolarized source can be modeled by the superposition of two linearly polarized plane waves, which are mutually orthogonal and normal to the propagation direction [7]. The polarization directions of the electric fields corresponding to the TE ($\hat{\mathbf{e}}_{TE}$) and TM ($\hat{\mathbf{e}}_{TM}$) modes are calculated as

$$\hat{\mathbf{e}}_{TE} = -\sin\phi\hat{\mathbf{p}}_X + \cos\phi\hat{\mathbf{p}}_Y, \tag{10.3}$$

$$\hat{\mathbf{e}}_{TM} = \sin\theta\cos\phi\hat{\mathbf{p}}_X + \sin\theta\cos\phi\hat{\mathbf{p}}_Y - \cos\theta\hat{\mathbf{p}}_Z, \tag{10.4}$$

where $\hat{\mathbf{p}}_X$ and $\hat{\mathbf{p}}_Y$ are the unit vectors along the P_X and P_Y axes, and $\hat{\mathbf{p}}_Z = \hat{\mathbf{p}}_X \times \hat{\mathbf{p}}_Y$.

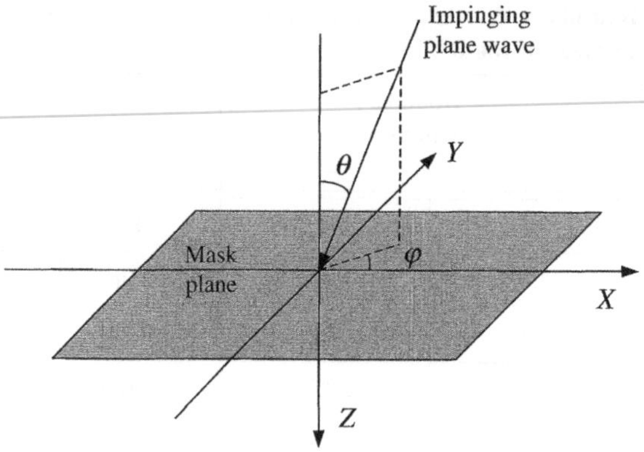

Figure 10.4 Each source point of the partially coherent illumination generates a plane wave impinging on the mask plane with incident azimuth angle ϕ and elevation angle θ.

According to Eqs. (2.7) and (2.8), when the incident elevation angle is small, the Hopkins approximation can be used to model the partially coherent illumination. In this case, the oblique impinging plane wave is assumed equal to the normal impinging wave, except for the corresponding frequency shift. As illustrated in Fig. 10.5b, based on the Hopkins approximation, the TE and TM modes of the unpolarized source can be approximated to have constant directions along P_X and P_Y axes [2, 4]. Therefore, the BL model of partially coherent illumination is approximated as the superposition of the BL model in coherent imaging system contributed by each source point. For the on-axis source point, the BL model parameters are described in Section 10.2.1. For the off-axis source point, the boundary layer parameters in Table 10.1 have also been proven to lead to accurate results in the 4× optical lithography system with partial coherence factor $\sigma \in [0.3, 0.6]$ [4].

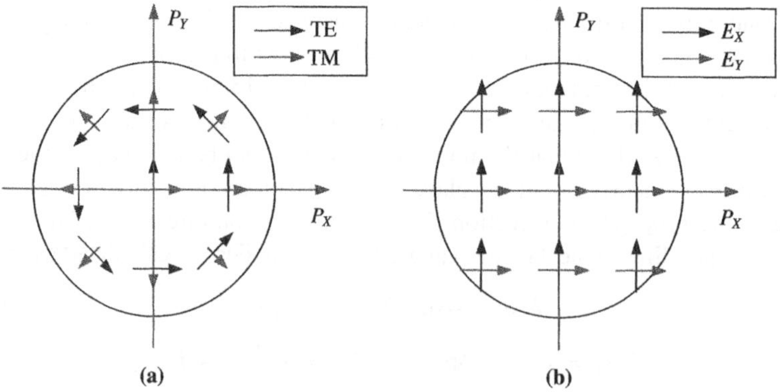

Figure 10.5 (a) Source polarization modes of TE and TM. (b) Approximated source polarization modes of E_X and E_Y (redrawn based on Fig. 4.11 in Ref. [4]).

10.3 LITHOGRAPHY PRELIMINARIES

In this section, OPC and alternating PSM inverse lithography methods are formulated in coherent imaging system. Let $M(x, y)$ be the input binary or phase-shifting mask to an optical lithography system $T\{\cdot\}$ with coherent illumination. The system $T\{\cdot\}$ includes two steps. The first step is the evaluation of near field of the thick mask, which is based on the BL model. The second step is the optical imaging system leading to the aerial image on the wafer, which is approximated by the Hopkins diffraction model described in Eq. (2.14). The output aerial image on the wafer is denoted as $I(x, y) = T\{M(x, y)\}$. Given a $N \times N$ desired output pattern $\tilde{I}(x, y)$, the goal of OPC and PSM designs is to find the optimized $M(x, y)$ called $\hat{M}(x, y)$ such that the distance

$$D = d\left(I(x, y), \tilde{I}(x, y)\right) = d\left(T\{M(x, y)\}, \tilde{I}(x, y)\right) \qquad (10.5)$$

is minimized, where $d(\cdot, \cdot)$ is the square of the l^2 norm criterion. The OPC and PSM optimization problems can thus be formulated as the search of $\hat{M}(x, y)$ over the $N \times N$ real space $\Re^{N \times N}$ such that

$$\hat{M}(x, y) = \arg \min_{M(x,y) \in \Re^{N \times N}} d\left(T\{M(x, y)\}, \tilde{I}(x, y)\right). \qquad (10.6)$$

The forward imaging process is illustrated in Fig. 10.6. The electric field propagating through the thick-mask pattern is affected by the 3D topography of the mask, forming the near field that is then influenced by the diffraction and mutual interference in the optical imaging system. Light that is transmitted through the optical system reaches the photoresist and forms the aerial image. In Fig. 10.6, the output of the convolution and the absolute-square operation is the intensity distribution of the aerial image.

Following the definitions above, the following notations are used:

1. The $M_{N \times N}$ matrix represents the mask pattern, with entry values equal to 0 or 1 for the binary mask, and -1, 0, or 1 for the phase-shifting mask. The $N^2 \times 1$ equivalent raster-scanned vector representation is denoted as $\underline{m}$.
2. The $\Gamma_{N \times N}(M)$ matrix represents the interior transmission area pattern of the near field corresponding to the mask M, with entry values equal to 0 or 1 for the binary mask, and -1, 0, or 1 for the phase-shifting mask. Its vector representation is denoted as $\underline{\gamma}$.

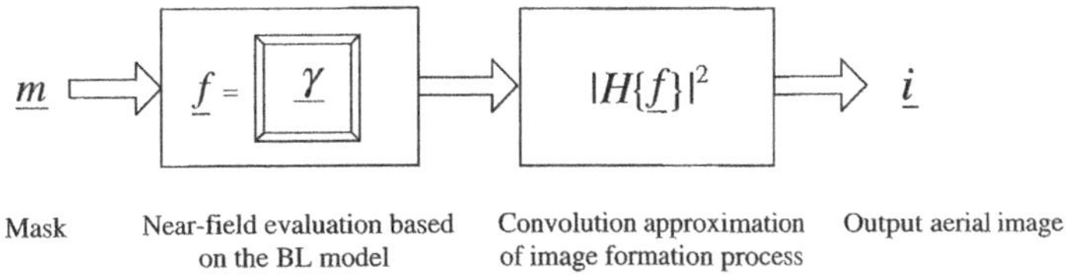

Figure 10.6 Approximated forward process model based on the BL model under coherent illumination, where the polarization of the electric field is assigned to be in the horizontal direction.

3. The $\Gamma \uparrow$ and $\Gamma \downarrow$ represent the shifted version of Γ by shifting Γ along the vertical direction (up and down) by one pixel, respectively. Their vector representations are denoted as $\underline{\gamma} \uparrow$ and $\underline{\gamma} \downarrow$.
4. The $F_{N\times N}(M)$ matrix represents the near field corresponding to the mask M, with complex entry values. Its vector representation is denoted as $\underline{f}$. Let the polarization of the impinging electric field **E** be in the horizontal direction. For the binary mask in the first type of optical lithography system, all the boundary layers are opaque, with transmission coefficient of 0. To represent all the features on the mask by integral number of pixels, the pixel size is assigned equal to the greatest common divisor between the boundary layer width and the minimum opening size, thus, the pixel size is set to be 24.8 nm. The minimum opening size is 248 nm = 10 × pixel size. Thus, the near field is the same as the interior transmission area. Therefore,

$$\underline{f}_p = \underline{\gamma}_p, \quad p = 1, 2, \ldots, N^2. \tag{10.7}$$

For the binary mask in the second type of optical lithography system, the normal boundary layers are opaque and the tangential boundary layers have complex transmission coefficient of $0.8i$. The pixel size is set to be 14.5 nm. To represent the minimum opening size by integral number of pixels, the minimum opening size is increased to be 203 nm = 14 × pixel size. Therefore,

$$\underline{f}_p = \begin{cases} 0.8j & : \left(\underline{\gamma}_{p-N} = 1 \text{ and } \underline{\gamma}_p = 0\right), \\ 0.8j & : \left(\underline{\gamma}_{p+N} = 1 \text{ and } \underline{\gamma}_p = 0\right), \\ \underline{\gamma}_p & : \text{otherwise.} \end{cases} \tag{10.8}$$

Equation (10.8) can be rewritten as

$$\underline{f}_p = 0.8j\left(1 - \underline{\gamma}_p\right)\underline{\gamma}_{p-N} + 0.8j\left(1 - \underline{\gamma}_p\right)\underline{\gamma}_{p+N} + \underline{\gamma}_p, \quad p = 1, 2, \ldots, N^2, \tag{10.9}$$

where $\underline{\gamma}_p = 0$, if $p < 1$ or $p > N^2$. For the phase-shifting mask in the first type of optical lithography system, the normal boundary layers are opaque. The tangential boundary layers of clear openings have transmission coefficient of 0. The tangential boundary layers of 180° phase-shifting openings have transmission coefficient of $-0.52j$. The pixel size is set to be 27.9 nm. The boundary layer width of the clear opening is approximated as 27.9 nm = pixel size. The boundary layer width of the 180° phase-shifting opening is 55.8 nm = 2 × pixel size. The minimum opening size is increased to be 306.9 nm = 11 × pixel size. Therefore,

$$\underline{f}_p = \begin{cases} -0.52j & : \left(\underline{\gamma}_{p-2N} = -1 \text{ and } \underline{\gamma}_p = 0\right), \\ -0.52j & : \left(\underline{\gamma}_{p+2N} = -1 \text{ and } \underline{\gamma}_p = 0\right), \\ \underline{\gamma}_p & : \text{otherwise.} \end{cases} \tag{10.10}$$

Equation (10.10) can be rewritten as

$$\underline{f}_p = \frac{0.52j}{2}\left(1-\underline{\gamma}_p\right)\left(1+\underline{\gamma}_p\right)\underline{\gamma}_{p-2N}\left(1-\underline{\gamma}_{p-2N}\right) + \frac{0.52j}{2}\left(1-\underline{\gamma}_p\right)\left(1+\underline{\gamma}_p\right)\underline{\gamma}_{p+2N}\left(1-\underline{\gamma}_{p+2N}\right) + \underline{\gamma}_p, \quad (10.11)$$

where $p = 1, 2, \ldots, N^2$ and $\underline{\gamma}_p = 0$, if $p < 1$ or $p > N^2$. For the phase-shifting mask in the second type of optical lithography system, the normal boundary layers are opaque. The tangential boundary layers of clear openings have transmission coefficient of $0.8j$. The tangential boundary layers of 180° phase-shifting openings have transmission coefficient of $-0.30j$. The pixel size is set to be 14.5 nm. The boundary layer width of the clear opening is 14.5 nm = pixel size. The boundary layer width of the 180° phase-shifting opening is approximated as 58 nm = 4 × pixel size. The minimum opening size is increased to be 261 nm = 18 × pixel size. Therefore,

$$\underline{f}_p = \begin{cases} 0.8j & : \left(\underline{\gamma}_{p-N} = 1 \text{ and } \underline{\gamma}_p = 0\right), \\ 0.8j & : \left(\underline{\gamma}_{p+N} = 1 \text{ and } \underline{\gamma}_p = 0\right), \\ -0.30j & : \left(\underline{\gamma}_{p-4N} = -1 \text{ and } \underline{\gamma}_p = 0\right), \\ -0.30j & : \left(\underline{\gamma}_{p+4N} = -1 \text{ and } \underline{\gamma}_p = 0\right), \\ \underline{\gamma}_p & : \text{otherwise.} \end{cases} \quad (10.12)$$

Equation (10.12) can be rewritten as

$$\underline{f}_p = \frac{0.8j}{2}\left(1-\underline{\gamma}_p\right)\left(1+\underline{\gamma}_p\right)\underline{\gamma}_{p-N}\left(1+\underline{\gamma}_{p-N}\right), + \frac{0.8j}{2}\left(1-\underline{\gamma}_p\right)\left(1+\underline{\gamma}_p\right)\underline{\gamma}_{p+N}\left(1+\underline{\gamma}_{p+N}\right), + \frac{0.3j}{2}\left(1-\underline{\gamma}_p\right)\left(1+\underline{\gamma}_p\right)\underline{\gamma}_{p-4N}\left(1-\underline{\gamma}_{p-4N}\right), + \frac{0.3j}{2}\left(1-\underline{\gamma}_p\right)\left(1+\underline{\gamma}_p\right)\underline{\gamma}_{p+4N}\left(1-\underline{\gamma}_{p+4N}\right) + \underline{\gamma}_p, \quad (10.13)$$

where $p = 1, 2, \ldots, N^2$ and $\underline{\gamma}_p = 0$, if $p < 1$ or $p > N^2$.

5. A convolution matrix H is a $N^2 \times N^2$ matrix with an equivalent two-dimensional filter h.
6. The desired $N \times N$ binary output pattern is denoted as $\tilde{I}$. It is the desired aerial image sought on the wafer. Its vector representation is denoted as $\underline{\tilde{i}}$.
7. For the OPC optimization, the initial interior transmission area $\tilde{\Gamma}$ of the optimization process is assigned equal to $\tilde{I}$. For the PSM optimization, the amplitude of $\tilde{\Gamma}$ is assigned equal to $\tilde{I}$. The phase of $\tilde{\Gamma}$ must be assigned *a priori*,

where phases in neighboring blocks are assigned alternately. For each $\tilde{\Gamma}$, the corresponding initial mask pattern is $\tilde{M}$.

8. The output aerial image is the $N \times N$ matrix denoted as

$$I = |H\{F\}|^2. \tag{10.14}$$

The equivalent vector is denoted as $\underline{i}$.

9. The optimized mask denoted as $\hat{M}$ minimizes the distance between I and $\tilde{I}$, that is,

$$\hat{M} = \arg\min_{M} d(|H\{F\}|^2, \tilde{I}). \tag{10.15}$$

Given the output aerial image $\underline{i} = |H\underline{f}|^2$, the pth entry in this vector can be represented as

$$\underline{i}_p = \left| \sum_{q=1}^{N^2} h_{pq} \underline{f}_q \right|^2, \quad p = 1, \ldots N^2, \tag{10.16}$$

where h_{pq} is the p, qth entry of the filter. The cost function is the square of l^2 norm of the difference between $\underline{i}$ and $\underline{\tilde{i}}$. Therefore,

$$d = F(\underline{\gamma}) = \left\| \underline{\tilde{i}} - \underline{i} \right\|_2^2 = \sum_{p=1}^{N^2} \left(\underline{\tilde{i}}_p - \underline{i}_p \right)^2, \tag{10.17}$$

where $\underline{i}_p$ is already represented in Eq. (10.16).

The performance of the OPC and PSM optimization algorithms is evaluated by the output pattern error, which is defined as d in Eq. (10.17). According to Eq. (10.17), the output pattern error results from the comparison between the desired pattern and the aerial image without threshold. Thus, Eq. (10.17) does not account for the photoresist effect. It has been proven that ignoring the photoresist effect may improve the aerial image contrast [69].

In the following, the sensitivity of the cost function F with respect to the change in the interior transmission area will be used to guide the optimization process. The sensitivity of the cost function F is ∇F. For the OPC optimization in the first type of optical lithography system,

$$\nabla F = -4H^T \left[(\underline{\tilde{i}} - \underline{i}) \odot H\underline{\gamma} \right], \tag{10.18}$$

where T is the conjugate transposition and $\odot$ is the element-by-element multiplication operation [45]. For the OPC optimization in the second type of optical lithography system,

$$\begin{aligned} \nabla F = & -4\mathrm{Re} \left\{ H^T[(\underline{\tilde{i}} - \underline{i}) \odot H\underline{f}] \odot (0.8j\underline{\gamma} \uparrow +0.8j\underline{\gamma} \downarrow +\underline{1}) \right. \\ & + H^T[(\underline{\tilde{i}} - \underline{i}) \odot H\underline{f}] \downarrow \odot 0.8j(\underline{1} - \underline{\gamma} \downarrow) \\ & \left. + H^T[(\underline{\tilde{i}} - \underline{i}) \odot H\underline{f}] \uparrow \odot 0.8j(\underline{1} - \underline{\gamma} \uparrow) \right\}, \end{aligned} \tag{10.19}$$

where $\mathrm{Re}\{\cdot\}$ denotes the real part of the argument. $\uparrow$ and $\downarrow$ are shifting operations by shifting the $N \times N$ equivalent matrix of the vector in the argument along vertical direction (up and down) by one pixel, respectively [45]. For the PSM optimization in the first type of optical lithography system,

$$\begin{aligned}\nabla F = &-4\mathrm{Re}\left\{H^T\left[(\underline{\tilde{i}}-\underline{i})\odot H\underline{f}\right]\odot\left[0.52j\underline{\gamma}\odot\underline{\gamma}\uparrow_2\odot(\underline{1}-\underline{\gamma}\uparrow_2)\right.\right.\\&+0.52j\underline{\gamma}\odot\underline{\gamma}\downarrow_2\odot(\underline{1}-\underline{\gamma}\downarrow_2)+\underline{1}\big]+H^T\left[(\underline{\tilde{i}}-\underline{i})\odot H\underline{f}\right]\downarrow_2\\&\odot\left[-0.26j\,(\underline{1}-\underline{\gamma}\downarrow_2)\odot(\underline{1}+\underline{\gamma}\downarrow_2)\odot(\underline{1}-2\underline{\gamma})\right]\\&+H^T\left[(\underline{\tilde{i}}-\underline{i})\odot H\underline{f}\right]\uparrow_2\odot\big[-0.26j\,(\underline{1}-\underline{\gamma}\uparrow_2)\odot(\underline{1}+\underline{\gamma}\uparrow_2)\\&\left.\left.\odot\ (\underline{1}-2\underline{\gamma})\right]\right\},\end{aligned} \tag{10.20}$$

where $\uparrow_2$ and $\downarrow_2$ are shifting operations by shifting the $N \times N$ equivalent matrix of the vector in the argument along vertical direction (up and down) by two pixels, respectively [48]. For the PSM optimization in the second type of optical lithography system,

$$\begin{aligned}\nabla F = &-4\mathrm{Re}\left\{H^T\left[(\underline{\tilde{i}}-\underline{i})\odot H\underline{f}\right]\odot\left[0.3j\underline{\gamma}\odot\underline{\gamma}\uparrow_4\odot(\underline{1}-\underline{\gamma}\uparrow_4)\right.\right.\\&+0.3j\underline{\gamma}\odot\underline{\gamma}\downarrow_4\odot(\underline{1}-\underline{\gamma}\downarrow_4)+0.8j\underline{\gamma}\odot\underline{\gamma}\uparrow\odot(\underline{1}-\underline{\gamma}\uparrow)+0.8j\underline{\gamma}\\&\odot\underline{\gamma}\downarrow\odot(\underline{1}-\underline{\gamma}\downarrow)+\underline{1}\big]+H^T\left[(\underline{\tilde{i}}-\underline{i})\odot H\underline{f}\right]\downarrow\odot\big[-0.4j\,(\underline{1}-\underline{\gamma}\downarrow)\\&\odot\ (\underline{1}+\underline{\gamma}\downarrow)\odot(\underline{1}+2\underline{\gamma})\big]+H^T\left[(\underline{\tilde{i}}-\underline{i})\odot H\underline{f}\right]\uparrow\\&\odot\left[-0.4j\,(\underline{1}-\underline{\gamma}\uparrow)\odot(\underline{1}+\underline{\gamma}\uparrow)\odot(\underline{1}+2\underline{\gamma})\right]\\&+H^T\left[(\underline{\tilde{i}}-\underline{i})\odot H\underline{f}\right]\downarrow_4\odot\big[-0.15j\,(\underline{1}-\underline{\gamma}\downarrow_4)\\&\odot\ (\underline{1}+\underline{\gamma}\downarrow_4)\odot(\underline{1}-2\underline{\gamma})\big]\\&+H^T\left[(\underline{\tilde{i}}-\underline{i})\odot H\underline{f}\right]\uparrow_4\odot\big[-0.15j\,(\underline{1}-\underline{\gamma}\uparrow_4)\\&\left.\left.\odot\ (\underline{1}+\underline{\gamma}\uparrow_4)\odot(\underline{1}-2\underline{\gamma})\right]\right\},\end{aligned} \tag{10.21}$$

where $\uparrow_4$ and $\downarrow_4$ are shifting operations by shifting the $N \times N$ equivalent matrix of the vector in the argument along vertical direction (up and down) by four pixels, respectively [48]. The derivations of Eq. (10.18) to (10.21) are shown in Appendix G.

10.4 OPC OPTIMIZATION

10.4.1 Topological Constraint

According to the BL model under coherent illumination summarized in Section 10.2.1, the interior transmission area has a one-to-one correspondence to the mask. Therefore, the following OPC mask design algorithm directly optimizes the interior transmission area, from which the mask can be easily reconstructed. The BL model constrains the minimum size of the openings on the binary mask [81, 82]. To meet the requirements, some topological constraints are imposed on the optimization process of the interior transmission area [33, 46, 98]. In the following, some definitions and operations for shape topologies are listed.

Definition 10.1 (White block and black block). Any pixel in the interior transmission area can have either a value 0 or 1. A white block is a square area with all pixels having values equal to 1, while a black block has all of its pixels equal to 0.

Definition 10.2 (Flipping-on and flipping-off operations). Turning the pixel value from 0 to 1 and from 1 to 0 is referred to as flipping-on and flipping-off a pixel, respectively. In general, flipping-on and flipping-off operations of a block means to turn the block to a white block and to a black block, respectively.

Definition 10.3 (Type I singular pixel). A type I singular pixel is one that has value of 1, and does not belong to any $L \times L$ white block on the interior transmission area pattern Γ, where L depends on the minimum opening size of the BL model.

Definition 10.4 (Type II singular pixel). A type II singular pixel is one that has value of 0, and does not belong to any 3×3 black block on the interior transmission area pattern Γ. Since the openings on the optimized binary mask contain the additional surrounding boundary layers compared to the corresponding interior transmission areas, the type II singular pixel introduces the mergence of adjacent openings on the mask.

Definition 10.5 (Cost sensitivity matrix of a block). The cost sensitivity function corresponding to a block G on the interior transmission area pattern, calculated by Eq. (10.18) or (10.19), is $\nabla F(G)$ defined as the cost sensitivity matrix of the block G.

Definition 10.6 (Changeable block). A $K \times K$ changeable block is a block whose cost sensitivity matrix contains K positive or negative values. If the cost sensitivity matrix contains K positive values, the block is defined as a positive changeable block. Likewise, if it contains K negative values, it is defined as a negative changeable block. Note that a block may be both positive and negative changeable blocks at the same time.

In this OPC optimization approach, only the positive or negative changeable blocks are considered to be flipped-off or flipped-on. These topological constraints guarantee that the features of the optimized binary mask are larger than the minimum opening size.

10.4.2 OPC Optimization Algorithm Based on BL Model Under Coherent Illumination

Following the topological constraints, the OPC optimization algorithm is shown in Table 10.2, where the parameters K in **Step 3** and L used in **Definition 10.3** depend on the minimum opening size assigned in Section 10.2.

Table 10.2 The Binary Mask Optimization Algorithm

Step 1	Initialization of the interior transmission area pattern: $\tilde{\Gamma} = \tilde{I}$. The corresponding initial mask pattern is $\tilde{M}$.
Step 2	Calculate the cost sensitivity function using Eq. (10.18) for the first type of optical lithography system or Eq. (10.19) for the second type of optical lithography system.
Step 3	Scan the cost sensitivity matrix from top to bottom and from left to right. Find the first encountered $K \times K$ changeable block G.
Step 4	Flip-on or flip-off G if it is a negative or positive changeable block.
Step 5	If (flipping operation has introduced type I or type II singular pixel) flag=1.
Step 6	If (flag==1) or (cost function F is increased) or (any pixel value $\neq 0$ or 1) restore G to its original values.
Step 7	Clear the cost sensitivity matrix of G: $\nabla F(G) = \mathbf{0}$.
Step 8	If $\nabla F \neq \mathbf{0}$ Go to step 3. Otherwise If no block is flipped in the current iteration End. Otherwise Go to step 2.

10.4.3 Simulations

To prove the efficiency of the OPC optimization algorithm, the method described in Table 10.2 is used to design a mask targeting the desired aerial image shown in Fig. 10.7. In Fig. 10.7, p is the pitch width. For the first type of optical lithography

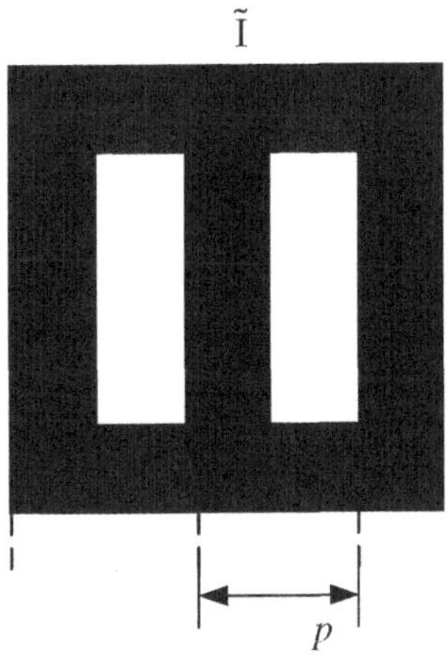

Figure 10.7 Desired pattern of the aerial image searched on the wafer.

system, $p = 223.2\,\text{nm}$ on the wafer, and the system parameters are $\text{NA} = 0.68$ and $\lambda = 248\,\text{nm}$. Since the system is a $4\times$ projection system, the pitch width of the initial interior transmission area pattern $\tilde{\Gamma}$ on the mask is $892.8\,\text{nm} = 4 \times p$. In the simulation, the mask pattern has the dimension of $2.23\,\mu\text{m} \times 2.23\,\mu\text{m}$. The pixel size is $24.8\,\text{nm} \times 24.8\,\text{nm}$. The convolution kernel is

$$h(\mathbf{r}) = \frac{J_1(2\pi r\text{NA}/\lambda)}{2\pi r\text{NA}/\lambda}, \tag{10.22}$$

which is assumed to vanish outside the area A_{h1} defined by $x, y \in [-1.5\,\mu\text{m}, 1.5\,\mu\text{m}]$. The parameters of the optimization algorithm are $K = L = 8$. The simulation results using the algorithm depicted in Table 10.2 for the first type of optical lithography system are shown in Fig. 10.8. The top row (from left to right) shows the initial mask pattern and the corresponding output aerial image, with output pattern error of 1200.1. The middle row (from left to right) shows the optimized binary mask $\tilde{M}'$ using the algorithm depicted in Table 10.2 based on thin-mask approximation and the corresponding output aerial image, with output pattern error of 1039.4. The bottom row (from left to right) shows the optimized binary mask based on BL model and the corresponding output aerial image, with output pattern error of 972.3. In the mask patterns, black and white represent 0 and 1, respectively. It is shown that optimization of the binary mask based on thin-mask approximation reduces the output pattern error by 13.4%. On the other hand, algorithm based on BL model reduces the output pattern error by 19.0%. Figure 10.9 illustrates the intersections of the aerial images on the 45th row. The solid, dashed, and dotted lines represent the intersections corresponding to the initial mask, OPC based on thin-mask assumption, and OPC based on thick-mask assumption, respectively.

For the second type of optical lithography system, $p = 137.8\,\text{nm}$ on the wafer, the system parameters are $\text{NA} = 0.85$ and $\lambda = 193\,\text{nm}$. The pitch width of the initial interior transmission area pattern $\tilde{\Gamma}$ on the mask is $551.0\,\text{nm} = 4 \times p$. In the simulation, the mask pattern has the dimension of $1.38\,\mu\text{m} \times 1.38\,\mu\text{m}$. The pixel size is $14.5\,\text{nm} \times 14.5\,\text{nm}$. The convolution kernel is assumed to vanish outside the area A_{h2} defined by $x, y \in [-1.0\,\mu\text{m}, 1.0\,\mu\text{m}]$. The parameters of the optimization algorithm are $K = L = 12$. The simulation results for the second type of optical lithography system are shown in Fig. 10.10. The top row (from left to right) shows the initial mask pattern and the corresponding output aerial image, with output pattern error of 1352.4. The middle row (from left to right) shows the optimized binary mask $\tilde{M}'$ using the algorithm depicted in Table 10.2 based on thin-mask approximation and the corresponding output aerial image, with output pattern error of 1135.9. The bottom row (from left to right) shows the optimized binary mask based on BL model and the corresponding output aerial image, with output pattern error of 1089.6. In the mask patterns, black and white represent 0 and 1, respectively. It is shown that optimization of the binary mask based on thin-mask approximation reduces the output pattern error by 16.0%. On the other hand, algorithm based on BL model reduces the output pattern error by 19.4%. Figure 10.11 illustrates the intersections of the aerial images on the 48th row. The solid, dashed, and dotted lines represent the intersections corresponding to the initial mask, OPC based on thin-mask

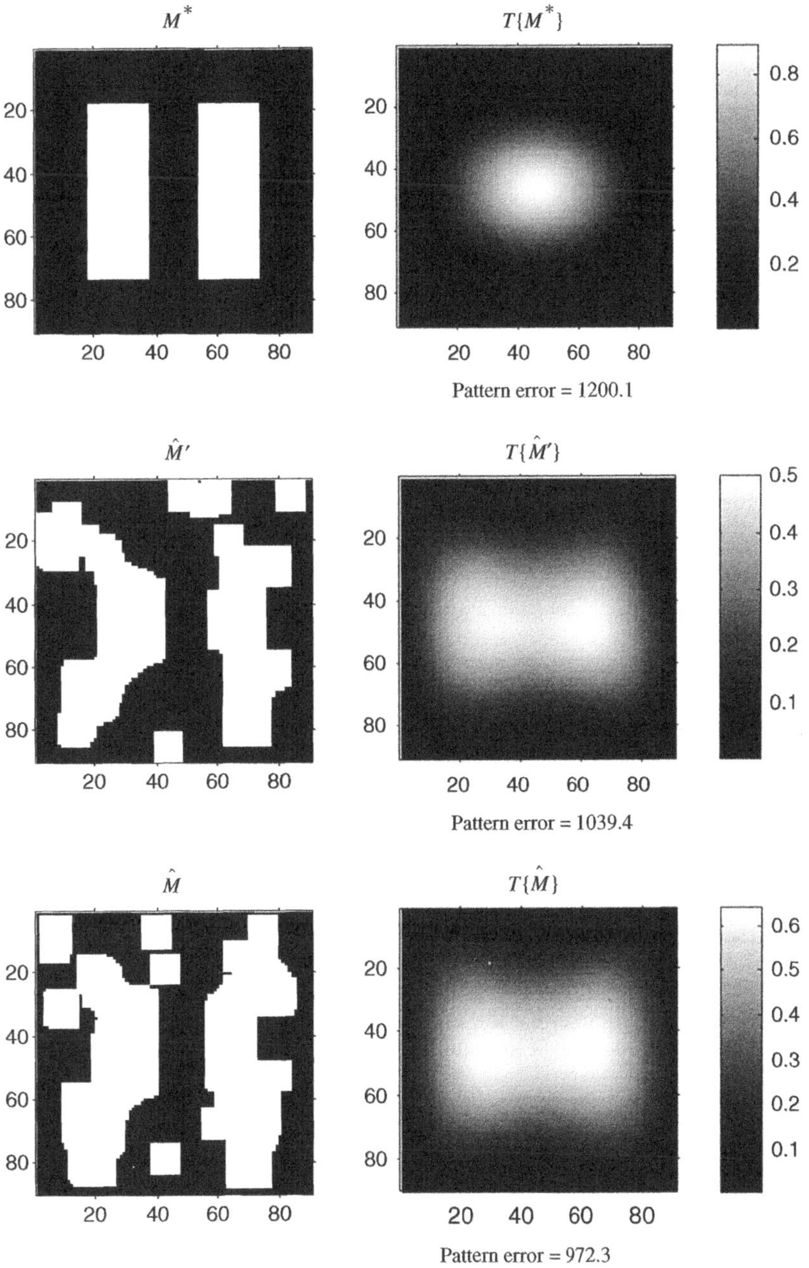

Figure 10.8 OPC optimization based on BL model for the first type of coherent optical lithography system. NA = 0.68 and $\lambda = 248$ nm. *Top row* (from left to right): The initial mask pattern and the corresponding output aerial image. *Middle row* (from left to right): The optimized binary mask based on thin-mask approximation and the corresponding output aerial image. *Bottom row* (from left to right): The optimized binary mask based on BL model and the corresponding output aerial image. In the mask patterns, black and white represent 0 and 1, respectively.

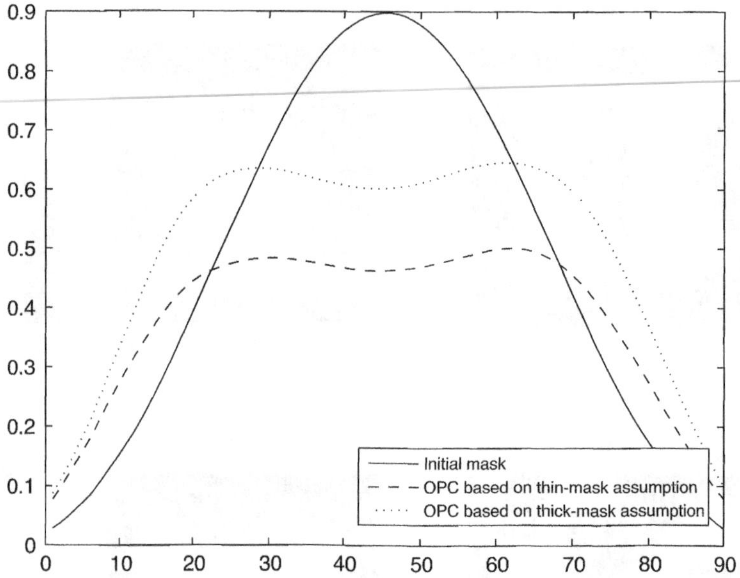

Figure 10.9 Intersections of the aerial images shown in Fig. 10.8 on the 45th row.

assumption, and OPC based on thick-mask assumption, respectively. As shown in Figs. 10.8 and 10.10, the described OPC optimization algorithm effectively reduces the output pattern errors and obtains more desirable aerial images. The performance differences between the optimizations of mask based on thin-mask approximation and on BL model show the necessity of the described algorithms taking into account the thick-mask effect. These results are consistent with those obtained in other simulations with different desired patterns. It is to be noted that the algorithms here result in asymmetric structures in the optimized mask patterns. These asymmetric structures introduce higher degree of freedom in the optimization process and lead to small output pattern errors. To obtain symmetric structures, the algorithms can be easily modified to optimize just the top half part of the mask patterns with respect to the midline. The symmetric pixels in the bottom half part are flipped in the same way.

10.5 PSM OPTIMIZATION

10.5.1 Topological Constraint

In this section, the OPC optimization algorithms described in Section 10.4 are generalized to the case of phase-shifting mask design. To achieve this goal, we first modify the topological constraints described in Section 10.4.1 to fit the PSM optimization. The modified definitions and operations for shape topologies are listed in the following.

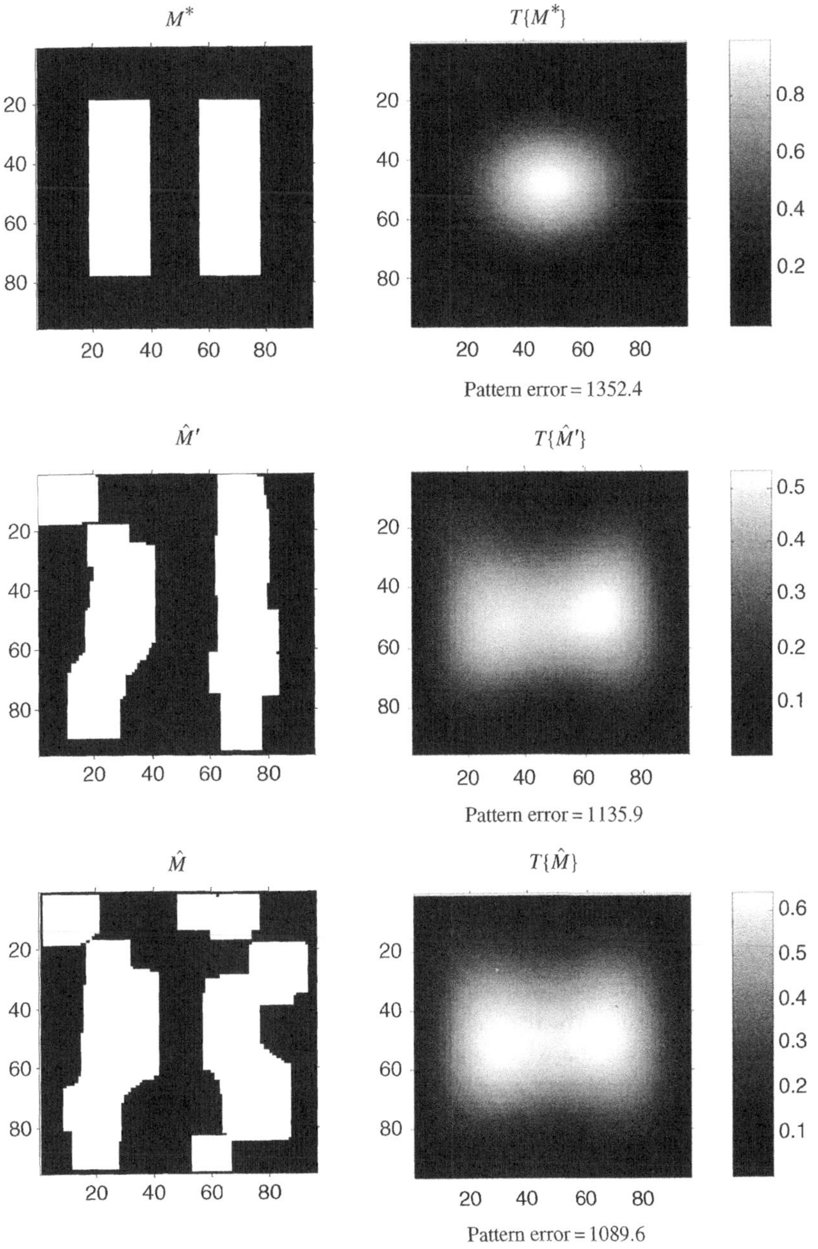

Figure 10.10 OPC optimization based on BL model for the second type of coherent optical lithography system. NA $= 0.85$ and $\lambda = 193$ nm. *Top row* (from left to right): The initial mask pattern and the corresponding output aerial image. *Middle row* (from left to right): The optimized binary mask based on thin-mask approximation and the corresponding output aerial image. *Bottom row* (from left to right): The optimized binary mask based on BL model and the corresponding output aerial image. In the mask patterns, black and white represent 0 and 1, respectively.

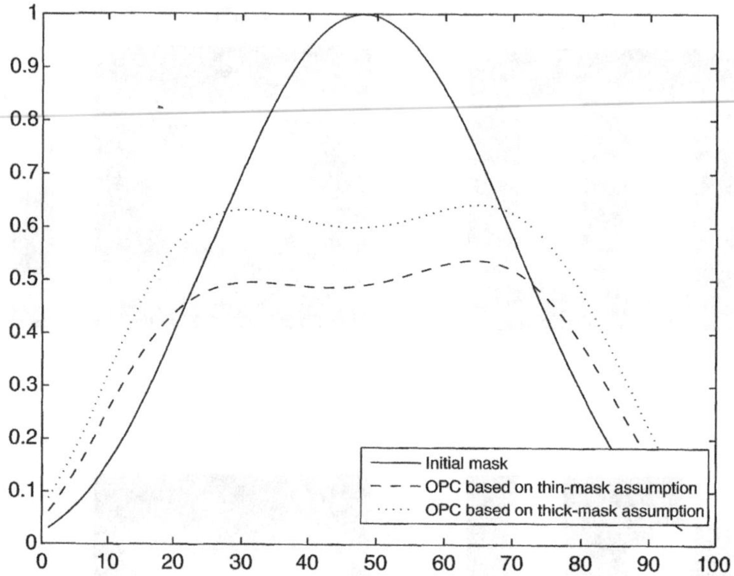

Figure 10.11 Intersections of the aerial images shown in Fig. 10.10 on the 48th row.

Definition 10.7 (White, gray, and black blocks). Any pixel in the interior transmission area can have a value of 1, 0, or -1. A white block is a square area with all pixel values equal to 1. A gray block has all of its pixels equal to 0. A black block has all of its pixels equal to -1.

Definition 10.8 (Flipping-on and flipping-off operations). Increasing or decreasing a pixel value by 1 is referred to as flipping-on or flipping-off a pixel. In general, consider a block with maximum pixel value equal to $p_{\max}$ and minimum pixel value $p_{\min}$. If $p_{\max} = 1$, flipping-on operation means to turn the block to a white block. Otherwise, flipping-on operation means to assign all the pixel values in this block equal to $p_{\max} + 1$. Similarly, if $p_{\min} = -1$, flipping-off operation means to turn the block to a black block. Otherwise, flipping-off operation means to assign all the pixel values in this block equal to $p_{\min} - 1$.

Definition 10.9 (Type I and II singular pixels). A type I or II singular pixel is one that has value of 1 (-1), and does not belong to any $L_1 \times L_1$ white (black) block on Γ, where L_1 depends on the minimum opening size.

Definition 10.10 (Type III singular pixel). A type III singular pixel is one that has value of 0, and does not belong to any $L_2 \times L_2$ gray block on Γ. Since the openings on the optimized PSM contain the additional surrounding boundary layers compared to the corresponding interior transmission areas, the type III singular pixel introduces the mergence of adjacent openings on the mask.

Definition 10.11 (Cost sensitivity matrix of a block). The cost sensitivity function corresponding to a block G on the interior transmission area pattern, calculated by Eq. (10.20) or (10.21) is $\nabla F(G)$ defined as the cost sensitivity matrix of the block G.

Definition 10.12 (Changeable block). A $K \times K$ changeable block is a block whose cost sensitivity matrix contains K positive or negative values. If the cost sensitivity matrix contains K positive values, the block is defined as a positive changeable block. Likewise, when it contains K negative values, it is defined as a negative changeable block. Note that a block may be both positive and negative changeable blocks at the same time.

In our PSM optimization approach, only the positive or negative changeable blocks are considered to be flipped-off or flipped-on. These topological constraints guarantee that the features of the optimized PSM are larger than the minimum opening size.

10.5.2 PSM Optimization Algorithm Based on BL Model Under Coherent Illumination

Following the topological constraints, the PSM optimization algorithm is shown in Table 10.3, where the parameters K in **Step 3** and L_1 used in **Definition 10.9** depend on the minimum opening size assigned in Section 10.2. In **Definition 10.10**, $L_2 = 5$ and 9 for the first and the second types of optical lithography systems, respectively.

10.5.3 Simulations

To prove the efficiency of the PSM optimization algorithm in the first type of optical lithography system, the method described in Table 10.3 is used to design a mask targeting the desired aerial image shown in Fig. 10.7. In Fig. 10.7, p is the pitch width. Note that the desired aerial image is symmetric with respect to the horizontal midline, which can be exploited to effectively reduce 50% computationally complexity of the algorithm. In the following simulations, the optimization process is carried out just in the top half part of the mask. The symmetric pixels with respect to the midline in the bottom half part are flipped in the same way. In general, the mask desired aerial image is not symmetric, and the algorithms are easily extended to the full-chip optimization. For the first type of optical lithography system, $p = 223.2$ nm on the wafer. The system parameters are NA $= 0.68$ and $\lambda = 248$ nm. Since the system is a $4\times$ projection system, the pitch width of the initial interior transmission area pattern $\tilde{\Gamma}$ on the mask is 892.8 nm $= 4 \times p$. In the simulation, the mask pattern has the dimension of 2.23 μm × 2.23 μm. The pixel size is 27.9 nm × 27.9 nm. The convolution kernel is

$$h(\mathbf{r}) = \frac{J_1(2\pi r \mathrm{NA}/\lambda)}{2\pi r \mathrm{NA}/\lambda}. \tag{10.23}$$

The convolution kernel is assumed to vanish outside the area A_{h1} defined by $x, y \in [-1.5\ \mu\text{m}, 1.5\ \mu\text{m}]$. The parameters of the optimization algorithm are $K = L_1 = 9$.

Table 10.3 The PSM Optimization Algorithm

Step 1	Initialization of the amplitude of interior transmission area pattern: $\|\tilde{\Gamma}\| = \tilde{I}$. The phase of $\tilde{\Gamma}$ must be assigned *a priori*, where phases in neighboring blocks are assigned alternately. The corresponding initial mask pattern is $\tilde{M}$.
Step 2	Calculate the cost sensitivity function using Eq. (10.20) for the first type of optical lithography system or Eq. (10.21) for the second type of optical lithography system.
Step 3	Scan the cost sensitivity matrix from top to bottom and from left to right. Find the first encountered $K \times K$ changeable block G.
Step 4	Flip-on or flip-off G if it is a negative or positive changeable block.
Step 5	If (flipping operation has introduced type I, II, or III singular pixel) flag=1.
Step 6	If (flag==1) or (cost function F is increased) or (any pixel value $\neq -1$ or 0 or 1) restore G to its original values.
Step 7	Clear the cost sensitivity matrix of G: $\nabla F(G) = \mathbf{0}$.
Step 8	If $\nabla F \neq \mathbf{0}$ Go to step 3. Otherwise If no block is flipped in the current iteration End. Otherwise Go to step 2.

The simulation results using the algorithm depicted in Table 10.3 for the first type of optical lithography system are shown in Fig. 10.12. The top row (from left to right) shows the initial mask pattern and the corresponding output aerial image, with output pattern error of 629.4. The middle row (from left to right) shows the optimized PSM $\tilde{M}'$ using the algorithm depicted in Table 10.3 based on thin-mask approximation and the corresponding output aerial image, with output pattern error of 675.8. The bottom row (from left to right) shows the optimized PSM based on BL model and the corresponding output aerial image, with output pattern error of 596.2. In the mask patterns, black, gray, and white represent -1, 0, and 1, respectively. It is shown that optimization of the PSM based on thin-mask approximation increases the output pattern error by 7.4%. On the other hand, algorithm based on BL model reduces the output pattern error by 5.3%. Figure 10.13 illustrates the intersections of the aerial images on the 40th row. The solid, dashed, and dotted lines represent the intersections corresponding to the initial mask, PSM based on thin-mask assumption, and PSM based on thick-mask assumption, respectively.

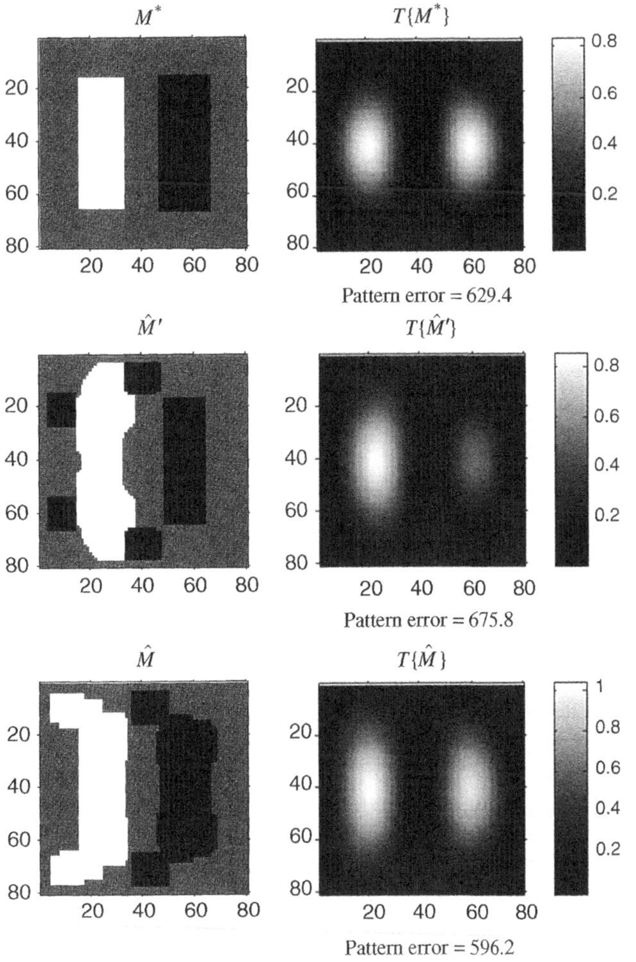

Figure 10.12 PSM optimization based on BL model for the first type of coherent optical lithography system. NA $= 0.68$ and $\lambda = 248$ nm. *Top row* (from left to right): The initial mask pattern and the corresponding output aerial image. *Middle row* (from left to right): The optimized PSM based on thin-mask approximation and the corresponding output aerial image. *Bottom row* (from left to right): The optimized PSM based on BL model and the corresponding output aerial image. In the mask patterns, black, gray, and white represent -1, 0, and 1 respectively.

For the second type of optical lithography system, $p = 137.8$ nm on the wafer. The system parameters are NA $= 0.85$ and $\lambda = 193$ nm. The pitch width of the initial interior transmission area pattern $\tilde{\Gamma}$ on the mask is 551 nm $= 4 \times p$. In the simulation, the initial mask pattern has the dimension of 2.2 μm × 2.2 μm. The pixel size is 14.5 nm × 14.5 nm, which is the same as the boundary width. The convolution kernel is assumed to vanish outside the area A_{h2} defined by $x, y \in [-1.0\ \mu\text{m}, 1.0\ \mu\text{m}]$. The parameters of the optimization algorithm are $K = L = 16$. The simulation results for

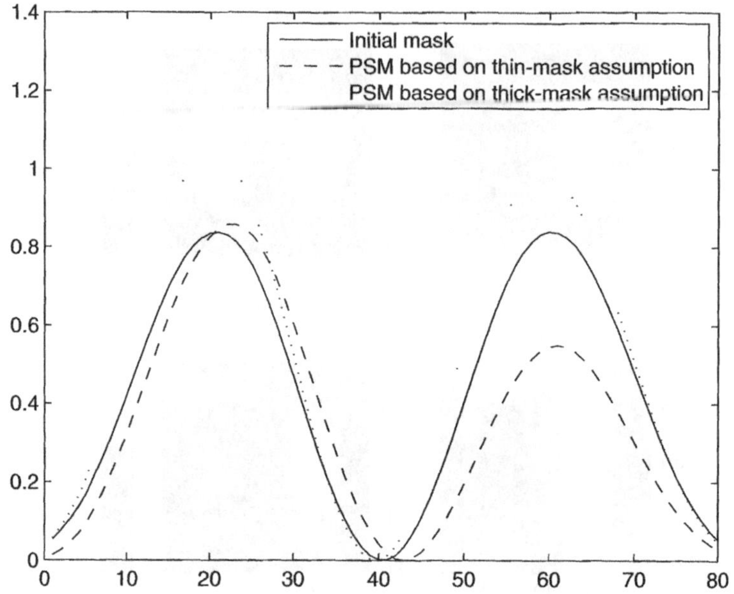

Figure 10.13 Intersection of the aerial image shown in Fig. 10.12 on the 40th row.

the second type of optical lithography system are shown in Fig. 10.14. The top row (from left to right) shows the initial mask pattern and the corresponding output aerial image, with output pattern error of 883.6. The middle row (from left to right) shows the optimized PSM $\tilde{M}'$ using the algorithm depicted in Table 10.3 based on thin-mask approximation and the corresponding output aerial image, with output pattern error of 1099.7. The bottom row (from left to right) shows the optimized PSM based on BL model and the corresponding output aerial image, with output pattern error of 720.8. In the mask patterns, black, gray, and white represent -1, 0, and 1, respectively. It is shown that optimization of the PSM based on thin-mask approximation increases the output pattern error by 24.5%. On the other hand, algorithm based on BL model reduces the output pattern error by 18.4%. Figure 10.15 illustrates the intersections of the aerial images on the 76th row. The solid dashed, and dotted lines represent the intersections corresponding to the initial mask, PSM based on thin-mask assumption, and PSM based on thick-mask assumption, respectively. As shown in Figs. 10.12 and 10.14, the described PSM optimization algorithm effectively reduces the output pattern errors and obtains more desirable aerial images. The performance differences between the optimizations of mask based on thin-mask approximation and on BL model show the necessity of the described algorithms taking into account the thick-mask effect. These results are consistent with those obtained in other simulations with different desired patterns.

The OPC and PSM optimization frameworks in this chapter are developed for coherent imaging systems and can be extended for partially coherent imaging systems. As described in Section 10.2.2, the BL model under partially coherent illumination

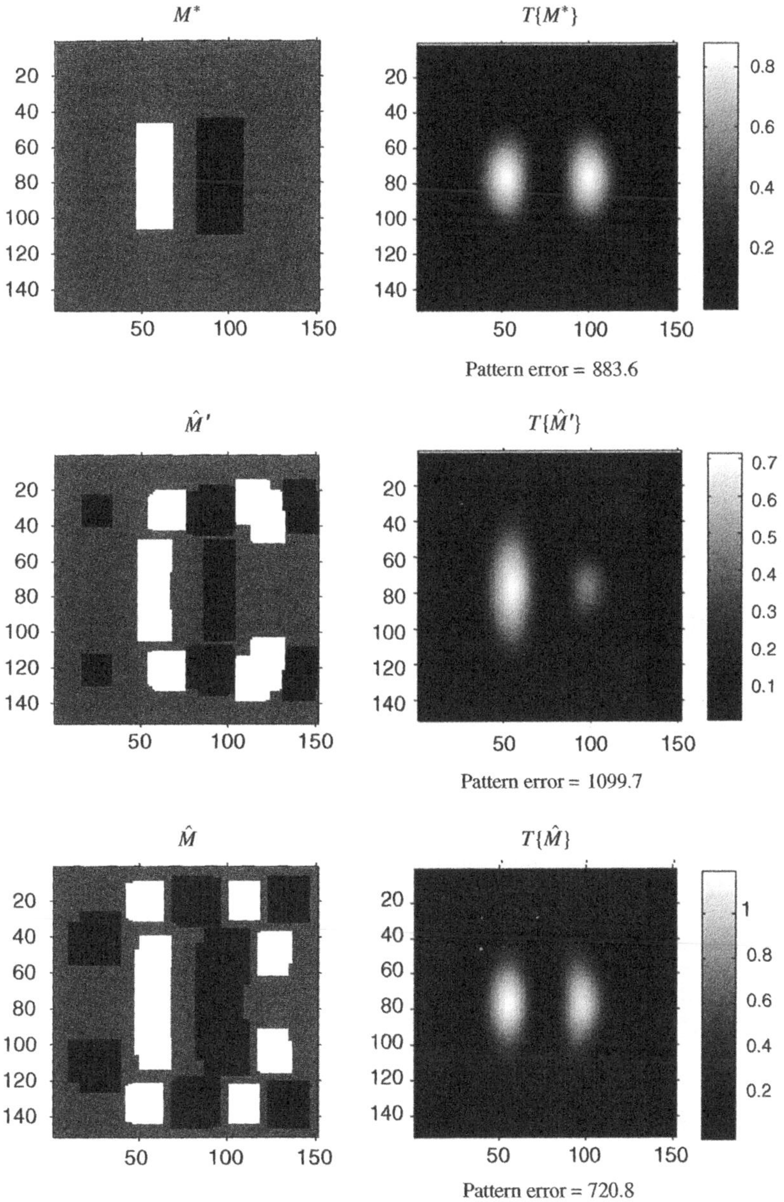

Figure 10.14 PSM optimization based on BL model for the second type of coherent optical lithography system. NA = 0.85 and $\lambda = 193$ nm. *Top row* (from left to right): The initial mask pattern and the corresponding output aerial image. *Middle row* (from left to right): The optimized PSM based on thin-mask approximation and the corresponding output aerial image. *Bottom row* (from left to right): The optimized PSM based on BL model and the corresponding output aerial image. In the mask patterns, black, gray, and white represent -1, 0, and 1, respectively.

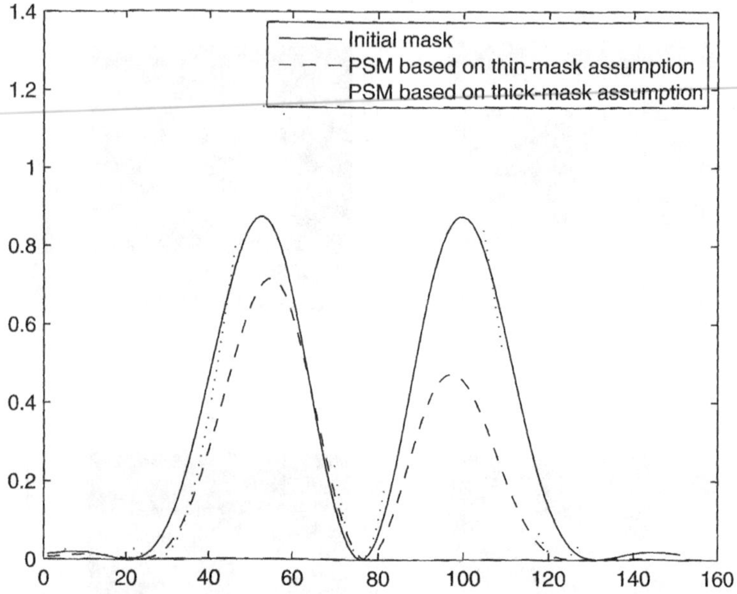

Figure 10.15 Intersection of the aerial image shown in Fig. 10.14 on the 76th row.

is approximated as the superposition of the BL model in coherent imaging system contributed by each source point. Thus, the partially coherent imaging system can be decomposed into the superposition of several coherent imaging systems based on SOCS model [74]. The near field on the exiting surface, contributed by each coherent component, can be evaluated by the BL model in coherent imaging system. The overall near field is the superposition of the near field resulting from all coherent components. Given the overall near field, the aerial image projected on the wafer under partially coherent illumination can be calculated using the SOCS model. Following the derivation in this chapter, the OPC and PSM optimization problems can be formulated in partially coherent imaging systems. However, this extension is beyond the scope of this book and hence not considered here.

10.6 SUMMARY

This chapter developed OPC and PSM optimizations in coherent imaging system under thick-mask assumption, where the BL model was applied to evaluate the near field of the thick mask. Two typical kinds of optical lithography systems were taken into account. Topological constraints were applied to limit the minimum feature size on the mask.

11

Conclusions and New Directions of Computational Lithography

11.1 CONCLUSION

In this book, we described a variety of gradient-based inverse lithography algorithms for OPC and PSM optimizations, as well as simultaneous source and mask optimization. Different optical lithography systems, ranging from coherent to partially coherent imaging systems, were taken into account in these algorithms. We also presented a set of regularization frameworks to reduce the output pattern errors and complexity of the optimized mask and source patterns. Subsequently, these algorithms were extended to OPC and PSM optimizations under the thick-mask assumption, which accounted for the polarization and diffraction effects due to the 3D topography of the mask.

In Chapter 2, the fundamentals of the optical lithography systems were summarized. The Abbe's model and the Hopkins diffraction model were discussed, both of which are based on the spatial discretization of the source into a number of spatially incoherent point sources. In the Abbe's model, the coherent images generated by every source point are incoherently added together to produce the final partially coherent image. On the other hand, the Hopkins diffraction model is a simplified and approximate version of the Abbe's model, which carries out the integration over the source first, and the result directly provides the aerial image of the partially coherent image system. Subsequently, three types of approximation models were summarized to represent the partially coherent imaging systems, such as the Fourier series expansion model, the SVD model, and the average coherent approximation model.

In Chapter 3, the RET approaches were first classified into rule-based, model-based, and hybrid RET approaches. Subsequently, this chapter focuses on the rule-based RETs, where rule-based OPC, PSM, and OAI approaches were described in detail. For all these rule-based RET methods, the rules to modify the masks and illuminations were summarized.

In Chapter 4, the definition and the classification of different optimization problems were summarized. Subsequently, this chapter focused on the unconstrained

Computational Lithography By Xu Ma and Gonzalo R. Arce

optimization, which was used to solve the inverse lithography optimization problems. Different types of unconstrained optimization algorithms were summarized, such as steepest descent method, Newton method, quasi-Newton method, conjugate gradient method, and so on. The steepest descent algorithm was applied to solve the gradient-based inverse lithography optimization problems.

In Chapter 5, OPC and two-phase PSM optimization algorithms based on coherent imaging system and thin-mask assumption were summarized and discussed. Subsequently, a generalized gradient-based inverse lithography optimization algorithm for PSM design was presented. This generalized approach enables the algorithm to search for a solution in the entire complex plane, and as such it is capable of generating adequate PSM for mask patterns having arbitrary Manhattan geometries, thus avoiding possible phase conflicts.

In Chapter 6, a set of regularization frameworks were discussed to bias the solution space of the optimization to sample solutions that had desired properties. The discretization penalty introduced in Ref. [68] was generalized to PSM optimization. In addition, we introduced the wavelet penalty to reduce the mask complexity. Because of the "localization property" of the wavelet penalty, regional weighting can be applied to different areas on the mask pattern. The advantages and trade-offs of the wavelet penalty and of the total variation penalty were discussed in the generalized PSM optimization.

In Chapter 7, the inverse lithography optimization algorithms were extended to the partially coherent imaging systems. Based on the Fourier series expansion model and the average coherent approximation model, two kinds of OPC optimization algorithms were described in the partially coherent imaging systems. The algorithm based on the Fourier series expansion model leads to lower output pattern errors. On the other hand, the algorithm based on the average coherent approximation model is capable of reducing the computational complexity. As a trade-off, the optimized output pattern errors are increased, which is caused by the inaccuracy of the average coherent approximation model. Based on the SVD model, a PSM optimization algorithm was developed for the partially coherent imaging system, where the first-order coherent approximation was used to represent the Hopkins diffraction model. This algorithm is most effective with small to medium partial coherence factors.

In Chapter 8, various techniques were developed to improve the performance of OPC and PSM optimization algorithms. A double patterning optimization method was presented as an alternative for the multiphase PSM. The double patterning method can be used to avoid the phase conflict and it results in much less pattern error; however, it requires more complicated processing and longer fabrication time. In addition, a post-processing based on 2D DCT was discussed to simultaneously reduce the complexity of the mask and the output pattern error. Finally, a photoresist tone reversing technique was exploited to design masks capable of projecting extremely sparse patterns, whose resolution limit is much higher than the traditional case without application of photoresist tone reversing.

In Chapter 9, computationally efficient and effective gradient-based simultaneous source–mask optimization algorithms were developed. Partially coherent illuminations were modeled by the Fourier series expansion model. Based on this model, the SMO algorithms were developed for both binary mask and PSM, where cost sensitivity was used to drive the cost function in the descent direction and topological constraints were applied in the optimization framework leading to desired manufacturability properties.

In Chapter 10, OPC and PSM optimizations for inverse lithography were studied taking into account the thick-mask effects under coherent illumination. The BL model was applied to evaluate the near field of the thick mask. Based on this model, the OPC and PSM optimization algorithms were described for two typical kinds of optical lithography systems. Topological constraints were applied in the optimization framework to limit the minimum feature size on the mask. Illustrative examples were presented.

11.2 NEW DIRECTIONS OF COMPUTATIONAL LITHOGRAPHY

11.2.1 OPC Optimization for the Next-Generation Lithography Technologies

OPC optimization algorithms developed for the traditional optical lithography systems will be generalized under extreme ultraviolet lithography (EUVL) and e-beam lithography (EBL) environments, which are significant departure from the deep ultraviolet lithography (DUVL) used today. EUVL systems use radiation with a wavelength ranging from 10 to 14 nm to carry out projection imaging, which is absorbed in virtually all materials, even gases [6]. Thus, EUVL systems are entirely reflective and operated under near-vacuum environment. Some prior works have been done to analyze the EUVL mask effects [18] and to optimize the EUVL phase mask [54]. On the other hand, EBL uses electron beam to directly write patterns on the wafer. The comprehensive simulation of EBL processes has been proposed using PROLITH/3D and TEMPTATION software tools [5]. It is expected that the OPC optimization algorithm for inverse lithography will be developed to compensate for the proximity effects in these next-generation lithography technologies.

11.2.2 Initialization Approach for the Inverse Lithography Optimization

Recently, a dynamic programming-based initialization scheme has been proposed to preassign phases for the alternative phase-shifting mask optimization [12]. This approach may be generalized to the multiphase mask optimization described in Section 5.4. The similar concept can also be extended to the assignment of photoresist distribution for the photoresist tone reversing method described in Section 8.3. In addition, based on this initialization approach, a joint optimization algorithm for the photoresist distribution and mask patterns could be developed.

11.2.3 Double Patterning and Double Exposure Methods in Partially Coherent Imaging System

In Section 8.1, a double patterning method was presented under coherent illumination. Poonawala and Milanfar have developed a double exposure method under coherent illumination [65, 66]. Both of these approaches aimed at avoiding the phase conflict of the two-phase PSM approach. In the future work, the double patterning and the double exposure methods may be extended to the partially coherent imaging systems. In addition, multipatterning and multiexposure methods may be investigated, where more than two exposure processes will be used to project image.

11.2.4 OPC and PSM Optimizations for Inverse Lithography Based on Rigorous Mask Models in Partially Coherent Imaging System

The gradient-based OPC and PSM optimizations based on the boundary layer model developed in Chapter 10 can be generalized into partially coherent systems. Approaches and regularization frameworks could be developed to further reduce the complexity of the algorithms and the optimized mask patterns. Other more accurate or computationally efficient rigorous mask models could also be proposed and applied in the OPC and PSM optimizations under the thick-mask assumption. In addition, 180° phase-shifting opening with 35 nm will be taken into account in the PSM optimization based on rigorous mask models.

11.2.5 Simultaneous Source and Mask Optimization for Inverse Lithography Based on Rigorous Mask Models

The SMO algorithms developed in Chapter 9 may be extended to the thick-mask cases. The SMO algorithms jointly optimize the source and mask patterns, where the source is depicted as a partially coherent illumination. Thus, the BL model in the partially coherent imaging system described in Section 10.2.2 can be applied to evaluate the near field on the mask exiting surface, generated by arbitrary partially coherent illumination.

11.2.6 Investigation of Factors Influencing the Complexity of the OPC and PSM Optimization Algorithms

As mentioned in Section 7.1.5, except for the computational operations in the algorithms itself, some other factors such as the treatment of boundary regions and hierarchy management can also affect the overall run time of the OPC and PSM optimization algorithms. Especially for the large-scale mask designs, these factors should be taken into account. Such considerations of the complexity of the optimization algorithms could be addressed.

Appendix A

Formula Derivation in Chapter 5

The derivation of Eq. (5.24) is as follows:

$$\frac{\partial F(\underline{\theta})}{\partial \underline{\theta}_m} = 2\sum_{i=1}^{N^2}\left[\tilde{\underline{z}}_i - \frac{1}{1+\exp\left[-a\sqrt{\left(\sum_{j=1}^{N^2} h_{ij}\cos\underline{\theta}_j\right)^2} + at_r\right]}\right]$$

$$\times \frac{1}{\left\{1+\exp\left[-a\sqrt{\left(\sum_{j=1}^{N^2} h_{ij}\cos\underline{\theta}_j\right)^2} + at_r\right]\right\}^2}$$

$$\times\exp\left[-a\sqrt{\left(\sum_{j=1}^{N^2} h_{ij}\cos\underline{\theta}_j\right)^2}\right] \times (-a) \times \frac{1}{2\sqrt{\left(\sum_{j=1}^{N^2} h_{ij}\cos\underline{\theta}_j\right)^2}}$$

$$\times\left[2\left(\sum_{j=1}^{N^2} h_{ij}\cos\underline{\theta}_j\right) h_{im}\left(-\sin\underline{\theta}_m\right)\right]$$

$$= 2\sum_{i=1}^{N^2}\left[\tilde{\underline{z}}_i - \frac{1}{1+\exp\left[-a\sqrt{\left(\sum_{j=1}^{N^2} h_{ij}\cos\underline{\theta}_j\right)^2} + at_r\right]}\right]$$

Computational Lithography By Xu Ma and Gonzalo R. Arce

$$\times \frac{1}{1+\exp\left[-a\sqrt{\left(\sum_{j=1}^{N^2} h_{ij}\cos\underline{\theta}_j\right)^2}+at_r\right]}$$

$$\times \frac{\exp\left[-a\sqrt{\left(\sum_{j=1}^{N^2} h_{ij}\cos\underline{\theta}_j\right)^2}\right]}{1+\exp\left[-a\sqrt{\left(\sum_{j=1}^{N^2} h_{ij}\cos\underline{\theta}_j\right)^2}+at_r\right]} \times (-a)$$

$$\times \frac{1}{2\sqrt{\left(\sum_{j=1}^{N^2} h_{ij}\cos\underline{\theta}_j\right)^2}} \times \left[2\left(\sum_{j=1}^{N^2} h_{ij}\cos\underline{\theta}_j\right) h_{im}\left(-\sin\underline{\theta}_m\right)\right]$$

$$= 2a\sum_{i=1}^{N^2} h_{im} \times \left(\underline{\tilde{z}}_i - \underline{z}_i\right) \times \underline{z}_i \times \left(1-\underline{z}_i\right)$$

$$\times \frac{\sum_{j=1}^{N^2} h_{ij}\cos\underline{\theta}_j}{\left|\sum_{j=1}^{N^2} h_{ij}\cos\underline{\theta}_j\right|} \times \sin\underline{\theta}_m. \tag{A.1}$$

Therefore,

$$\nabla F(\underline{\theta}) = \underline{d} = 2a\left(H^T\left[(\underline{\tilde{z}} - \underline{z}) \odot \underline{z} \odot (\underline{1} - \underline{z}) \odot \underline{\text{sig}}\right]\right) \odot \sin\underline{\theta}, \tag{A.2}$$

where

$$\underline{\text{sig}} = (\text{sig}_1, \text{sig}_2, \ldots, \text{sig}_{N^2}) \tag{A.3}$$

and

$$\text{sig}_i = \begin{cases} 0 \;:\; \sum_{j=1}^{N^2} h_{ij}\cos \le 0, \\ 1 \;:\; \sum_{j=1}^{N^2} h_{ij}\cos > 0, \end{cases} \qquad i = 1, \ldots, N^2. \tag{A.4}$$

The derivation of Eq. (5.36) is as follows:

$$\frac{\partial F(\underline{\theta},\underline{\phi})}{\partial \underline{\theta}_m}$$

$$= 2\sum_{i=1}^{N^2}\left[\tilde{\underline{z}}_i - \frac{1}{1+\exp\left[-a\sqrt{\left(\sum_{j=1}^{N^2} h_{ij}\underline{r}_j\cos\underline{\theta}_j\right)^2+\left(\sum_{j=1}^{N^2} h_{ij}\underline{r}_j\sin\underline{\theta}_j\right)^2}+at_r\right]}\right]$$

$$\times\frac{1}{\left\{1+\exp\left[-a\sqrt{\left(\sum_{j=1}^{N^2} h_{ij}\underline{r}_j\cos\underline{\theta}_j\right)^2+\left(\sum_{j=1}^{N^2} h_{ij}\underline{r}_j\sin\underline{\theta}_j\right)^2}+at_r\right]\right\}^2}$$

$$\times\exp\left[-a\sqrt{\left(\sum_{j=1}^{N^2} h_{ij}\underline{r}_j\cos\underline{\theta}_j\right)^2+\left(\sum_{j=1}^{N^2} h_{ij}\underline{r}_j\sin\underline{\theta}_j\right)^2}\right]\times(-a)$$

$$\times\frac{1}{\sqrt{\left(\sum_{j=1}^{N^2} h_{ij}\underline{r}_j\cos\underline{\theta}_j\right)^2+\left(\sum_{j=1}^{N^2} h_{ij}\underline{r}_j\sin\underline{\theta}_j\right)^2}}$$

$$\times\left[\left(\sum_{j=1}^{N^2} h_{ij}\frac{1+\cos\underline{\phi}_j}{2}\cos\underline{\theta}_j\right)h_{im}\frac{1+\cos\underline{\phi}_m}{2}(-\sin\underline{\theta}_m)\right.$$

$$\left.+\left(\sum_{j=1}^{N^2} h_{ij}\frac{1+\cos\underline{\phi}_j}{2}\sin\underline{\theta}_j\right)h_{im}\frac{1+\cos\underline{\phi}_m}{2}\cos\underline{\theta}_m\right]. \tag{A.5}$$

Thus,

$$\frac{\partial F(\underline{\theta},\underline{\phi})}{\partial \underline{\theta}_m} =$$

$$2\sum_{i=1}^{N^2}\left[\tilde{\underline{z}}_i - \frac{1}{1+\exp\left[-a\sqrt{\left(\sum_{j=1}^{N^2} h_{ij}\underline{r}_j\cos\underline{\theta}_j\right)^2+\left(\sum_{j=1}^{N^2} h_{ij}\underline{r}_j\sin\underline{\theta}_j\right)^2}+at_r\right]}\right]$$

$$\times \frac{1}{1+\exp\left[-a\sqrt{\left(\sum_{j=1}^{N^2} h_{ij}\underline{r}_j\cos\underline{\theta}_j\right)^2+\left(\sum_{j=1}^{N^2} h_{ij}\underline{r}_j\sin\underline{\theta}_j\right)^2}+at_r\right]}$$

$$\times \frac{\exp\left[-a\sqrt{\left(\sum_{j=1}^{N^2} h_{ij}\underline{r}_j\cos\underline{\theta}_j\right)^2+\left(\sum_{j=1}^{N^2} h_{ij}\underline{r}_j\sin\underline{\theta}_j\right)^2}\right]}{1+\exp\left[-a\sqrt{\left(\sum_{j=1}^{N^2} h_{ij}\underline{r}_j\cos\underline{\theta}_j\right)^2+\left(\sum_{j=1}^{N^2} h_{ij}\underline{r}_j\sin\underline{\theta}_j\right)^2}+at_r\right]} \times (-a)$$

$$\times \frac{1}{\sqrt{\left(\sum_{j=1}^{N^2} h_{ij}\underline{r}_j\cos\underline{\theta}_j\right)^2+\left(\sum_{j=1}^{N^2} h_{ij}\underline{r}_j\sin\underline{\theta}_j\right)^2}}$$

$$\times \left[\left(\sum_{j=1}^{N^2} h_{ij}\frac{1+\cos\underline{\phi}_j}{2}\cos\underline{\theta}_j\right) h_{im}\frac{1+\cos\underline{\phi}_m}{2}\left(-\sin\underline{\theta}_m\right)\right.$$

$$\left.+\left(\sum_{j=1}^{N^2} h_{ij}\frac{1+\cos\underline{\phi}_j}{2}\sin\underline{\theta}_j\right) h_{im}\frac{1+\cos\underline{\phi}_m}{2}\cos\underline{\theta}_m\right]$$

$$= 2a \times \frac{1+\cos\underline{\phi}_m}{2} \times \sum_{i=1}^{N^2} h_{im} \times \left(\underline{\tilde{z}}_i-\underline{z}_i\right) \times \underline{z}_i \times \left(1-\underline{z}_i\right)$$

$$\times \frac{1}{\sqrt{\left(\sum_{j=1}^{N^2} h_{ij}\underline{r}_j\cos\underline{\theta}_j\right)^2+\left(\sum_{j=1}^{N^2} h_{ij}\underline{r}_j\sin\underline{\theta}_j\right)^2}}$$

$$\times \left[\left(\sum_{j=1}^{N^2} h_{ij}\frac{1+\cos\underline{\phi}_j}{2}\cos\underline{\theta}_j\right)\sin\underline{\theta}_m\right.$$

$$\left.+\left(\sum_{j=1}^{N^2} h_{ij}\frac{1+\cos\underline{\phi}_j}{2}\sin\underline{\theta}_j\right)\left(-\cos\underline{\theta}_m\right)\right]. \quad \text{(A.6)}$$

Therefore,

$$\nabla F(\underline{\theta},\underline{\phi})_{\underline{\theta}} = 2a \times \frac{\underline{1}+\cos\underline{\phi}}{2} \odot \sin\underline{\theta} \odot \left\{H^T\left[\left(\underline{\tilde{z}}-\underline{z}\right)\odot\underline{z}\odot\left(\underline{1}-\underline{z}\right)\right.\right.$$

$$\left.\left.\odot H\left(\underline{m}_{\mathrm{R}}\right)\odot T\left(\underline{m}\right)\right]\right\} - 2a \times \frac{\underline{1}+\cos\underline{\phi}}{2} \odot \cos\underline{\theta}$$

$$\odot\left\{H^T\left[\left(\underline{\tilde{z}}-\underline{z}\right)\odot\underline{z}\odot\left(\underline{1}-\underline{z}\right)\odot H\left(\underline{m}_{\mathrm{I}}\right)\odot T\left(\underline{m}\right)\right]\right\}, \quad \text{(A.7)}$$

where $T(\underline{m}) = [(H\underline{m}_{\mathrm{R}})^2 + (H\underline{m}_{\mathrm{I}})^2]^{-\frac{1}{2}}$. $\underline{m}_{\mathrm{R}}$ and $\underline{m}_{\mathrm{I}}$ are the real and the imaginary part of $\underline{m}$.

The derivation of Eq. (5.37) is as follows:

$$\frac{\partial F\left(\underline{\theta},\underline{\phi}\right)}{\partial \underline{\phi}_m} =$$

$$\sum_{i=1}^{N^2}\left[\tilde{\underline{z}}_i - \frac{1}{1+\exp\left[-a\sqrt{\left(\sum_{j=1}^{N^2} h_{ij}\underline{r}_j\cos\underline{\theta}_j\right)^2 + \left(\sum_{j=1}^{N^2} h_{ij}\underline{r}_j\sin\underline{\theta}_j\right)^2} + at_r\right]}\right]$$

$$\times \frac{1}{\left\{1+\exp\left[-a\sqrt{\left(\sum_{j=1}^{N^2} h_{ij}\underline{r}_j\cos\underline{\theta}_j\right)^2 + \left(\sum_{j=1}^{N^2} h_{ij}\underline{r}_j\sin\underline{\theta}_j\right)^2} + at_r\right]\right\}^2}$$

$$\times \exp\left[-a\sqrt{\left(\sum_{j=1}^{N^2} h_{ij}\underline{r}_j\cos\underline{\theta}_j\right)^2 + \left(\sum_{j=1}^{N^2} h_{ij}\underline{r}_j\sin\underline{\theta}_j\right)^2}\right] \times a$$

$$\times \frac{1}{\sqrt{\left(\sum_{j=1}^{N^2} h_{ij}\underline{r}_j\cos\underline{\theta}_j\right)^2 + \left(\sum_{j=1}^{N^2} h_{ij}\underline{r}_j\sin\underline{\theta}_j\right)^2}}$$

$$\times \left[\left(\sum_{j=1}^{N^2} h_{ij}\frac{1+\cos\underline{\phi}_j}{2}\cos\underline{\theta}_j\right) h_{im}\cos\underline{\theta}_m\sin\underline{\phi}_m\right.$$

$$\left. + \left(\sum_{j=1}^{N^2} h_{ij}\frac{1+\cos\underline{\phi}_j}{2}\sin\underline{\theta}_j\right) h_{im}\sin\underline{\theta}_m\sin\underline{\phi}_m\right]. \tag{A.8}$$

Thus,

$$\frac{\partial F\left(\underline{\theta},\underline{\phi}\right)}{\partial \underline{\phi}_m} =$$

$$\sum_{i=1}^{N^2}\left[\tilde{\underline{z}}_i - \frac{1}{1+\exp\left[-a\sqrt{\left(\sum_{j=1}^{N^2} h_{ij}\underline{r}_j\cos\underline{\theta}_j\right)^2 + \left(\sum_{j=1}^{N^2} h_{ij}\underline{r}_j\sin\underline{\theta}_j\right)^2} + at_r\right]}\right]$$

$$\times \frac{1}{1+\exp\left[-a\sqrt{\left(\sum_{j=1}^{N^2} h_{ij}\underline{r}_j\cos\underline{\theta}_j\right)^2+\left(\sum_{j=1}^{N^2} h_{ij}\underline{r}_j\sin\underline{\theta}_j\right)^2}+at_r\right]}$$

$$\times \frac{\exp\left[-a\sqrt{\left(\sum_{j=1}^{N^2} h_{ij}\underline{r}_j\cos\underline{\theta}_j\right)^2+\left(\sum_{j=1}^{N^2} h_{ij}\underline{r}_j\sin\underline{\theta}_j\right)^2}\right]}{1+\exp\left[-a\sqrt{\left(\sum_{j=1}^{N^2} h_{ij}\underline{r}_j\cos\underline{\theta}_j\right)^2+\left(\sum_{j=1}^{N^2} h_{ij}\underline{r}_j\sin\underline{\theta}_j\right)^2}+at_r\right]} \times a$$

$$\times \frac{1}{\sqrt{\left(\sum_{j=1}^{N^2} h_{ij}\underline{r}_j\cos\underline{\theta}_j\right)^2+\left(\sum_{j=1}^{N^2} h_{ij}\underline{r}_j\sin\underline{\theta}_j\right)^2}}$$

$$\times \left[\left(\sum_{j=1}^{N^2} h_{ij}\frac{1+\cos\underline{\phi}_j}{2}\cos\underline{\theta}_j\right) h_{im}\cos\underline{\theta}_m\sin\underline{\phi}_m\right.$$

$$\left.+\left(\sum_{j=1}^{N^2} h_{ij}\frac{1+\cos\underline{\phi}_j}{2}\sin\underline{\theta}_j\right) h_{im}\sin\underline{\theta}_m\sin\underline{\phi}_m\right]$$

$$= a \times \sin\underline{\phi}_m \times \sum_{i=1}^{N^2} h_{im} \times \left(\tilde{\underline{z}}_i - \underline{z}_i\right) \times \underline{z}_i \times \left(1-\underline{z}_i\right)$$

$$\times \frac{1}{\sqrt{\left(\sum_{j=1}^{N^2} h_{ij}\underline{r}_j\cos\underline{\theta}_j\right)^2+\left(\sum_{j=1}^{N^2} h_{ij}\underline{r}_j\sin\underline{\theta}_j\right)^2}}$$

$$\times \left[\left(\sum_{j=1}^{N^2} h_{ij}\frac{1+\cos\underline{\phi}_j}{2}\cos\underline{\theta}_j\right)\cos\underline{\theta}_m\right.$$

$$\left.+\left(\sum_{j=1}^{N^2} h_{ij}\frac{1+\cos\underline{\phi}_j}{2}\sin\underline{\theta}_j\right)\sin\underline{\theta}_m\right]. \tag{A.9}$$

Therefore,

$$\nabla F\left(\underline{\theta},\underline{\phi}\right)_{\underline{\phi}} = a\times\sin\underline{\phi}\odot\cos\underline{\theta}\odot\left\{H^T\left[\left(\tilde{\underline{z}}-\underline{z}\right)\odot\underline{z}\odot\left(\underline{1}-\underline{z}\right)\odot H\left(\underline{m}_{\mathrm{R}}\right)\right.\right.$$
$$\left.\left.\odot T\left(\underline{m}\right)\right]\right\} + a\times\sin\underline{\phi}\odot\sin\underline{\theta}\odot\left\{H^T\left[\left(\tilde{\underline{z}}-\underline{z}\right)\odot\underline{z}\right.\right.$$
$$\left.\left.\odot\left(\underline{1}-\underline{z}\right)\odot H\left(\underline{m}_{\mathrm{I}}\right)\odot T\left(\underline{m}\right)\right]\right\}. \tag{A.10}$$

Appendix B

Manhattan Geometry

The definition of the Manhattan geometry is adopted from Ref. [58]. The a pattern has a Manhattan geometry, if at least one partition of the pattern includes a set of objects satisfying the following conditions: First, all objects are rectangles, and any two rectangles that share an entire edge are merged and considered as a single rectangular object. Second, all edges of objects have either vertical or horizontal orientation. Third, all edges have length $\geq d$, where d is the minimum feature size of the pattern. A typical Manhattan geometry pattern is illustrated in Fig. B.1.

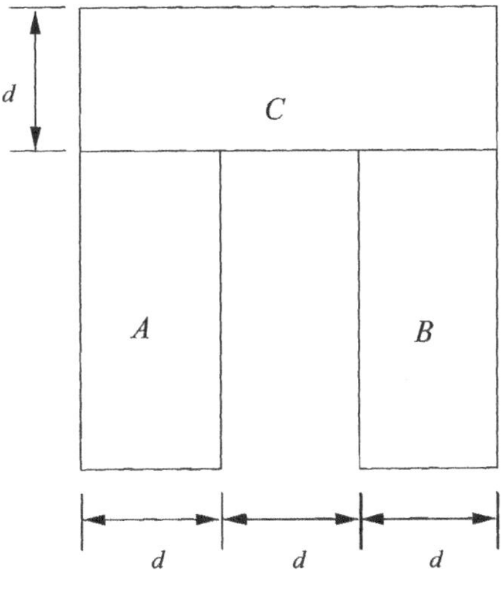

Figure B.1 Manhattan geometry.

Computational Lithography By Xu Ma and Gonzalo R. Arce

Appendix C

Formula Derivation in Chapter 6

The derivation of Eq. (6.33) is as follows:

$$
\begin{aligned}
&\frac{\partial E_{\text{detail}}}{\partial \underline{\phi}_{(2(i-1)+p)(2(j-1)+q)}} \\
&\quad = \frac{\partial\left(\left(1+\cos\underline{\phi}_{(2(i-1)+p)(2(j-1)+q)}/2\right) \mathrm{e}^{j\underline{\theta}_{(2(i-1)+p)(2(j-1)+q)}}\right)}{\partial \underline{\phi}_{(2(i-1)+p)(2(j-1)+q)}} \\
&\qquad \times \left(3\underline{m}^{*}_{(2(i-1)+p)(2(j-1)+q)} - \underline{m}^{*}_{(2(i-1)+p_1)(2(j-1)+q)}\right. \\
&\qquad \left. - \underline{m}^{*}_{(2(i-1)+p)(2(j-1)+q_1)} - \underline{m}^{*}_{(2(i-1)+p_1)(2(j-1)+q_1)}\right) \\
&\qquad + \frac{\partial\left(\left(1+\cos\underline{\phi}_{(2(i-1)+p)(2(j-1)+q)}/2\right) \mathrm{e}^{-j\underline{\theta}_{(2(i-1)+p)(2(j-1)+q)}}\right)}{\partial \underline{\phi}_{(2(i-1)+p)(2(j-1)+q)}} \\
&\qquad \times \left(3\underline{m}_{(2(i-1)+p)(2(j-1)+q)} - \underline{m}_{(2(i-1)+p_1)(2(j-1)+q)}\right. \\
&\qquad \left. - \underline{m}_{(2(i-1)+p)(2(j-1)+q_1)} - \underline{m}_{(2(i-1)+p_1)(2(j-1)+q_1)}\right) \\
&\quad = -\frac{1}{2}\sin\underline{\phi}_{(2(i-1)+p)(2(j-1)+q)} \mathrm{e}^{j\underline{\theta}_{(2(i-1)+p)(2(j-1)+q)}} \\
&\qquad \times \left(3\underline{m}^{*}_{(2(i-1)+p)(2(j-1)+q)} - \underline{m}^{*}_{(2(i-1)+p_1)(2(j-1)+q)}\right. \\
&\qquad \left. - \underline{m}^{*}_{(2(i-1)+p)(2(j-1)+q_1)} - \underline{m}^{*}_{(2(i-1)+p_1)(2(j-1)+q_1)}\right) \\
&\qquad - \frac{1}{2}\sin\underline{\phi}_{(2(i-1)+p)(2(j-1)+q)} \mathrm{e}^{-j\underline{\theta}_{(2(i-1)+p)(2(j-1)+q)}} \\
&\qquad \times \left(3\underline{m}_{(2(i-1)+p)(2(j-1)+q)} - \underline{m}_{(2(i-1)+p_1)(2(j-1)+q)}\right. \\
&\qquad \left. - \underline{m}_{(2(i-1)+p)(2(j-1)+q_1)} - \underline{m}_{(2(i-1)+p_1)(2(j-1)+q_1)}\right). \qquad \text{(C.1)}
\end{aligned}
$$

Therefore,

$$
\begin{aligned}
&\frac{\partial E_{\text{detail}}}{\partial \underline{\phi}_{(2(i-1)+p)(2(j-1)+q)}} \\
&\quad = -\sin \underline{\phi}_{(2(i-1)+p)(2(j-1)+q)} \times \operatorname{Re}\Big[\mathrm{e}^{-j\underline{\theta}_{(2(i-1)+p)(2(j-1)+q)}} \\
&\qquad \times \big(3\underline{m}_{(2(i-1)+p)(2(j-1)+q)} - \underline{m}_{(2(i-1)+p_1)(2(j-1)+q)} \\
&\qquad - \underline{m}_{(2(i-1)+p)(2(j-1)+q_1)} - \underline{m}_{(2(i-1)+p_1)(2(j-1)+q_1)}\big) \Big] .
\end{aligned} \tag{C.2}
$$

The derivation of Eq. (6.34) is as follows:

$$
\begin{aligned}
&\frac{\partial E_{\text{detail}}}{\partial \underline{\theta}_{(2(i-1)+p)(2(j-1)+q)}} \\
&\quad = \frac{\partial \left(\left(1 + \cos \underline{\phi}_{(2(i-1)+p)(2(j-1)+q)} / 2\right) \mathrm{e}^{j\underline{\theta}_{(2(i-1)+p)(2(j-1)+q)}} \right)}{\partial \underline{\theta}_{(2(i-1)+p)(2(j-1)+q)}} \\
&\qquad \times \left(3\underline{m}^*_{(2(i-1)+p)(2(j-1)+q)} - \underline{m}^*_{(2(i-1)+p_1)(2(j-1)+q)} \right. \\
&\qquad \left. - \underline{m}^*_{(2(i-1)+p)(2(j-1)+q_1)} - \underline{m}^*_{(2(i-1)+p_1)(2(j-1)+q_1)} \right) \\
&\qquad + \frac{\partial \left(\left(1 + \cos \underline{\phi}_{(2(i-1)+p)(2(j-1)+q)} / 2\right) \mathrm{e}^{-j\underline{\theta}_{(2(i-1)+p)(2(j-1)+q)}} \right)}{\partial \underline{\theta}_{(2(i-1)+p)(2(j-1)+q)}} \\
&\qquad \times \big(3\underline{m}_{(2(i-1)+p)(2(j-1)+q)} - \underline{m}_{(2(i-1)+p_1)(2(j-1)+q)} \\
&\qquad - \underline{m}_{(2(i-1)+p)(2(j-1)+q_1)} - \underline{m}_{(2(i-1)+p_1)(2(j-1)+q_1)}\big) \\
&\quad = j\mathrm{e}^{j\underline{\theta}_{(2(i-1)+p)(2(j-1)+q)}} \frac{1 + \cos \underline{\phi}_{(2(i-1)+p)(2(j-1)+q)}}{2} \\
&\qquad \times \left(3\underline{m}^*_{(2(i-1)+p)(2(j-1)+q)} - \underline{m}^*_{(2(i-1)+p_1)(2(j-1)+q)} \right. \\
&\qquad \left. - \underline{m}^*_{(2(i-1)+p)(2(j-1)+q_1)} - \underline{m}^*_{(2(i-1)+p_1)(2(j-1)+q_1)} \right) \\
&\qquad - j\mathrm{e}^{j\underline{\theta}_{(2(i-1)+p)(2(j-1)+q)}} \frac{1 + \cos \underline{\phi}_{(2(i-1)+p)(2(j-1)+q)}}{2} \\
&\qquad \times \big(3\underline{m}_{(2(i-1)+p)(2(j-1)+q)} - \underline{m}_{(2(i-1)+p_1)(2(j-1)+q)} \\
&\qquad - \underline{m}_{(2(i-1)+p)(2(j-1)+q_1)} - \underline{m}_{(2(i-1)+p_1)(2(j-1)+q_1)}\big) .
\end{aligned} \tag{C.3}
$$

Therefore,

$$\frac{\partial E_{\text{detail}}}{\partial \underline{\theta}_{(2(i-1)+p)(2(j-1)+q)}}$$
$$= \left(1 + \cos\underline{\phi}_{(2(i-1)+p)(2(j-1)+q)}\right) \times \text{Re}\Big[(-j)\text{e}^{-j\underline{\theta}_{(2(i-1)+p)(2(j-1)+q)}}$$
$$\times \big(3\underline{m}_{(2(i-1)+p)(2(j-1)+q)} - \underline{m}_{(2(i-1)+p_1)(2(j-1)+q)}$$
$$-\underline{m}_{(2(i-1)+p)(2(j-1)+q_1)} - \underline{m}_{(2(i-1)+p_1)(2(j-1)+q_1)}\big)\Big]. \tag{C.4}$$

Appendix D

Formula Derivation in Chapter 7

The derivation of Eq. (7.11) is as follows:

$$\frac{\partial F(\underline{\theta})}{\partial \underline{\theta}_k} = 2\sum_{i=1}^{N^2}\left\{\tilde{\underline{z}}_i - \frac{1}{1+\exp\left[-a\sum_{\mathbf{m}}\Gamma_{\mathbf{m}}\left|\sum_{j=1}^{N^2} h_{ij}^{\mathbf{m}}\frac{1+\cos\underline{\theta}_j}{2}\right|^2 + at_r\right]}\right\}$$

$$\times \frac{1}{\left\{1+\exp\left[-a\sum_{\mathbf{m}}\Gamma_{\mathbf{m}}\left|\sum_{j=1}^{N^2} h_{ij}^{\mathbf{m}}\frac{1+\cos\underline{\theta}_j}{2}\right|^2 + at_r\right]\right\}^2}$$

$$\times \exp\left[-a\sum_{\mathbf{m}}\Gamma_{\mathbf{m}}\left|\sum_{j=1}^{N^2} h_{ij}^{\mathbf{m}}\frac{1+\cos\underline{\theta}_j}{2}\right|^2 + at_r\right] \times (-a)$$

$$\times \sum_{\mathbf{m}}\Gamma_{\mathbf{m}}\left[\left(\sum_{j=1}^{N^2} h_{ij}^{\mathbf{m}}\frac{1+\cos\underline{\theta}_j}{2}\right) h_{ik}^{\mathbf{m}*}\left(-\frac{\sin\underline{\theta}_k}{2}\right)\right.$$

$$\left. + \left(\sum_{j=1}^{N^2} h_{ij}^{\mathbf{m}*}\frac{1+\cos\underline{\theta}_j}{2}\right) h_{ik}^{\mathbf{m}}\left(-\frac{\sin\underline{\theta}_k}{2}\right)\right]$$

$$= \sum_{i=1}^{N^2}\left\{\tilde{\underline{z}}_i - \frac{1}{1+\exp\left[-a\sum_{\mathbf{m}}\Gamma_{\mathbf{m}}\left|\sum_{j=1}^{N^2} h_{ij}^{\mathbf{m}}\frac{1+\cos\underline{\theta}_j}{2}\right|^2 + at_r\right]}\right\}$$

$$\times \frac{1}{1+\exp\left[-a\sum_{\mathbf{m}}\Gamma_{\mathbf{m}}\left|\sum_{j=1}^{N^2} h_{ij}^{\mathbf{m}}\frac{1+\cos\underline{\theta}_j}{2}\right|^2+at_r\right]}$$

$$\times \frac{\exp\left[-a\sum_{\mathbf{m}}\Gamma_{\mathbf{m}}\left|\sum_{j=1}^{N^2} h_{ij}^{\mathbf{m}}\frac{1+\cos\underline{\theta}_j}{2}\right|^2+at_r\right]}{1+\exp\left[-a\sum_{\mathbf{m}}\Gamma_{\mathbf{m}}\left|\sum_{j=1}^{N^2} h_{ij}^{\mathbf{m}}\frac{1+\cos\underline{\theta}_j}{2}\right|^2+at_r\right]} \times a$$

$$\times \sum_{\mathbf{m}}\Gamma_{\mathbf{m}}\left[\left(\sum_{j=1}^{N^2} h_{ij}^{\mathbf{m}}\frac{1+\cos\underline{\theta}_j}{2}\right)h_{ik}^{\mathbf{m}*}\sin\underline{\theta}_k\right.$$

$$\left.+\left(\sum_{j=1}^{N^2} h_{ij}^{\mathbf{m}*}\frac{1+\cos\underline{\theta}_j}{2}\right)h_{ik}^{\mathbf{m}}\sin\underline{\theta}_k\right]. \tag{D.1}$$

Thus,

$$\frac{\partial F(\underline{\theta})}{\partial \underline{\theta}_k} = a\times\sin\underline{\theta}_k\times\sum_{\mathbf{m}}\Gamma_{\mathbf{m}}\sum_{i=1}^{N^2}(\tilde{\underline{z}}_i-\underline{z}_i)\times\underline{z}_i\times(1-\underline{z}_i)$$

$$\times\left[\left(\sum_{j=1}^{N^2} h_{ij}^{\mathbf{m}}\frac{1+\cos\underline{\theta}_j}{2}\right)h_{ik}^{\mathbf{m}*}+\left(\sum_{j=1}^{N^2} h_{ij}^{\mathbf{m}*}\frac{1+\cos\underline{\theta}_j}{2}\right)h_{ik}^{\mathbf{m}}\right]. \tag{D.2}$$

Using Eqs. (7.5) and (7.8), the gradient above can be written as

$$\nabla F(\underline{\theta}) = a\times\sin\underline{\theta}\odot\left\{\sum_{\mathbf{m}}\Gamma_{\mathbf{m}}(H^{\mathbf{m}})^{*T}[(\tilde{\underline{z}}-\underline{z})\odot\underline{z}\odot(\underline{1}-\underline{z})\right.$$

$$\left.\odot(H^{\mathbf{m}})^*(\underline{m})]\right\}+a\times\sin\underline{\theta}\odot\left\{\sum_{\mathbf{m}}\Gamma_{\mathbf{m}}(H^{\mathbf{m}})^T[(\tilde{\underline{z}}-\underline{z})\right.$$

$$\left.\odot\underline{z}\odot(\underline{1}-\underline{z})\odot(H^{\mathbf{m}})(\underline{m})]\right\}. \tag{D.3}$$

The derivation of Eq. (7.20) is as follows:

$$\frac{\partial F(\underline{\theta})}{\partial \underline{\theta}_k} = 2\sum_{i=1}^{N^2}\left\{\tilde{\underline{z}}_i-\frac{1}{1+\exp\left[-a\left(\left|\sum_{j=1}^{N^2} h_{\mathrm{C}ij}\underline{m}_j\right|^2+\sum_{j=1}^{N^2}\left|h_{\mathrm{I}ij}\right|^2\left|\underline{m}_j\right|^2\right)+at_r'\right]}\right\}$$

$$\times \frac{1}{\left\{1+\exp\left[-a\left(\left|\sum_{j=1}^{N^2} h_{\mathrm{C}ij}\underline{m}_j\right|^2+\sum_{j=1}^{N^2}|h_{\mathrm{I}ij}|^2|\underline{m}_j|^2\right)+at_r'\right]\right\}^2}$$

$$\times \exp\left[-a\left(\left|\sum_{j=1}^{N^2} h_{\mathrm{C}ij}\underline{m}_j\right|^2+\sum_{j=1}^{N^2}|h_{\mathrm{I}ij}|^2|\underline{m}_j|^2\right)+at_r'\right]\times(-a)$$

$$\times\left[\left(\sum_{j=1}^{N^2}2h_{\mathrm{C}ij}\frac{1+\cos\underline{\theta}_j}{2}\right)h_{\mathrm{C}ik}\left(-\frac{\sin\underline{\theta}_k}{2}\right)+2\left(\frac{1+\cos\underline{\theta}_k}{2}\right)h_{\mathrm{I}ik}^2\left(-\frac{\sin\underline{\theta}_k}{2}\right)\right]. \tag{D.4}$$

Thus,

$$\frac{\partial F(\theta)}{\partial\underline{\theta}_k}=2\sum_{i=1}^{N^2}\left\{\tilde{\underline{z}}_i-\frac{1}{1+\exp\left[-a\left(\left|\sum_{j=1}^{N^2} h_{\mathrm{C}ij}\underline{m}_j\right|^2+\sum_{j=1}^{N^2}|h_{\mathrm{I}ij}|^2|\underline{m}_j|^2\right)+at_r'\right]}\right\}$$

$$\times\frac{1}{1+\exp\left[-a\left(\left|\sum_{j=1}^{N^2} h_{\mathrm{C}ij}\underline{m}_j\right|^2+\sum_{j=1}^{N^2}|h_{\mathrm{I}ij}|^2|\underline{m}_j|^2\right)+at_r'\right]}$$

$$\times\frac{\exp\left[-a\left(\left|\sum_{j=1}^{N^2} h_{\mathrm{C}ij}\underline{m}_j\right|^2+\sum_{j=1}^{N^2}|h_{\mathrm{I}ij}|^2|\underline{m}_j|^2\right)+at_r'\right]}{1+\exp\left[-a\left(\left|\sum_{j=1}^{N^2} h_{\mathrm{C}ij}\underline{m}_j\right|^2+\sum_{j=1}^{N^2}|h_{\mathrm{I}ij}|^2|\underline{m}_j|^2\right)+at_r'\right]}\times a$$

$$\times\left[\left(\sum_{j=1}^{N^2}2h_{\mathrm{C}ij}\frac{1+\cos\underline{\theta}_j}{2}\right)h_{\mathrm{C}ik}\frac{\sin\underline{\theta}_k}{2}+2\left(\frac{1+\cos\underline{\theta}_k}{2}\right)h_{\mathrm{I}ik}^2\frac{\sin\underline{\theta}_k}{2}\right]$$

$$=2a\times\sin\underline{\theta}_k\times\sum_{i=1}^{N^2}(\tilde{\underline{z}}_i-\underline{z}_i)\times\underline{z}_i\times(1-\underline{z}_i)$$

$$\times\left[\left(\sum_{j=1}^{N^2}h_{\mathrm{C}ij}\frac{1+\cos\underline{\theta}_j}{2}\right)h_{\mathrm{C}ik}+\left(\frac{1+\cos\underline{\theta}_k}{2}\right)h_{\mathrm{I}ik}^2\right]. \tag{D.5}$$

Using Eqs. (7.8) and (7.18), the gradient above can be written as

$$\nabla F(\underline{\theta}) = \underline{d}_{\theta} = 2a \times \sin\underline{\theta} \odot \left\{ H_{\mathrm{C}}^{T} \left[(\underline{\tilde{z}} - \underline{z}) \odot \underline{z} \odot (\underline{1} - \underline{z}) \odot (H_{\mathrm{C}}(\underline{m})) \right] \right\}$$
$$+ 2a \times \sin\underline{\theta} \odot \left\{ H_{\mathrm{I}}^{2T} \left[(\underline{\tilde{z}} - \underline{z}) \odot \underline{z} \odot (\underline{1} - \underline{z}) \odot (\underline{m}) \right] \right\}. \tag{D.6}$$

Appendix E

Formula Derivation in Chapter 8

The derivation of Eq. (8.9) is as follows:

$$
\begin{aligned}
&\frac{\partial F}{\partial \underline{\theta}_{pm}} \\
&= 2\sum_{i=1}^{N^2}\left\{\tilde{z}_i - \frac{1}{2}\left[\tanh\left(\underline{z}_{1i} + \underline{z}_{2i} - 1\right) + 1\right]\right\} \times \frac{1}{2}\mathrm{sech}^2\left(\underline{z}_{1i} + \underline{z}_{2i} - 1\right) \\
&\quad \times \frac{1}{\left\{1 + \exp\left[-a\sqrt{\left(\sum_{j=1}^{N^2} h_{ij}\underline{r}_{pj}\cos\underline{\theta}_{pj}\right)^2 + \left(\sum_{j=1}^{N^2} h_{ij}\underline{r}_{pj}\sin\underline{\theta}_{pj}\right)^2} + at_r\right]\right\}^2} \\
&\quad \times \exp\left[-a\sqrt{\left(\sum_{j=1}^{N^2} h_{ij}\underline{r}_{pj}\cos\underline{\theta}_{pj}\right)^2 + \left(\sum_{j=1}^{N^2} h_{ij}\underline{r}_{pj}\sin\underline{\theta}_{pj}\right)^2}\right] \times (-a) \\
&\quad \times \frac{1}{\sqrt{\left(\sum_{j=1}^{N^2} h_{ij}\underline{r}_{pj}\cos\underline{\theta}_{pj}\right)^2 + \left(\sum_{j=1}^{N^2} h_{ij}\underline{r}_{pj}\sin\underline{\theta}_{pj}\right)^2}} \\
&\quad \times \left[\left(\sum_{j=1}^{N^2} h_{ij}\frac{1+\cos\underline{\phi}_{pj}}{2}\cos\underline{\theta}_{pj}\right) h_{im}\frac{1+\cos\underline{\phi}_{pm}}{2}(-\sin\underline{\theta}_{pm})\right. \\
&\quad \left. + \left(\sum_{j=1}^{N^2} h_{ij}\frac{1+\cos\underline{\phi}_{pj}}{2}\sin\underline{\theta}_{pj}\right) h_{im}\frac{1+\cos\underline{\phi}_{pm}}{2}\cos\underline{\theta}_{pm}\right],
\end{aligned}
\tag{E.1}
$$

where $p = 1$ or 2 and

$$\underline{z}_{1i} = \frac{1}{1 + \exp\left[-a \left| \sum_{k=1}^{N^2} h_{ik} \frac{1+\cos\underline{\phi}_{1k}}{2} e^{j\underline{\theta}_{1k}} \right| + at_r \right]}, \quad i = 1, \ldots, N^2, \qquad \text{(E.2)}$$

$$\underline{z}_{2i} = \frac{1}{1 + \exp\left[-a \left| \sum_{k=1}^{N^2} h_{ik} \frac{1+\cos\underline{\phi}_{2k}}{2} e^{j\underline{\theta}_{2k}} \right| + at_r \right]}, \quad i = 1, \ldots, N^2. \qquad \text{(E.3)}$$

Thus,

$$\frac{\partial F}{\partial \underline{\theta}_{pm}}$$

$$= 2\sum_{i=1}^{N^2} \left\{ \tilde{\underline{z}}_i - \frac{1}{2}\left[\tanh\left(\underline{z}_{1i} + \underline{z}_{2i} - 1\right) + 1\right] \right\} \times \frac{1}{2}\operatorname{sech}^2\left(\underline{z}_{1i} + \underline{z}_{2i} - 1\right)$$

$$\times \frac{1}{1 + \exp\left[-a\sqrt{\left(\sum_{j=1}^{N^2} h_{ij}\underline{r}_{pj}\cos\underline{\theta}_{pj}\right)^2 + \left(\sum_{j=1}^{N^2} h_{ij}\underline{r}_{pj}\sin\underline{\theta}_{pj}\right)^2} + at_r\right]}$$

$$\times \frac{\exp\left[-a\sqrt{\left(\sum_{j=1}^{N^2} h_{ij}\underline{r}_{pj}\cos\underline{\theta}_{pj}\right)^2 + \left(\sum_{j=1}^{N^2} h_{ij}\underline{r}_{pj}\sin\underline{\theta}_{pj}\right)^2}\right]}{1 + \exp\left[-a\sqrt{\left(\sum_{j=1}^{N^2} h_{ij}\underline{r}_{pj}\cos\underline{\theta}_{pj}\right)^2 + \left(\sum_{j=1}^{N^2} h_{ij}\underline{r}_{pj}\sin\underline{\theta}_{pj}\right)^2} + at_r\right]} \times a$$

$$\times \frac{1}{\sqrt{\left(\sum_{j=1}^{N^2} h_{ij}\underline{r}_{pj}\cos\underline{\theta}_{pj}\right)^2 + \left(\sum_{j=1}^{N^2} h_{ij}\underline{r}_{pj}\sin\underline{\theta}_{pj}\right)^2}}$$

$$\times \left[\left(\sum_{j=1}^{N^2} h_{ij} \frac{1+\cos\underline{\phi}_{pj}}{2}\cos\underline{\theta}_{pj}\right) h_{im}\frac{1+\cos\underline{\phi}_{pm}}{2}\sin\underline{\theta}_{pm}\right.$$

$$\left. + \left(\sum_{j=1}^{N^2} h_{ij} \frac{1+\cos\underline{\phi}_{pj}}{2}\sin\underline{\theta}_{pj}\right) h_{im}\frac{1+\cos\underline{\phi}_{pm}}{2}\left(-\cos\underline{\theta}_{pm}\right)\right]$$

$$= a \times \frac{1+\cos\underline{\phi}_{pm}}{2} \times \sum_{i=1}^{N^2} h_{im} \times \left\{ \tilde{\underline{z}}_i - \frac{1}{2}\left[\tanh\left(\underline{z}_{1i} + \underline{z}_{2i} - 1\right) + 1\right] \right\}$$

$$\times \text{sech}^2\left(\underline{z}_{1i}+\underline{z}_{2i}-1\right)\times \underline{z}_p\times\left(\underline{1}-\underline{z}_p\right)$$

$$\times\frac{1}{\sqrt{\left(\sum_{j=1}^{N^2} h_{ij}\underline{r}_{pj}\cos\underline{\theta}_{pj}\right)^2+\left(\sum_{j=1}^{N^2} h_{ij}\underline{r}_{pj}\sin\underline{\theta}_{pj}\right)^2}}$$

$$\times\left[\left(\sum_{j=1}^{N^2} h_{ij}\frac{1+\cos\underline{\phi}_{pj}}{2}\cos\underline{\theta}_{pj}\right)\sin\underline{\theta}_{pm}\right.$$

$$\left.+\left(\sum_{j=1}^{N^2} h_{ij}\frac{1+\cos\underline{\phi}_{pj}}{2}\sin\underline{\theta}_{pj}\right)\left(-\cos\underline{\theta}_{pm}\right)\right]. \tag{E.4}$$

Therefore,

$$\begin{aligned}\nabla F_{\underline{\theta}_p} &= a\times\frac{\underline{1}+\cos\underline{\phi}_p}{2}\odot\sin\underline{\theta}_p\odot\left\{H^T\left[\left(\underline{\tilde{z}}-\underline{z}\right)\odot\text{sech}^2\left(\underline{z}_1+\underline{z}_2-\underline{1}\right)\right.\right.\\&\left.\left.\odot\underline{z}_p\odot\left(\underline{1}-\underline{z}_p\right)\odot H\left(\underline{m}_{p\text{R}}\right)\odot T\left(\underline{m},p\right)\right]\right\}\\&-a\times\frac{\underline{1}+\cos\underline{\phi}_p}{2}\odot\cos\underline{\theta}_p\odot\left\{H^T\left[\left(\underline{\tilde{z}}-\underline{z}\right)\odot\text{sech}^2\left(\underline{z}_1+\underline{z}_2-\underline{1}\right)\right.\right.\\&\left.\left.\odot\underline{z}_p\odot\left(\underline{1}-\underline{z}\right)_p\odot H\left(\underline{m}_{p\text{I}}\right)\odot T\left(\underline{m},p\right)\right]\right\},\end{aligned} \tag{E.5}$$

where $T(\underline{m})=[(H\underline{m}_\text{R})^2+(H\underline{m}_\text{I})^2]^{-\frac{1}{2}}$. $\underline{m}_\text{R}$ and $\underline{m}_\text{I}$ are the real and imaginary parts of $\underline{m}$, respectively.

The derivation of Eq. (8.10) is as follows:

$$\frac{\partial F(\underline{\theta},\underline{\phi})}{\partial\underline{\phi}_m}$$

$$=\sum_{i=1}^{N^2}\left\{\underline{\tilde{z}}_i-\frac{1}{2}\left[\tanh\left(\underline{z}_{1i}+\underline{z}_{2i}-1\right)+1\right]\right\}\times\frac{1}{2}\text{sech}^2\left(\underline{z}_{1i}+\underline{z}_{2i}-1\right)$$

$$\times\frac{1}{\left\{1+\exp\left[-a\sqrt{\left(\sum_{j=1}^{N^2} h_{ij}\underline{r}_{pj}\cos\underline{\theta}_{pj}\right)^2+\left(\sum_{j=1}^{N^2} h_{ij}\underline{r}_{pj}\sin\underline{\theta}_{pj}\right)^2}+at_r\right]\right\}^2}$$

$$\times\exp\left[-a\sqrt{\left(\sum_{j=1}^{N^2} h_{ij}\underline{r}_{pj}\cos\underline{\theta}_{pj}\right)^2+\left(\sum_{j=1}^{N^2} h_{ij}\underline{r}_{pj}\sin\underline{\theta}_{pj}\right)^2}\right]\times a$$

$$\times \frac{1}{\sqrt{\left(\sum_{j=1}^{N^2} h_{ij}\underline{r}_{pj}\cos\underline{\theta}_{pj}\right)^2 + \left(\sum_{j=1}^{N^2} h_{ij}\underline{r}_{pj}\sin\underline{\theta}_{pj}\right)^2}}$$

$$\times \left[\left(\sum_{j=1}^{N^2} h_{ij}\frac{1+\cos\underline{\phi}_{pj}}{2}\cos\underline{\theta}_{pj}\right) h_{im}\cos\underline{\theta}_{pm}\sin\underline{\phi}_{pm}\right.$$

$$\left. + \left(\sum_{j=1}^{N^2} h_{ij}\frac{1+\cos\underline{\phi}_{pj}}{2}\sin\underline{\theta}_{pj}\right) h_{im}\sin\underline{\theta}_{pm}\sin\underline{\phi}_{pm}\right]. \tag{E.6}$$

Thus,

$$\frac{\partial F(\underline{\theta}, \underline{\phi})}{\partial \underline{\phi}_m}$$

$$= \sum_{i=1}^{N^2} \left\{\tilde{\underline{z}}_i - \frac{1}{2}\left[\tanh\left(\underline{z}_{1i} + \underline{z}_{2i} - 1\right) + 1\right]\right\} \times \frac{1}{2}\mathrm{sech}^2\left(\underline{z}_{1i} + \underline{z}_{2i} - 1\right)$$

$$\times \frac{1}{1+\exp\left[-a\sqrt{\left(\sum_{j=1}^{N^2} h_{ij}\underline{r}_{pj}\cos\underline{\theta}_{pj}\right)^2 + \left(\sum_{j=1}^{N^2} h_{ij}\underline{r}_{pj}\sin\underline{\theta}_{pj}\right)^2} + at_r\right]}$$

$$\times \frac{\exp\left[-a\sqrt{\left(\sum_{j=1}^{N^2} h_{ij}\underline{r}_{pj}\cos\underline{\theta}_{pj}\right)^2 + \left(\sum_{j=1}^{N^2} h_{ij}\underline{r}_{pj}\sin\underline{\theta}_{pj}\right)^2}\right]}{1+\exp\left[-a\sqrt{\left(\sum_{j=1}^{N^2} h_{ij}\underline{r}_{pj}\cos\underline{\theta}_{pj}\right)^2 + \left(\sum_{j=1}^{N^2} h_{ij}\underline{r}_{pj}\sin\underline{\theta}_{pj}\right)^2} + at_r\right]} \times a$$

$$\times \frac{1}{\sqrt{\left(\sum_{j=1}^{N^2} h_{ij}\underline{r}_{pj}\cos\underline{\theta}_{pj}\right)^2 + \left(\sum_{j=1}^{N^2} h_{ij}\underline{r}_{pj}\sin\underline{\theta}_{pj}\right)^2}}$$

$$\times \left[\left(\sum_{j=1}^{N^2} h_{ij}\frac{1+\cos\underline{\phi}_{pj}}{2}\cos\underline{\theta}_{pj}\right) h_{im}\cos\underline{\theta}_{pm}\sin\underline{\phi}_{pm}\right.$$

$$\left. + \left(\sum_{j=1}^{N^2} h_{ij}\frac{1+\cos\underline{\phi}_{pj}}{2}\sin\underline{\theta}_{pj}\right) h_{im}\sin\underline{\theta}_{pm}\sin\underline{\phi}_{pm}\right]$$

$$= \frac{a}{2} \times \sin\underline{\phi}_{pm} \times \sum_{i=1}^{N^2} \left\{ \underline{\tilde{z}}_i - \frac{1}{2} \left[\tanh\left(\underline{z}_{1i} + \underline{z}_{2i} - 1 \right) + 1 \right] \right\}$$

$$\times \text{sech}^2\left(\underline{z}_{1i} + \underline{z}_{2i} - 1 \right) \times \underline{z}_p \times \left(\underline{1} - \underline{z}_p \right)$$

$$\times \frac{1}{\sqrt{\left(\sum_{j=1}^{N^2} h_{ij} \underline{r}_{pj} \cos\underline{\theta}_{pj} \right)^2 + \left(\sum_{j=1}^{N^2} h_{ij} \underline{r}_{pj} \sin\underline{\theta}_{pj} \right)^2}}$$

$$\times \left[\left(\sum_{j=1}^{N^2} h_{ij} \frac{1 + \cos\underline{\phi}_{pj}}{2} \cos\underline{\theta}_{pj} \right) h_{im} \cos\underline{\theta}_{pm} \right.$$

$$\left. + \left(\sum_{j=1}^{N^2} h_{ij} \frac{1 + \cos\underline{\phi}_{pj}}{2} \sin\underline{\theta}_{pj} \right) h_{im} \sin\underline{\theta}_{pm} \right], \tag{E.7}$$

where $p = 1$ or 2. Therefore,

$$\nabla F_{\underline{\phi}_p} = \frac{a}{2} \times \sin\underline{\phi}_p \odot \cos\underline{\theta}_p \odot \left\{ H^T \left[(\underline{\tilde{z}} - \underline{z}) \odot \text{sech}^2(\underline{z}_1 + \underline{z}_2 - \underline{1}) \right.\right.$$

$$\left.\left. \odot \underline{z}_p \odot (\underline{1} - \underline{z}_p) \odot H(\underline{m}_{p\mathrm{R}}) \odot T(\underline{m}, p) \right] \right\}$$

$$+ \frac{a}{2} \times \sin\underline{\phi}_p \odot \sin\underline{\theta}_p \odot \left\{ H^T \left[(\underline{\tilde{z}} - \underline{z}) \odot \text{sech}^2(\underline{z}_1 + \underline{z}_2 - \underline{1}) \right.\right.$$

$$\left.\left. \odot \underline{z}_p \odot (\underline{1} - \underline{z}_p) \odot H(\underline{m}_{p\mathrm{I}}) \odot T(\underline{m}, p) \right] \right\}. \tag{E.8}$$

Appendix F

Formula Derivation in Chapter 9

The derivation of Eq. (9.17) is as follows:

$$\frac{\partial F}{\partial \Gamma_{\mathbf{m}}} = 2\sum_{p=1}^{N^2} \left(\tilde{\underline{i}}_p - \underline{i}_p\right) \times (-1) \times \left|\sum_{q=1}^{N^2} h_{pq}^{\mathbf{m}} \underline{m}_q\right|^2 . \tag{F.1}$$

Therefore,

$$\frac{\partial F}{\partial \Gamma_{\mathbf{m}}} = -2\left(\tilde{\underline{i}} - \underline{i}\right)^T \left|H^{\mathbf{m}}(\underline{m})\right|^2 . \tag{F.2}$$

The derivation of Eq. (9.18) is as follows:

$$\frac{\partial F(\underline{\theta})}{\partial \underline{m}_k} = 2\sum_{p=1}^{N^2} \left\{ \tilde{\underline{i}}_p - \sum_{\mathbf{m}} \Gamma_{\mathbf{m}} \left|\sum_{q=1}^{N^2} h_{pq}^{\mathbf{m}} \underline{m}_q\right|^2 \right\} \times (-1)$$

$$\times \sum_{\mathbf{m}} \Gamma_{\mathbf{m}} \left[\left(\sum_{j=1}^{N^2} h_{ij}^{\mathbf{m}} \underline{m}_j\right) h_{ik}^{\mathbf{m}*} + \left(\sum_{j=1}^{N^2} h_{ij}^{\mathbf{m}*} \underline{m}_j\right) h_{ik}^{\mathbf{m}} \right] . \tag{F.3}$$

Therefore,

$$\frac{\partial F}{\partial M} = -4\mathrm{Re}\left\{ \sum_{\mathbf{m}} \Gamma_{\mathbf{m}} (H^{\mathbf{m}})^T \left[(\tilde{\underline{i}} - \underline{i}) \odot H^{\mathbf{m}}(\underline{m}) \right] \right\} . \tag{F.4}$$

Appendix G

Formula Derivation in Chapter 10

The derivation of Eq. (10.18) is as follows:

$$\frac{\partial F}{\partial \underline{\gamma}_m} = 2\sum_{u=1}^{N^2}\left(\tilde{\underline{i}}_u - \left|\sum_{v=1}^{N^2} h_{uv} f_v\right|^2\right) \times (-1)$$

$$\times 2\mathrm{Re}\left\{\left(\sum_{v=1}^{N^2} h_{uv} f_v\right) h_{um} \times \frac{\partial \underline{f}_m}{\underline{\gamma}_m}\right\}. \tag{G.1}$$

For the OPC optimization in the first type of optical lithography system, all the entries of $\underline{f}$ have real values. In addition, according to Eq. (10.7), $\underline{f}_m = \underline{\gamma}_m$, where $m = 1, 2, \ldots, N^2$. Thus,

$$\nabla F = -4H^T\left[\left(\tilde{\underline{i}} - \underline{i}\right) \odot H\underline{\gamma}\right]. \tag{G.2}$$

The derivation of Eq. (10.19) is as follows:

$$\frac{\partial F}{\partial \underline{\gamma}_m} = 2\sum_{u=1}^{N^2}\left(\tilde{\underline{i}}_u - \left|\sum_{v=1}^{N^2} h_{uv} f_v\right|^2\right) \times (-1) \times 2\mathrm{Re}\left\{\left(\sum_{v=1}^{N^2} h_{uv} f_v\right)\right.$$

$$\times h_{um} \times \frac{\partial f_m^*}{\partial \underline{\gamma}_m} + \left(\sum_{v=1}^{N^2} h_{uv} f_v\right) \times h_{u,m-N} \times \frac{\partial f_{m-N}^*}{\partial \underline{\gamma}_m}$$

$$\left. + \left(\sum_{v=1}^{N^2} h_{uv} f_v\right) \times h_{u,m+N} \times \frac{\partial f_{m+N}^*}{\partial \underline{\gamma}_m}\right\}$$

$$= 2\sum_{u=1}^{N^2}\left(\tilde{\underline{i}}_u - \left|\sum_{v=1}^{N^2} h_{uv} f_v\right|^2\right) \times (-1) \times 2\mathrm{Re}\left\{\left(\sum_{v=1}^{N^2} h_{uv} f_v\right)\right.$$

$$\times h_{um}(0.8j\gamma_{m-N} + 0.8j\gamma_{m+N} + 1) + \left(\sum_{v=1}^{N^2} h_{uv} f_v\right) h_{u,m-N}$$

$$\times [-0.8j(1-\gamma_{m-N})] + \left(\sum_{v=1}^{N^2} h_{uv} f_v\right) h_{u,m+N}$$

$$\times [-0.8j(1-\gamma_{m+N})] \Bigg\}. \tag{G.3}$$

Therefore,

$$\begin{aligned} \nabla F = -4\mathrm{Re}\Big\{ & H^T \left[\left(\tilde{\underline{i}} - \underline{i}\right) \odot H\underline{f}\right] \odot \left(0.8j\underline{\gamma}\uparrow + 0.8j\underline{\gamma}\downarrow + \underline{1}\right) \\ & + H^T \left[\left(\tilde{\underline{i}} - \underline{i}\right) \odot H\underline{f}\right]\downarrow \odot 0.8j\left(\underline{1} - \underline{\gamma}\downarrow\right) \\ & + H^T \left[\left(\tilde{\underline{i}} - \underline{i}\right) \odot H\underline{f}\right]\uparrow \odot 0.8j\left(\underline{1} - \underline{\gamma}\uparrow\right) \Big\}. \end{aligned} \tag{G.4}$$

The derivation of Eq. (10.20) is as follows:

$$\begin{aligned} \frac{\partial F}{\partial \underline{\gamma}_m} &= 2\sum_{u=1}^{N^2} \left(\tilde{\underline{i}}_u - \left|\sum_{v=1}^{N^2} h_{uv} f_v\right|^2\right) \times (-1) \times 2\mathrm{Re}\left\{\left(\sum_{v=1}^{N^2} h_{uv} f_v\right)\right. \\ &\quad \times h_{um} \times \frac{\partial f_m^*}{\partial \underline{\gamma}_m} + \left(\sum_{v=1}^{N^2} h_{uv} f_v\right) \times h_{u,m-2N} \times \frac{\partial f_{m-2N}^*}{\partial \underline{\gamma}_m} \\ &\quad \left. + \left(\sum_{v=1}^{N^2} h_{uv} f_v\right) \times h_{u,m+2N} \times \frac{\partial f_{m+2N}^*}{\partial \underline{\gamma}_m} \right\} \\ &= 2\sum_{u=1}^{N^2} \left(\tilde{\underline{i}}_u - \left|\sum_{v=1}^{N^2} h_{uv} f_v\right|^2\right) \times (-1) \times 2\mathrm{Re}\left\{\left(\sum_{v=1}^{N^2} h_{uv} f_v\right)\right. \\ &\quad \times h_{um}\left[0.52j\gamma_m\gamma_{m-2N}(1-\gamma_{m-2N}) + 0.52j\gamma_m\gamma_{m+2N}(1-\gamma_{m+2N}) + 1\right] \\ &\quad + \left(\sum_{v=1}^{N^2} h_{uv} f_v\right) h_{u,m-2N}\left[-\frac{0.52j}{2}(1-\gamma_{m-2N})(1+\gamma_{m-2N})(1-2\gamma_m)\right] \\ &\quad \left. + \left(\sum_{v=1}^{N^2} h_{uv} f_v\right) h_{u,m+2N}\left[-\frac{0.52j}{2}(1-\gamma_{m+2N})(1+\gamma_{m+2N})(1-2\gamma_m)\right]\right\}. \end{aligned} \tag{G.5}$$

Therefore,

$$\nabla F = -4\mathrm{Re}\Big\{ H^T \left[(\underline{\tilde{i}} - \underline{i}) \odot H\underline{f} \right] \odot \left[0.52 j\underline{\gamma} \odot \underline{\gamma} \uparrow_2 \odot (\underline{1} - \underline{\gamma} \uparrow_2) \right.$$
$$+ 0.52 j\underline{\gamma} \odot \underline{\gamma} \downarrow_2 \odot (\underline{1} - \underline{\gamma} \downarrow_2) + \underline{1} \big] + H^T \left[(\underline{\tilde{i}} - \underline{i}) \odot H\underline{f} \right] \downarrow_2$$
$$\odot \left[-0.26 j (\underline{1} - \underline{\gamma} \downarrow_2) \odot (\underline{1} + \underline{\gamma} \downarrow_2) \odot (\underline{1} - 2\underline{\gamma}) \right]$$
$$+ H^T \left[(\underline{\tilde{i}} - \underline{i}) \odot H\underline{f} \right] \uparrow_2 \odot \left[-0.26 j (\underline{1} - \underline{\gamma} \uparrow_2) \odot (\underline{1} + \underline{\gamma} \uparrow_2) \right.$$
$$\left. \odot (\underline{1} - 2\underline{\gamma}) \right] \Big\}. \tag{G.6}$$

The derivation of Eq. (10.21) is as follows:

$$\frac{\partial F}{\partial \underline{\gamma}_m} = 2\sum_{u=1}^{N^2} \left(\tilde{i}_u - \left| \sum_{v=1}^{N^2} h_{uv} f_v \right|^2 \right) \times (-1) \times 2\mathrm{Re}\left\{ \left(\sum_{v=1}^{N^2} h_{uv} f_v \right) \right.$$
$$\times h_{um} \times \frac{\partial f_m^*}{\partial \underline{\gamma}_m} + \left(\sum_{v=1}^{N^2} h_{uv} f_v \right) \times h_{u,m-N} \times \frac{\partial f_{m-N}^*}{\partial \underline{\gamma}_m}$$
$$+ \left(\sum_{v=1}^{N^2} h_{uv} f_v \right) \times h_{u,m+N} \times \frac{\partial f_{m+N}^*}{\partial \underline{\gamma}_m}$$
$$+ \left(\sum_{v=1}^{N^2} h_{uv} f_v \right) \times h_{u,m-4N} \times \frac{\partial f_{m-4N}^*}{\partial \underline{\gamma}_m}$$
$$\left. + \left(\sum_{v=1}^{N^2} h_{uv} f_v \right) \times h_{u,m+4N} \times \frac{\partial f_{m+4N}^*}{\partial \underline{\gamma}_m} \right\}$$
$$= 2\sum_{u=1}^{N^2} \left(\tilde{i}_u - \left| \sum_{v=1}^{N^2} h_{uv} f_v \right|^2 \right) \times (-1) \times 2\mathrm{Re}\left\{ \left(\sum_{v=1}^{N^2} h_{uv} f_v \right) \right.$$
$$\times h_{um}[0.3 j\gamma_m \gamma_{m-4N}(1 - \gamma_{m-4N}) + 0.3 j\gamma_m \gamma_{m+4N}(1 - \gamma_{m+4N})$$
$$+ 0.8 j\gamma_m \gamma_{m-N}(1 - \gamma_{m-N}) + 0.8 j\gamma_m \gamma_{m+N}(1 - \gamma_{m+N}) + 1]$$
$$+ \left(\sum_{v=1}^{N^2} h_{uv} f_v \right) h_{u,m-N} \left[-\frac{0.8 j}{2}(1 - \gamma_{m-N})(1 + \gamma_{m-N})(1 - 2\gamma_m) \right]$$
$$+ \left(\sum_{v=1}^{N^2} h_{uv} f_v \right) h_{u,m+N} \left[-\frac{0.8 j}{2}(1 - \gamma_{m+N})(1 + \gamma_{m+N})(1 - 2\gamma_m) \right]$$

$$+\left(\sum_{v=1}^{N^2} h_{uv} f_v\right) h_{u,m-4N}\left[-\frac{0.3j}{2}(1-\gamma_{m-4N})(1+\gamma_{m-4N})(1-2\gamma_m)\right]$$

$$+\left(\sum_{v=1}^{N^2} h_{uv} f_v\right) h_{u,m+4N}\left[-\frac{0.3j}{2}(1-\gamma_{m+4N})(1+\gamma_{m+4N})(1-2\gamma_m)\right]\Bigg\}. \quad \text{(G.7)}$$

Therefore,

$$\begin{aligned}
\nabla F = -4\mathrm{Re}\Big\{ & H^T\left[(\tilde{\underline{i}}-\underline{i})\odot H\underline{f}\right]\odot\left[0.3j\underline{\gamma}\odot\underline{\gamma}\uparrow_4\odot(\underline{1}-\underline{\gamma}\uparrow_4)\right.\\
& +0.3j\underline{\gamma}\odot\underline{\gamma}\downarrow_4\odot(\underline{1}-\underline{\gamma}\downarrow_4)+0.8j\underline{\gamma}\odot\underline{\gamma}\uparrow\odot(\underline{1}-\underline{\gamma}\uparrow)+0.8j\underline{\gamma}\\
& \left.\odot\underline{\gamma}\downarrow\odot(\underline{1}-\underline{\gamma}\downarrow)+\underline{1}\right]+H^T\left[(\tilde{\underline{i}}-\underline{i})\odot H\underline{f}\right]\downarrow\odot\left[-0.4j\,(\underline{1}-\underline{\gamma}\downarrow)\right.\\
& \left.\odot(\underline{1}+\underline{\gamma}\downarrow)\odot(\underline{1}+2\underline{\gamma})\right]+H^T\left[(\tilde{\underline{i}}-\underline{i})\odot H\underline{f}\right]\uparrow\\
& \odot\left[-0.4j\,(\underline{1}-\underline{\gamma}\uparrow)\odot(\underline{1}+\underline{\gamma}\uparrow)\odot(\underline{1}+2\underline{\gamma})\right]\\
& +H^T\left[(\tilde{\underline{i}}-\underline{i})\odot H\underline{f}\right]\downarrow_4\odot\left[-0.15j\,(\underline{1}-\underline{\gamma}\downarrow_4)\right.\\
& \left.\odot(\underline{1}+\underline{\gamma}\downarrow_4)\odot(\underline{1}-2\underline{\gamma})\right]\\
& +H^T\left[(\tilde{\underline{i}}-\underline{i})\odot H\underline{f}\right]\uparrow_4\odot\left[-0.15j\,(\underline{1}-\underline{\gamma}\uparrow_4)\right.\\
& \left.\odot(\underline{1}+\underline{\gamma}\uparrow_4)\odot(\underline{1}-2\underline{\gamma})\right]\Big\}.
\end{aligned} \quad \text{(G.8)}$$

Appendix H

Software Guide

Chapter 5 and 6	
GPSM_wa	Generalized PSM optimization with discretization penalty and wavelet penalty in coherent imaging system.
GPSM_tv	Generalized PSM optimization with discretization penalty and total variation penalty in coherent imaging system.
OPC_tv	OPC optimization with discretization penalty and total variation penalty in coherent imaging system.
PSM_tv	Two-phase PSM optimization with discretization penalty and total variation penalty in coherent imaging system.
Chapter 7	
OPC_acaa	OPC optimization using the average coherent approximation model in partially coherent imaging system.
OPC_fse	OPC optimization using the Fourier series expansion model in partially coherent imaging system.
PSM_svd	Two-phase PSM optimization using the singular value decomposition model in partially coherent imaging system.
SOCS	Calculate the transmission cross-coefficient.
Chapter 8	
double_pattern	Double patterning optimization using two generalized PSMs in coherent imaging system.
proc_dct	Post-processing based on the two dimensional discrete cosine transform.
PSM_dct	Two-phase PSM optimization with the 2D DCT post-processing in partially coherent imaging system.
resisttone	Photoresist tone reversing method in partially coherent imaging system.
Chapter 9	
smo_OPC	Simultaneous source and binary mask optimization.
smo_OPC_mask	Binary mask optimization based on the SMO algorithm without source optimization.
smo_PSM	Simultaneous source and phase-shifting mask optimization.
smo_PSM_mask	Phase-shifting mask optimization based on the SMO algorithm without source optimization.

Computational Lithography By Xu Ma and Gonzalo R. Arce

Chapter 10

check_OPC	Check whether the topology of the binary mask pattern satisfies the topological constraint.
check_PSM	Check whether the topology of the phase-shifting mask pattern satisfies the topological constraint.
OPC_3D1	OPC optimization based on the boundary layer model in the first kind of coherent imaging system.
OPC_3D2	OPC optimization based on the boundary layer model in the second kind of coherent imaging system.
PSM_3D1	PSM optimization based on the boundary layer model in the first kind of coherent imaging system.
PSM_3D2	PSM optimization based on the boundary layer model in the second kind of coherent imaging system.

GPSM_wa

Purpose Generalized PSM optimization with discretization penalty and wavelet penalty in coherent imaging system.

Syntax GPSM_wa(N,pz,ra,phase_n,s_phi,s_theta,a,t_r,t_m, gamma_r_D,gamma_a_D,gamma_r_WA, gamma_a_WA,scale,epsilon,maxloop).

Description GPSM_wa performs the generalized gradient-based phase-shifting mask optimization with a N × N desired pattern in coherent imaging system. This algorithm generates the optimized four-phase or two-phase PSMs and includes discretization penalty and localized wavelet penalty. Different regions on the mask can be assigned with different weights of localized wavelet penalty. If all the regional weights are equal to 1, the localized wavelet penalty reduces to global wavelet penalty. The optimization iteration is terminated when either the tolerable output pattern error (epsilon) or the maximum iteration number (maxloop) is reached. The input parameters are

- N: Dimension of the mask.
- pz: Desired output pattern.
- ra: Initial phase pattern of the mask.
- phase_n: Number of discrete phase levels of the optimized mask. In the algorithm, phase_n can be 2 or 4.
- s_phi: Step size of the mask amplitude optimization.
- s_theta: Step size of the mask phase optimization.
- a: Steepness of the sigmoid function.
- t_r: Process threshold of the sigmoid function.
- t_m: Global threshold of the mask.
- gamma_r_D: Weight of the discretization penalty corresponding to mask amplitude.
- gamma_a_D: Weight of the discretization penalty corresponding to mask phase.
- gamma_r_WA: Weight of the wavelet penalty corresponding to mask amplitude.

GPSM_wa (*Continued*)

gamma_a_WA: Weight of the wavelet penalty corresponding to mask phase.
scale: Regional weights of the localized wavelet penalty.
epsilon: Tolerable output pattern error.
maxloop: Maximum iteration number.

Example

GPSM_wa(80,pz_f,ra_f,2,2,0.01,80,0.5,0.5,0,0,0,0,scale1,28,120); The result is shown in Fig. 5.16.
GPSM_wa(80,pz_f,ra_f,2,2,0.01,80,0.5,0.5,0.001,0.0001,0,0,scale1, 32,100); The result is shown in Fig. 6.10.
GPSM_wa(80,pz_u,ra_u,4,2,0.01,80,0.5,0.5,0.01,0.001,0.2,0.001, scale1,44,200); The result is shown in Fig. 6.15.
GPSM_wa(80,pz_f,ra_f,2,2,0.01,80,0.5,0.5,0.001,0.0001,0.03,0.001, scale1,36,230); The result is shown in Fig. 6.16.
GPSM_wa(80,pz_u,ra_u,4,2,0.01,80,0.5,0.5,0.01,0.001,0.2,0.001, scale2,38,200); The result is shown in Fig. 6.17.
In the above examples, pz_f is a 80×80 desired pattern of four horizontal bars. ra_f is a 80×80 initial phase pattern corresponding to pz_f. pz_u is a 80×80 desired pattern of U-junction. ra_u is a 80×80 initial phase pattern corresponding to pz_u. scale1 is a 80×80 regional weights matrix with all of the entries equal to 1. scale2 is a 80×80 regional weights matrix, where the gap between the vertical bars has a regional weight of 1.6. The other regions have a regional weight of 0.7. All these matrices are provided at ftp://ftp.wiley.com/public/sci_tech_med/computational_lithography.

Algorithm

The algorithms are described in Sections 5.4.1, 6.1.3, 6.2.2, and 6.2.3.

GPSM_tv

Purpose

Generalized PSM optimization with discretization penalty and total variation penalty in coherent imaging system.

Syntax

GPSM_tv(N,pz,ra,phase_n,s_phi,s_theta,a,t_r,t_m, gamma_r_D,gamma_a_D,gamma_r_TV, gamma_a_TV,epsilon,maxloop).

Description

GPSM_tv performs the generalized gradient-based phase-shifting mask optimization with a $N \times N$ desired pattern in coherent imaging system. This algorithm generates the optimized four-phase or two-phase PSMs and includes discretization penalty and total variation penalty. The optimization iteration is terminated when either the tolerable output pattern error (epsilon) or the maximum iteration number (maxloop) is reached. The input parameters are

N: Dimension of the mask.
pz: Desired output pattern.
ra: Initial phase pattern of the mask.
phase_n: Number of discrete phase levels of the optimized mask. In the algorithm, phase_n can be 2 or 4.
s_phi: Step size of the mask amplitude optimization.
s_theta: Step size of the mask phase optimization.
a: Steepness of the sigmoid function.

GPSM_tv (*Continued*)

t_r: Process threshold of the sigmoid function.
t_m: Global threshold of the mask.
gamma_r_D: Weight of the discretization penalty corresponding to mask amplitude.
gamma_a_D: Weight of the discretization penalty corresponding to mask phase.
gamma_r_TV: Weight of the total variation penalty corresponding to mask amplitude.
gamma_a_TV: Weight of the total variation penalty corresponding to mask phase.
epsilon: Tolerable output pattern error.
maxloop: Maximum iteration number.

GPSM_wa

Example GPSM_tv(80,pz_u,ra_u,4,2,0.01,80,0.5,0.5,0,0,0,0,18,25); The result is shown in Fig. 5.13.
GPSM_tv(80,pz_u,ra_u,4,2,0.01,80,0.5,0.5,0.045,0.001,0,0,29,18); The result is shown in Fig. 6.9.
GPSM_tv(80,pz_u,ra_u,4,2,0.01,80,0.5,0.5,0.01,0.001,0.1,0.001, 44,27); The result is shown in Fig. 6.13.
In the above examples, pz_u is a 80×80 desired pattern of U-junction. ra_u is a 80×80 initial phase pattern corresponding to pz_u. Both of these matrices are provided at ftp://ftp.wiley.com/public/sci_tech_med/computational_lithography.

Algorithm The algorithms are described in Sections 5.4.1, 6.1.3, and 6.2.1.

OPC_tv

Purpose OPC optimization with discretization penalty and total variation penalty in coherent imaging system.

Syntax OPC_tv(N,N_filter,k,pz,s,a,t_r,t_m,gamma_D, gamma_TV,epsilon,maxloop).

Description OPC_tv performs the gradient-based binary mask optimization with a $N \times N$ desired pattern in coherent imaging system. This algorithm generates the optimized binary masks and includes discretization penalty and total variation penalty. The optimization iteration is terminated when either the tolerable output pattern error (epsilon) or the maximum iteration number (maxloop) is reached. The input parameters are

N: Dimension of the mask.
N_filter: Dimension of the amplitude impulse response.
k: Process constant.
pz: Desired output pattern.
s: Step size.
a: Steepness of the sigmoid function.
t_r: Process threshold of the sigmoid function.
t_m: Global threshold of the mask.
gamma_D: Weight of the discretization penalty.
gamma_TV: Weight of the total variation penalty.
epsilon: Tolerable output pattern error.
maxloop: Maximum iteration number.

OPC_tv (*Continued*)

Example

OPC_tv(50,15,5,pz_t,0.2,90,0.5,0.5,0,0,4,350); The result is shown in Fig. 5.3.
OPC_tv(80,11,14,pz_f,0.5,80,0.5,0.5,0,0,4,90); The result is shown in Fig. 5.5.
OPC_tv(50,15,5,pz_t,0.2,90,0.5,0.5,0.025,0,4,350); The result is shown in Fig. 6.2.
OPC_tv(80,11,14,pz_f,0.5,80,0.5,0.5,0.01,0,0,70); The result is shown in Fig. 6.3.
OPC_tv(80,11,14,pz_f,0.5,80,0.5,0.5,0.01,0.025,6,70); The result is shown in Fig. 6.11.
In the above examples, pz_t is a 50×50 desired pattern of two vertical bars. pz_f is a 80×80 desired pattern of four horizontal bars. Both of these matrices are provided at ftp://ftp.wiley.com/public/sci_tech_med/computational_lithography.

Algorithm The algorithms are described in Sections 5.2.1, 6.1.1, and 6.2.1.

PSM_tv

Purpose Two-phase PSM optimization with discretization penalty and total variation penalty in coherent imaging system.

Syntax PSM_tv(N,N_filter,k,pz,r,s,a,t_r,t_m,gamma_D, gamma_TV,epsilon,maxloop).

Description PSM_tv performs the gradient-based phase-shifting mask optimization with a $N \times N$ desired pattern in coherent imaging system. This algorithm generates the optimized two-phase PSMs and includes discretization penalty and total variation penalty. The optimization iteration is terminated when either the tolerable output pattern error (epsilon) or the maximum iteration number (maxloop) is reached. The input parameters are

- N: Dimension of the mask.
- N_filter: Dimension of the amplitude impulse response.
- k: Process constant.
- pz: Desired output pattern.
- r: Initial phase pattern of the mask.
- s: Step size.
- a: Steepness of the sigmoid function.
- t_r: Process threshold of the sigmoid function.
- t_m: Global threshold of the mask.
- gamma_D: Weight of the discretization penalty.
- gamma_TV: Weight of the total variation penalty.
- epsilon: Tolerable output pattern error.
- maxloop: Maximum iteration number.

Example

PSM_tv(50,15,5,pz_t,r_t,1,90,0.5,0.5,0,0,10,160); The result is shown in Fig. 5.7.
PSM_tv(80,11,14,pz_f,r_f,0.5,80,0.5,0.5,0,0,10,230); The result is shown in Fig. 5.9.
PSM_tv(50,15,5,pz_t,r_t,1,90,0.5,0.5,0.0175,0,12,150); The result is shown in Fig. 6.5.
PSM_tv(80,11,14,pz_f,r_f,0.5,80,0.5,0.5,0.0025,

PSM_tv (*Continued*)

0,12,1200); The result is shown in Fig. 6.6.
PSM_tv(80,11,14,pz_f,r_f,0.5,80,0.5,0.5,0.0025,
0.008,0,59); The result is shown in Fig. 6.12.
In the above examples, pz_t is a 50 × 50 desired pattern of two vertical bars. r_t is a 50 × 50 initial phase pattern corresponding to pz_t. pz_f is a 80 × 80 desired pattern of four horizontal bars. r_f is a 80 × 80 initial phase pattern corresponding to pz_f. All of these matrices are provided at ftp://ftp.wiley.com/public/sci_tech_med/computational_lithography.

Algorithm The algorithms are described in Sections 5.3.1, 6.1.2, and 6.2.1.

OPC_acaa

Purpose OPC optimization using the average coherent approximation model in partially coherent imaging system.

Syntax OPC_acaa(N,pz,N_filter,pixel,k,NA,lamda,order,sigma_large_inner, sigma_large_outer,step,a,t_r,tr_approx,t_m,gamma_D,gamma_WA, epsilon,maxloop).

Description OPC_acaa performs the gradient-based binary mask optimization using the average coherent approximation model in partially coherent imaging system. The desired output pattern is represented by a N × N matrix. This algorithm generates the optimized binary masks and includes discretization penalty and global wavelet penalty. The optimization iteration is terminated when either the tolerable output pattern error (epsilon) or the maximum iteration number (maxloop) is reached. The input parameters are

- N: Dimension of the mask.
- pz: Desired output pattern.
- N_filter: Dimension of the amplitude impulse response.
- pixel: Pixel size (nm).
- k: Process constant.
- NA: Numerical aperture.
- lamda: Wavelength (nm).
- order: Order of Bessel function used in amplitude impulse response.
- sigma_large_inner: Inner partial coherence factor.
- sigma_large_outer: Outer partial coherence factor.
- step: Step size.
- a: Steepness of the sigmoid function.
- t_r: Process threshold of the sigmoid function.
- tr_approx: Process threshold for the average coherent approximation model.
- t_m: Global threshold of the mask.
- gamma_D: Weight of the discretization penalty.
- gamma_WA: Weight of the wavelet penalty.
- epsilon: Tolerable output pattern error.
- maxloop: Maximum iteration number.

Example OPC_acaa(184,pz_90,21,5.625,0.29,1.25,193,1,0.8,
0.975,0.5,25,0.19,0.09,0.5,0.025,0.025,0,200); The result is shown in the first row in Fig. 7.8.
OPC_acaa(184,pz_90,21,5.625,0.29,1.25,193,1,0.5,

OPC_acaa (*Continued*)

0.6,0.5,25,0.19,0.095,0.5,0.025,0.025,0,200); The result is shown in the second row in Fig. 7.8.
OPC_acaa(184,pz_90,21,5.625,0.29,1.25,193,1,0.3, 0.4,0.5,25,0.19,0.17,0.5,0.025,0.025,0,200); The result is shown in the third row in Fig. 7.8.
In the above examples, pz_90 is a 184 × 184 desired pattern with a pitch equal to 90 nm. This matrix is provided at ftp://ftp.wiley.com/public/sci_tech_med/computational_lithography.

Algorithm The algorithms are described in Section 7.1.3.

OPC_fse

Purpose OPC optimization using the Fourier series expansion model in partially coherent imaging system.

Syntax OPC_fse(N,pz,N_filter,pixel,k,NA,lamda,order,sigma_large_inner, sigma_large_outer,step,a,t_r,t_m,gamma_D,gamma_WA,epsilon, maxloop).

Description OPC_fse performs the gradient-based binary mask optimization using the Fourier series expansion model in partially coherent imaging system. The desired output pattern is represented by a N × N matrix. This algorithm generates the optimized binary masks and includes discretization penalty and global wavelet penalty. The optimization iteration is terminated when either the tolerable output pattern error (epsilon) or the maximum iteration number (maxloop) is reached. The input parameters are

N: Dimension of the mask.
pz: Desired output pattern.
N_filter: Dimension of the amplitude impulse response.
pixel: Pixel size (nm).
k: Process constant.
NA: Numerical aperture.
lamda: Wavelength (nm).
order: Order of Bessel function used in amplitude impulse response.
sigma_large_inner: Inner partial coherence factor.
sigma_large_outer: Outer partial coherence factor.
step: Step size.
a: Steepness of the sigmoid function.
t_r: Process threshold of the sigmoid function.
t_m: Global threshold of the mask.
gamma_D: Weight of the discretization penalty.
gamma_WA: Weight of the wavelet penalty.
epsilon: Tolerable output pattern error.
maxloop: Maximum iteration number.

Example OPC_fse(184,pz_90,21,5.625,0.29,1.25,193,1,0.8, 0.975,2,25,0.19,0.5,0.025,0.025,1426,43); The result is shown in the first row in Fig. 7.3.
OPC_fse(184,pz_90,21,5.625,0.29,1.25,193,1,0.5,0.6, 2,25,0.19,0.5,0.025,0.025,1582,60); The result is shown in the second row in Fig. 7.3.

OPC_fse (*Continued*)

OPC_fse(184,pz_90,21,5.625,0.29,1.25,193,1,0.3,0.4,2,25,0.19,0.5,0.025,0.025,1512,120); The result is shown in the third row in Fig. 7.3.
In the above examples, pz_90 is a 184 × 184 desired pattern with a pitch equal to 90 nm. This matrix is provided at ftp://ftp.wiley.com/public/sci_tech_med/computational_lithography.

Algorithm The algorithms are described in Section 7.1.1.

PSM_svd

Purpose Two-phase PSM optimization using the singular value decomposition model in partially coherent imaging system.

Syntax PSM_svd(N_mask,pz,r,pixel,k,NA,lamda,sigma,order,step,a,t_r,t_m,gamma_D,epsilon,maxloop).

Description PSM_svd performs the gradient-based phase-shifting mask optimization using the singular value decomposition model in partially coherent imaging system. The desired output pattern is represented by a N × N matrix. This algorithm generates the optimized two-phase PSM and includes discretization penalty. The optimization iteration is terminated when either the tolerable output pattern error (epsilon) or the maximum iteration number (maxloop) is reached. The input parameters are

- N: Dimension of the mask.
- pz: Desired output pattern.
- r: Initial phase pattern of the mask.
- pixel: Pixel size (nm).
- k: Process constant.
- NA: Numerical aperture.
- lamda: Wavelength (nm).
- sigma: Partial coherence factor.
- order: Order of Bessel function used in amplitude impulse response.
- step: Step size.
- a: Steepness of the sigmoid function.
- t_r: Process threshold for the first order coherent approximation.
- t_m: Global threshold of the mask.
- gamma_D: Weight of the discretization penalty.
- epsilon: Tolerable output pattern error.
- maxloop: Maximum iteration number.

Example PSM_svd(51,pz_t2,r_t2,11,0.29,1.35,193,0.3,1,0.2,200,0.003,0.33,0.1,9,300); The result is shown in Fig. 7.13.
PSM_svd(51,pz_t2,r_t2,11,0.29,1.35,193,0.6,1,0.2,200,0.003,0.33,0.1,9,300); The result is shown in Fig. 7.15.
In the above examples, pz_t2 is a 51 × 51 desired pattern of two vertical bars. r_t2 is a 51 × 51 initial phase pattern corresponding to pz_t2. Both of the matrices are provided at ftp://ftp.wiley.com/public/sci_tech_med/computational_lithography.

Algorithm The algorithms are described in Sections 7.2.1 and 7.2.2.

SOCS

Purpose Calculate the transmission cross-coefficient.

Syntax TCC=SOCS(N,pixel,k,NA,lamda,midway,sigma,order).

Description Given a partially coherent imaging system, SOCS calculates and returns the transmission cross-coefficient. The input parameters are

- N: Dimension of the mask.
- pixel: Pixel size (nm).
- k: Process constant.
- NA: Numerical aperture.
- lamda: Wavelength (nm).
- midway: Middle point of the mask.
- sigma: Partial coherence factor.
- order: Order of Bessel function used in amplitude impulse response.

The returned parameter is

- TCC: Transmission cross-coefficient.

Example SOCS(51,11,0.29,1.35,193,26,0.3,1); This example is used in Section 7.2.3.

Algorithm The algorithms are described in Section 2.1.2.

double_pattern

Purpose Double patterning optimization using two generalized PSMs in coherent imaging system.

Syntax double_pattern(s_phi_one,s_theta_one,s_phi_two,s_theta_two,a, t_r,t_m, gamma_r_D_one,gamma_a_D_one,gamma_r_WA_one, gamma_a_WA_one, gamma_r_D_two,gamma_a_D_two, gamma_r_WA_two,gamma_a_WA_two,epsilon,maxloop).

Description double_pattern performs the gradient-based double patterning optimization using two generalized PSMs in coherent imaging system. The desired output pattern is a U-junction represented by a 80×80 matrix. This algorithm generates a pair of optimized two-phase PSMs and includes discretization and wavelet penalties. The optimization iteration is terminated when either the tolerable output pattern error (epsilon) or the maximum iteration number (maxloop) is reached. The input parameters are

- s_phi_one: Step size of the first mask amplitude optimization.
- s_theta_one: Step size of the first mask phase optimization.
- s_phi_two: Step size of the second mask amplitude optimization.
- s_theta_two: Step size of the second mask phase optimization.
- a: Steepness of the sigmoid function.
- t_r: Process threshold.
- t_m: Global threshold of the mask.
- gamma_r_D_one: Weight of the discretization penalty corresponding to the first mask amplitude.
- gamma_a_D_one: Weight of the discretization penalty corresponding to the first mask phase.

double_pattern (*Continued*)

gamma_r_WA_one: Weight of the wavelet penalty corresponding to the first mask amplitude.
gamma_a_WA_one: Weight of the wavelet penalty corresponding to the first mask phase.
gamma_r_D_two: Weight of the discretization penalty corresponding to the second mask amplitude.
gamma_a_D_two: Weight of the discretization penalty corresponding to the second mask phase.
gamma_r_WA_two: Weight of the wavelet penalty corresponding to the second mask amplitude.
gamma_a_WA_two: Weight of the wavelet penalty corresponding to the second mask phase.
epsilon: Tolerable output pattern error.
maxloop: Maximum iteration number.

Example double_pattern(4,0.01,4,0.01,80,0.5,0.5,0.015,0.001, 0.5,0.003,0.015,0.001,0.5,0.003,10,100); The result is shown in Fig. 8.4.

Algorithm The algorithms are described in Section 8.1.

proc_dct

Purpose Post-processing based on the two-dimensional discrete cosine transform.

Syntax m_trinary_new=proc_dct(N_mask,pz,m,t_r,t_r_real,t_m,TCC, threshold).

Description proc_dct performs the 2D DCT post-processing on a given mask m, and returns the simplified mask pattern. The input parameters are

N_mask: Dimension of the mask.
pz: Desired output pattern.
m: Original mask.
t_r: Process threshold for the first order coherent approximation.
t_r_real: Process threshold for the entire partially coherent imaging system.
t_m: Global threshold of the mask.
TCC: Transmission cross-coefficient.
threshold: Threshold to cut off the high-frequency components of the DCT spectrum. The number of maintained low-frequency components is $(threshold - 1) \times (threshold - 2)/2$

The returned parameter is

m_trinary_new: Simplified mask after the post-processing.

Example proc_dct(51,pz_t2,m,0.003,t_r_real,0.33,TCC,18); The result is shown in Fig. 8.9.

In the above examples, pz_t2 is a 51 × 51 desired pattern of two vertical bars. This matrix is provided at ftp://ftp.wiley.com/public/sci_tech_med/computational_lithography. m is the mask pattern shown in the left figure of Fig. 8.7.

Algorithm The algorithms are described in Section 8.2.

PSM_dct

Purpose Two-phase PSM optimization with the 2D DCT post-processing in partially coherent imaging system.

Syntax PSM_dct(N_mask,pz,r,pixel,k,NA,lamda,sigma,order, threshold,step,a,t_r,t_m,gamma_D,epsilon,maxloop).

Description PSM_dct performs the gradient-based phase-shifting mask optimization using the SVD model in partially coherent imaging system. Then, it performs the 2D DCT post-processing on the optimized mask. The desired output pattern is represented by a N_mask × N_mask matrix. This algorithm generates the simplified two-phase PSMs and includes discretization penalty. The optimization iteration is terminated when either the tolerable output pattern error (epsilon) or the maximum iteration number (maxloop) is reached. The input parameters are

- N_mask: Dimension of the mask.
- pz: Desired output pattern.
- r: Initial phase pattern of the mask.
- pixel: Pixel size (nm).
- k: Process constant.
- NA: Numerical aperture.
- lamda: Wavelength (nm).
- sigma: Partial coherence factor.
- order: Order of Bessel function used in amplitude impulse response.
- threshold: Threshold to cut off the high-frequency components of the DCT spectrum. The number of maintained low-frequency components is $(threshold - 1) \times (threshold - 2)/2$
- step: Step size.
- a: Steepness of the sigmoid function.
- t_r: Process threshold for the first order coherent approximation.
- t_m: Global threshold of the mask.
- gamma_D: Weight of the discretization penalty.
- epsilon: Tolerable output pattern error.
- maxloop: Maximum iteration number.

Example PSM_dct(51,pz_t2,r_t2,11,0.29,1.35,193,0.3,1,18,0.2,200,0.003,0.33, 0.1,9,300); The result is shown in Fig. 8.9.
PSM_dct(51,pz_t2,r_t2,11,0.29,1.35,193,0.6,1,38,0.2,200,0.003,0.33, 0.1,9,300); The result is shown in Fig. 8.10.
In the above examples, pz_t2 is a 51 × 51 desired pattern of two vertical bars. r_t2 is a 51 × 51 initial phase pattern corresponding to pz_t2. Both of the matrices are provided at ftp://ftp.wiley.com/public/sci_tech_med/computational_lithography.

Algorithm The algorithms are described in Section 8.2.

resisttone

Purpose Photoresist tone reversing method in partially coherent imaging system.

Syntax resisttone(N_mask,desire_pattern,distribution,pixel,k, NA,lamda,sigma,order,threshold,flag,step,a,t_r, t_m,gamma_D,gamma_WA,epsilon,maxloop).

resisttone (*Continued*)

Description resisttone performs the gradient-based phase-shifting mask optimization using the SVD model and photoresist tone reversing method. The desired output pattern is represented by a N_mask × N_maxk matrix. The photoresist tone reversing method is applied to project extreme dense patterns. This algorithm generates the optimized two-phase PSMs and includes discretization and wavelet penalties. In addition, the user may choose whether to invoke the 2D DCT post-processing to simplify the optimized mask pattern or not. The optimization iteration is terminated when either the tolerable output pattern error (epsilon) or the maximum iteration number (maxloop) is reached. The input parameters are

- N_mask: Dimension of the mask.
- desire_pattern: Desired output pattern.
- distribution: Photoresist distribution.
- pixel: Pixel size (nm).
- k: Process constant.
- NA: Numerical aperture.
- lamda: Wavelength (nm).
- sigma: Partial coherence factor.
- order: Order of Bessel function used in amplitude impulse response.
- threshold: Threshold to cut off the high-frequency components of the DCT spectrum. The number of maintained low-frequency components is $(threshold - 1) \times (threshold - 2)/2$.
- flag: The mark controlling whether to invoke the 2D DCT post-processing. If $flag = 0$, then the algorithm does **NOT** invoke the post-processing. If $flag = 1$, then the algorithm invokes the post-processing.
- step: Step size.
- a: Steepness of the sigmoid function.
- t_r: Process threshold for the first order coherent approximation.
- t_m: Global threshold of the mask.
- gamma_D: Weight of the discretization penalty.
- gamma_WA: Weight of the wavelet penalty.
- epsilon: Tolerable output pattern error.
- maxloop: Maximum iteration number.

Example resisttone(51,pz_f2,distribution1,11,0.29,1.35,193,0.3,1,15,0,2,200, 0.01,0.33,0.1,0.1,9,100); The result is shown in Fig. 8.12.
resisttone(51,pz_f2,distribution1,11,0.29,1.35,193,0.3,1,15,1,2,200, 0.01,0.33,0.1,0.1,9,300); The result is shown in Fig. 8.13.
resisttone(51,pz_r,distribution2,11,0.29,1.35,193,0.3,1,49,0,0.2,200, 0.01,0.33,0.1,0,8,100); The result is shown in Fig. 8.15.
resisttone(51,pz_r,distribution2,11,0.29,1.35,193,0.3,1,49,1,0.2,200, 0.01,0.33,0.1,0,8,100); The result is shown in Fig. 8.16.
In the above examples, pz_f2 is a 51 × 51 desired pattern of four vertical bars. distribution1 is the photoresist distribution pattern corresponding to pz_f2. pz_r is a 51 × 51 desired pattern of contact hole. distribution2 is the photoresist distribution pattern corresponding to pz_r. All of these matrices are provided at ftp://ftp.wiley.com/public/sci_tech_med/computational_lithography.

Algorithm The algorithms are described in Section 8.3.

smo_OPC

Purpose Simultaneous source and binary mask optimization.

Syntax smo_OPC(N,pz,N_filter,pixel,k,NA,lamda,order, sigma_large_inner,sigma_large_outer).

Description smo_OPC performs the simultaneous source and binary mask optimization with an annular original illumination. The desired output pattern is represented by a $N \times N$ matrix. The algorithm generates the optimized binary masks and source patterns. The optimization iteration is terminated when the cost function cannot be reduced any more. The input parameters are

N: Dimension of the mask.
pz: Desired output pattern.
N_filter: Dimension of the amplitude impulse response.
pixel: Pixel size (nm).
k: Process constant.
NA: Numerical aperture.
lamda: Wavelength (nm).
order: Order of Bessel function used in amplitude impulse response.
sigma_large_inner: Inner partial coherence factor.
sigma_large_outer: Outer partial coherence factor.

Example smo_OPC(80,pz_smo1,21,15,0.29,1.25,193,1,0.4,0.5); The result is shown in the third row of Fig. 9.2.
In the above example, pz_smo1 is a 80×80 desired pattern, which is provided at ftp://ftp.wiley.com/public/sci_tech_med/computational_lithography.

Algorithm The algorithms are described in Section 9.3.

smo_OPC_mask

Purpose Binary mask optimization based on the SMO algorithm without source optimization.

Syntax smo_OPC_mask(N,pz,N_filter,pixel,k,NA,lamda,order, sigma_large_inner,sigma_large_outer).

Description smo_OPC_mask performs the binary mask optimization based on the SMO algorithm with an annular illumination. The desired output pattern is represented by a $N \times N$ matrix. The algorithm generates the optimized binary masks without source optimization. The optimization iteration is terminated when the cost function cannot be reduced any more. The input parameters are

N: Dimension of the mask.
pz: Desired output pattern.
N_filter: Dimension of the amplitude impulse response.
pixel: Pixel size (nm).
k: Process constant.
NA: Numerical aperture.
lamda: Wavelength (nm).
order: Order of Bessel function used in amplitude impulse response.
sigma_large_inner: Inner partial coherence factor.
sigma_large_outer: Outer partial coherence factor.

smo_OPC_mask (*Continued*)

Example smo_OPC_mask(80,pz_smo1,21,15,0.29,1.25,193,1,0.4,0.5); The result is shown in the second row of Fig. 9.2.
In the above example, pz_smo1 is a 80×80 desired pattern, which is provided at ftp://ftp.wiley.com/public/sci_tech_med/computational_lithography.

Algorithm The algorithms are described in Section 9.3.

smo_PSM

Purpose Simultaneous source and phase-shifting mask optimization.

Syntax smo_PSM(N,pz,N_filter,pixel,k,NA,lamda,order,sigma).

Description smo_PSM performs the simultaneous source and phase-shifting mask optimization with a circular original illumination. The desired output pattern is represented by a $N \times N$ matrix. The algorithm generates the optimized phase-shifting masks and source patterns. The optimization iteration is terminated when the cost function cannot be reduced any more. The input parameters are

N: Dimension of the mask.
pz: Desired output pattern.
N_filter: Dimension of the amplitude impulse response.
pixel: Pixel size (nm).
k: Process constant.
NA: Numerical aperture.
lamda: Wavelength (nm).
order: Order of Bessel function used in amplitude impulse response.
sigma: Partial coherence factor.

Example smo_PSM(80,pz_smo2,21,15,0.29,1.25,193,1,0.4); The result is shown in the third row of Fig. 9.3.
In the above example, pz_smo2 is a 80×80 desired pattern, which is provided at ftp://ftp.wiley.com/public/sci_tech_med/computational_lithography.

Algorithm The algorithms are described in Section 9.3.

smo_PSM_mask

Purpose Phase-shifting mask optimization based on the SMO algorithm without source optimization.

Syntax smo_PSM_mask(N,pz,N_filter,pixel,k,NA,lamda,order,sigma).

Description smo_PSM_mask performs the phase-shifting mask optimization based on the SMO algorithm with a circular illumination. The desired output pattern is represented by a $N \times N$ matrix. The algorithm generates the optimized phase-shifting masks without source optimization. The optimization iteration is terminated when the cost function cannot be reduced any more. The input parameters are

smo_PSM_mask (***Continued***)

N: Dimension of the mask.
pz: Desired output pattern.
N_filter: Dimension of the amplitude impulse response.
pixel: Pixel size (nm).
k: Process constant.
NA: Numerical aperture.
lamda: Wavelength (nm).
order: Order of Bessel function used in amplitude impulse response.
sigma: Partial coherence factor.

Example smo_PSM_mask(80,pz_smo2,21,15,0.29,1.25,193,1,0.4); The result is shown in the second row of Fig. 9.3.
In the above example, pz_smo2 is a 80×80 desired pattern, which is provided at ftp://ftp.wiley.com/public/sci_tech_med/computational_lithography.

Algorithm The algorithms are described in Section 9.3.

check_OPC

Purpose Check whether the topology of the binary mask pattern satisfies the topological constraint.

Syntax flag=check_OPC(m_dummy,N_dummy,singular1, singular2).

Description Given a binary mask pattern, check_OPC checks whether the mask satisfies the topological constraint described in Section 10.4.1. The input parameters are

m_dummy: The binary mask to be checked.
N_dummy: Dimension of the mask.
singular1: The minimum dimension used to remove the Type I singular pixel described in Section 10.4.1.
singular2: The minimum dimension used to remove the Type II singular pixel described in Section 10.4.1.

The returned parameter is

flag: If $flag = 1$, then the binary mask does **NOT** satisfy the topological constraint. If $flag = 0$, then the binary mask satisfies the topological constraint.

Example check_OPC(m_dummy,90,8,3); This example is used in Section 10.4.3.

Algorithm The algorithms are described in Section 10.4.1.

check_PSM

Purpose Check whether the topology of the phase-shifting mask pattern satisfies the topological constraint.

Syntax flag=check_PSM(m_dummy,N_dummy,singular12, boundary_clear,boundary_shift).

check_PSM (*Continued*)

Description Given a phase-shifting mask pattern, check_PSM checks whether the mask satisfies the topological constraint described in Section 10.5.1. The input parameters are

m_dummy: The phase-shifting mask to be checked.
N_dummy: Dimension of the mask.
singular12: The minimum dimension used to remove the Type I and II singular pixels described in Section 10.5.1.
boundary_clear: The boundary width of clear openings.
boundary_shift: The boundary width of 180° phase-shifting openings.

The returned parameter is

flag: If *flag* = 1, then the binary mask does **NOT** satisfy the topological constraint. If *flag* = 0, then the binary mask satisfies the topological constraint.

Example check_PSM(m_dummy,80,9,1,2); This example is used in Section 10.5.3.

Algorithm The algorithms are described in Section 10.5.1.

OPC_3D1

Purpose OPC optimization based on the boundary layer model in the first kind of coherent imaging system.

Syntax OPC_3D1(N,pz,N_filter,order).

Description OPC_3D1 performs the binary mask optimization based on the boundary layer model, which takes into account the thick-mask effects. The desired output pattern is represented by a $N \times N$ matrix. The optical lithography system is the first kind of coherent imaging system described in Section 10.2. The optimization iteration is terminated when the cost function cannot be reduced any more. The input parameters are

N: Dimension of the mask.
pz: Desired output pattern.
N_filter: Dimension of the amplitude impulse response.
order: Order of Bessel function used in amplitude impulse response.

Example OPC_3D1(90,pz_3D1,121,1); The result is shown in Fig. 10.8.
In the above example, pz_3D1 is a 90×90 desired pattern, which is provided at ftp://ftp.wiley.com/public/sci_tech_med/computational_lithography.

Algorithm The algorithms are described in Section 10.4.2.

OPC_3D2

Purpose OPC optimization based on the boundary layer model in the second kind of coherent imaging system.

Syntax OPC_3D2(N,pz,N_filter,order).

OPC_3D2 (*Continued*)

Description	OPC_3D2 performs the binary mask optimization based on the boundary layer model, which takes into account the thick mask effects. The desired output pattern is represented by a N × N matrix. The optical lithography system is the second kind of coherent imaging system described in Section 10.2. The optimization iteration is terminated when the cost function cannot be reduced any more. The input parameters are: N: Dimension of the mask. pz: Desired output pattern. N_filter: Dimension of the amplitude impulse response. order: Order of Bessel function used in amplitude impulse response.
Example	OPC_3D2(95,pz_3D2,139,1); The result is shown in Fig. 10.10. In the above example, pz_3D2 is a 95 × 95 desired pattern, which is provided at ftp://ftp.wiley.com/public/sci_tech_med/computational_lithography.
Algorithm	The algorithms are described in Section 10.4.2.

PSM_3D1

Purpose	PSM optimization based on the boundary layer model in the first kind of coherent imaging system.
Syntax	PSM_3D1(N,pz,N_filter,order).
Description	PSM_3D1 performs the phase-shifting mask optimization based on the boundary layer model, which takes into account the thick-mask effects. The desired output pattern is represented by a N × N matrix. The optical lithography system is the first kind of coherent imaging system described in Section 10.2. The optimization iteration is terminated when the cost function cannot be reduced any more. The input parameters are N: Dimension of the mask. pz: Desired output pattern. N_filter: Dimension of the amplitude impulse response. order: Order of Bessel function used in amplitude impulse response.
Example	PSM_3D1(80,pz_3D3,109,1); The result is shown in Fig. 10.12. In the above example, pz_3D3 is a 80 × 80 desired pattern, which is provided at ftp://ftp.wiley.com/public/sci_tech_med/computational_lithography.
Algorithm	The algorithms are described in Section 10.5.2.

PSM_3D2

Purpose	PSM optimization based on the boundary layer model in the second kind of coherent imaging system.
Syntax	PSM_3D2(N,pz,N_filter,order).
Description	PSM_3D2 performs the phase-shifting mask optimization based on the boundary layer model, which takes into account the thick-mask effects. The

PSM_3D2

desired output pattern is represented by a N × N matrix. The optical lithography system is the second kind of coherent imaging system described in Section 10.2. The optimization iteration is terminated when the cost function cannot be reduced any more. The input parameters are

N: Dimension of the mask.
pz: Desired output pattern.
N_filter: Dimension of the amplitude impulse response.
order: Order of Bessel function used in amplitude impulse response.

Example PSM_3D2(151,pz_3D4,139,1); The result is shown in Fig. 10.14.
In the above example, pz_3D4 is a 151 × 151 desired pattern, which is provided at ftp://ftp.wiley.com/public/sci_tech_med/computational_lithography.

Algorithm The algorithms are described in Section 10.5.2.

References

1. International technology roadmap for semiconductors. Technical report, http://public.itrs.net.
2. K. Adam. Domain decomposition methods for the electromagnetic simulation of scattering from three-dimensional structures with applications in lithography. PhD dissertation, Electrical Engineering and Computer Sciences, University of California, Berkeley, 2001.
3. K. Adam and A. R. Neureuther. Domain decomposition methods for the rapid electromagnetic simulation of photomask scattering. *Journal of Microlithography, Microfabrication, and Microsystems*, 1:253, 2002.
4. J. T. Azpiroz. Analysis and modeling of photomask near-fields in sub-wavelength deep ultraviolet lithography with optical proximity corrections. PhD dissertation, Department of Electrical Engineering, University of California, Los Angeles, 2004.
5. S. Babin, I. Yu. Kuzmin, and C. A. Mack. Comprehensive simulation of e-beam lithography processes using prolith/3D and temptation software tools. *Proceedings of SPIE*, Vol. 4186, p. 503, 2001.
6. J. E. Bjorkholm. EUV lithography: the successor to optical lithography? *Intel Technology Journal*, Vol. 2, Issue 3, Q3, 1998.
7. M. Born and E. Wolf. *Principles of Optics*, SPIE Press, 2001.
8. Y. A. Borodovsky. Lithographic enhancement method and apparatus for randomly spaced structures. US patent 5,424,154, Jun. 1995.
9. T. Brunner. Impact of lens aberrations on optical lithography. *IBM Journal of Research and Development*, 41(1/2):57–67, 1997.
10. M. Burkhardt, C. Progler, A. Yen, and G. Wells. Illuminator design for the printing of regular contact patterns. *Microelectronic Engineering*, 41:91–95, 1998.
11. S. A. Campbell. *The Science and Engineering of Microelectronic Fabrication*, Publishing House of Electronics Industry, 2003.
12. S. H. Chan, A. K. Wong, and E. Y. Lam. Initialization for robust inverse synthesis of phase-shifting masks in optical projection lithography. *Optics Express*, 16(19):14746–14760, 2008.
13. N. Cobb. Fast optical and process proximity correction algorithms for integrated circuit manufacturing. PhD dissertation, University of California at Berkeley, 1998.
14. N. Cobb and A. Zakhor. Fast sparse aerial image calculation for OPC. In *BACUS Symposium on Photomask Technology, Proceedings of SPIE*, Vol. 2440, pp. 313–327, 1995.
15. W. J. Cook, W. H. Cunningham, W. R. Pulleyblank, and A. Schrijver. *Combinatorial Optimization*, John Wiley & Sons, 1997.

Computational Lithography By Xu Ma and Gonzalo R. Arce

16. Y. Dai and Y. Yuan. A nonlinear conjugate gradient method with a strong global convergence property. *SIAM Journal on Optimization*, 10:177–182, 1999.
17. P. S. Davids and S. B. Bollepalli. Generalized inverse problem for partially coherent projection lithography. *Proceedings of SPIE*, Vol. 6924, p. 69240X, 2008.
18. V. Domnenko, T. Schmoeller, and T. Klimpel. Analysis of EUVL mask effects under partially coherent illumination. *Proceedings of SPIE*, Vol. 7271, p. 727141, 2009.
19. P. G. Engeldrum. Color gamut limits of halftone printing with and without the paper spread function. *Journal of Imaging Science and Technology*, 40(3):239–244, 1996.
20. A. Erdmann, P. Evanschitzky, G. Citarella, T. Fühner, and P. D. Bisschop, Rigorous mask modeling using waveguide and FDTD methods: An assess- ment for typical hyper NA imaging problems, in Photomask and Next- Generation Lithography Mask Technology XIII, Proc. of SPIE, vol. 6283, pp. 628319, 2006.
21. A. Erdmann, R. Farkas, T. Fuhner, B. Tollkuhn, and G. Kokai. Towards automatic mask and source optimization for optical lithography. *Optical Microlithography, Proceedings of SPIE*, Vol. 5377, pp. 646–657, 2004.
22. S. Farsiu, D. Robinson, M. Elad, and P. Milanfar. Fast and robust multi-frame super-resolution. *IEEE Transactions on Image Processing*, 13:1327–1344, 2004.
23. R. Fletcher and C. M. Reeves. Function minimization by conjugate gradients. *Computer Journal*, 7:149–154, 1964.
24. T. S. Gau, R. G. Liu, C. K. Chen, C. M. Lai, F. J. Liang, and C. C. Hsiar. The customized illumination aperture filter for low k1 photolithography process. *Proceedings of SPIE*, Vol. 4000, p. 271, 2000.
25. J. W. Goodman. *Introduction to Fourier Optics*, McGraw-Hill, 1968.
26. Y. Granik. Illuminator optimization methods in microlithography. *Optical Microlithography, Proceedings of SPIE*, Vol. 5754, pp. 217–229, 2005.
27. W. W. Hager and H. Zhang. A new conjugate gradient method with guaranteed descent and an efficient line search. *SIAM Journal on Optimization*, 16:170–192, 2005.
28. H. Hopkins. The concept of partial coherence in optics. *Proceeding of the Royal Society of London*, A208: pp. 263–377, 1951.
29. H. Hopkins. The concept of partial coherence in optics. *Proceedings of the Royal Society of London*, A217: 408–432, 1953.
30. J. Iba, K. Hashimoto, R. A. Ferguson, T. Yanagisawa, and D. J. Samuels. Electrical characterization of across field lithographic performance for 256 mbit dram technologies. *Proceedings of SPIE*, Vol. 2512, pp. 218–225, 1995.
31. J. Kirk and C. Progler. Pinholes and pupil fills. *Microlithography World*, 6(4):25–28, 1997.
32. J. Kirk and C. Progler. Pupil illumination: *in-situ* measurement of partial coherence. In L. van den Hove, editor, *Proceedings of SPIE*, Vol. 3334, pp. 281–288, 1998.
33. L. Lam, S. W. Lee, and C. Y. Suen. Thinning methodologies: a comprehensive survey. *IEEE Transactions on Pattern Analysis and Machine Intelligence*, 14:869–885, 1992.
34. D. L. Lau and G. R. Arce. *Modern Digital Halftoning*, Marcel Dekker, 2001.
35. M. D. Levenson, N. S. Viswanathan, and R. A. Simpson. Improving resolution in photolithography with a phase-shifting mask. *IEEE Transactions on Electron Devices*, ED-29:1828–1836, 1982.
36. H. J. Levinson. *Principles of Lithography*, SPIE Press, 2001.

37. L. Liebmann, S. Mansfield, A. Wong, M. Lavin, W. Leipold, and T. Dunham. TCAD development for lithography resolution enhancement. *IBM Journal of Research and Development*, 45(5): 651–665, 2001.
38. Y. Liu and A. Zakhor. Binary and phase shifting mask design for optical lithography. *IEEE Transactions on Semiconductor Manufacturing*, 5(2):138–152, 1992.
39. K. Lucas, H. Tanabe, and A. J. Strojwas. Efficient and rigorous three-dimensional model for optical lithography simulation. *Journal of the Optical Society of America A*, 13(11):2187–2199, 1996.
40. X. Ma and G. R. Arce. Generalized inverse lithography methods for phase-shifting mask design. *Proceedings of SPIE*, Vol. 6520, p. 65200U, 2007.
41. X. Ma and G. R. Arce. Generalized inverse lithography methods for phase-shifting mask design. *Optics Express*, 15(23):15066–15079, 2007.
42. X. Ma and G. R. Arce. Binary mask optimization for inverse lithography with partially coherent illumination. *Proceedings of SPIE*, Vol. 7140, p. 714087, 2008.
43. X. Ma and G. R. Arce. Binary mask optimization for inverse lithography with partially coherent illumination. *Journal of the Optical Society of America A*, 25(12):2960–2970, 2008.
44. X. Ma and G. R. Arce. PSM design for inverse lithography with partially coherent illumination. *Optics Express*, 16(24):20126–20141, 2008.
45. X. Ma and G. R. Arce. Binary mask optimization for forward lithography based on the boundary layer model in coherent systems. *Journal of the Optical Society of America A*, 26(7):1687–1695, 2009.
46. X. Ma and G. R. Arce. Pixel-based simultaneous source and mask optimization for resolution enhancement in optical lithography. *Optics Express*, 17(7):5783–5793, 2009.
47. X. Ma and G. R. Arce. PSM design for inverse lithography with partially coherent illumination. *Proceedings of SPIE*, Vol. 7274, p. 727437, 2009.
48. X. Ma and G. R. Arce. 3D PSM optimization based on boundary layer model. in review, *Journal of the Optical Society of America A*, 2010.
49. C. Mack. *Fundamental Principles of Optical Lithography: The Science of Microfabrication*, John Wiley & Sons, 2008.
50. S. M. Mansfield, L. W. Liebmann, A. F. Molless, and A. K. Wong. Lithographic comparison of assist feature design strategies. In C. Progler, editor, *Proceedings of SPIE*, Vol. 4000, pp. 63–76, 2000.
51. K. H. Nakagawa, F. Chen, R. Socha, T. Laidig, K. E. Wampler, D. V. D. Broeke, M. Dusa, and R. Caldwell. Halftone biasing OPC technology: an approach for achieving fine bias control on raster-scan systems. *Proceedings of SPIE*, Vol. 3748, pp. 315–323, 1999.
52. G. L. Nemhauser and L. A. Wolsey. *Integer and Combinatorial Optimization*, John Wiley & Sons, 1988.
53. J. Nocedal and S. J. Wright. *Numerical Optimization*, Springer Press, 2006.
54. A. M. Nugrowati, A. S. van de Nes, S. F. Pereira, and J. J. M. Braat. EUV phase mask engineering based on image optimisation. *Microelectronic Engineering*, 83:684–687, 2006.
55. L. Pang, Y. Liu, and D. Abrams. Inverse lithography technology (ILT): What is the impact to the photomask industry? *Proceedings of SPIE*, Vol. 6283, p. 62830X, 2006.
56. C. H. Papadimitriou and K. Steiglitz. *Combinatorial Optimization: Algorithms and Complexity*, Prentice Hall, 1982.

57. J. Park, C. Park, S. Rhie, Y. Kim, M. Yoo, J. Kong, H. Kim, and S. Yoo. An efficient rule-based OPC approach using a drc tool for 0.18 μm ASIC. *International Symposium on Quality Electronic Design, IEEE Proceedings of the 1st International Symposium on Quality of Electronic Design*, pp. 81–85, 2000.

58. Y. C. Pati and T. Kailath. Phase-shifting masks for microlithography: automated design and mask requirements. *Journal of the Optical Society of America A*, 11:2438–2452, 1994.

59. M. C. Peckerar and C. R. K. Marrian. Error measure comparison of currently employed dose-modulation schemes for e-beam proximity effect control. *Electron-Beam, X-Ray, EUV and Ion-Beam Submicrometer Lithographies for Manufacturing, Proceedings of SPIE*, Vol. 2437, pp. 222–238, 1995.

60. M. C. Peckerar and C. R. K. Marrian. Proximity correction algorithms and a co-processor based on regularized optimization. I. Description of the algorithm. *Journal of Vacuum Science and Technology B*, 13:2518–2525, 1995.

61. C. Pierrat and A. Wong. MEF revisited: Low k1 effects versus mask topography effects. *Proceedings of SPIE*, Vol. 5040, pp. 193–202, 2003.

62. C. Pierrat, A. Wong, and S. Vaidya. Phase-shifting mask topography effects on lithographic image quality. In *Technical Digest: International Electron Devices Meeting*, Vol. 41, pp. 53–56, 1992.

63. T. V. Pistor. Electromagnetic simulation and modeling with application in lithography. PhD dissertation, University of California at Berkeley, 2001.

64. T. V. Pistor, A. R. Neureuther, and R. J. Socha. Modeling oblique incidence effects in photomasks. *Proceedings of SPIE*, Vol. 4000, pp. 228–237, 2000.

65. A. Poonawala, Y. Borodovsky, and P. Milanfar. ILT for double exposure lithography with conventional and novel materials. *Proceedings of SPIE Advanced Lithography Symposium*, Vol. 6520, p. 65202Q, 2007.

66. A. Poonawala and P. Milanfar. Double exposure mask synthesis using inverse lithography. *Journal of Microlithography, Microfabrication, and Microsystems*, Vol. 6, Issue 4, pp. 043001, 2007.

67. A. Poonawala and P. Milanfar. OPC and PSM design using inverse lithography: a nonlinear optimization approach. *Proceedings of SPIE Microlithography Symposium*, Vol. 6154, pp. 1159–1172, 2006.

68. A. Poonawala and P. Milanfar. Mask design for optical microlithography: an inverse imaging problem. *IEEE Transactions on Image Processing*, 16(3):774–788, 2007.

69. A. Poonawala. Mask design for single and double exposure optical microlithography: an inverse imaging approach. PhD dissertation, Computer Engineering, University of California Santa Cruz, 2007.

70. C. Progler, W. Conley, B. Socha, and Y. Ham. Layout and source dependent phase mask transmission tuning. *Proceedings of SPIE*, Vol. 5754, p. 315, 2005.

71. M. D. Prouty and A. R. Neureuther. Optical imaging with phase shift masks. *Proceedings of SPIE*, Vol. 470, pp. 228–232, 1984.

72. S. Robert, X. Shi, and L. David. Simultaneous source mask optimization (SMO). *Proceedings of SPIE*, Vol. 5853, p. 180, 2005.

73. A. E. Rosenbluth, S. Bukofsky, C. Fonseca, and M. Hibbs. Optimum mask and source patterns to print a given shape. *Journal of Microlithography, Microfabrication, and Microsystems*, 1:13–30, 2002.

74. B. E. A. Saleh and M. Rabbani. Simulation of partially coherent imagery in the space and frequency domains and by modal expansion. *Applied Optics*, 21:2770–2777, 1982.

75. B. Salik, J. Rosen, and A. Yariv. Average coherent approximation for partially coherent optical systems. *Journal of the Optical Society of America A*, 13:2086–2090, 2001.

76. F. Schellenberg. *Resolution Enhancement Techniques in Optical Lithography*, SPIE Press, 2004.

77. F. Schellenberg. Resolution enhancement technology: the past, the present, and extensions for the future, optical microlithography. *Proceedings of SPIE*, Vol. 5377, pp. 1–20, 2004.

78. G. J. Schneider, J. Murakowski, S. Venkataraman, and D. W. Prather. Combination lithography for photonic crystal circuits. *Journal of Vacuum Science & Technology B*, 22(1):146–151, 2004.

79. S. Sherif, B. Saleh, and R. Leone. Binary image synthesis using mixed integer programming. *IEEE Transactions on Image Processing*, 4(9):1252–1257, 1995.

80. R. Shi, Y. Cai, X. Hong, W. Wu, and C. Yang. Important works about rules in rules-based optical proximity correction. *Chinese Journal of Semiconductors*, 23(7):701–706, 2002.

81. J. Tirapu-Azpiroz, P. Burchard, and E. Yablonovitch. Boundary layer model to account for thick mask effects in photolithography. *Optical Microlithography, Proceedings of SPIE*, Vol. 5040, pp. 1611–1619, 2003.

82. J. Tirapu-Azpiroz and E. Yablonovitch. Fast evaluation of photomask near-fields in subwavelength 193 nm lithography. *Optical Microlithography, Proceedings of SPIE*, Vol. 5377, pp. 1528–1535, 2004.

83. C. Vogel. *Computational Methods for Inverse Problems*. SIAM Press, 2002.

84. D. L. White, R. A. Cirelli, S. J. Spector, M. I. Blakey, and O. R. Wood. Phase-mask effects by dark-field lithography. In C. Progler, editor, *Proceedings of SPIE*, Vol. 4000, pp. 366–372, 2000.

85. R. Wilson. *Fourier Series and Optical Transform Techniques in Contemporary Optics*, John Wiley & Sons, 1995.

86. G. Wojcik, J. J. Mould, R. Ferguson, R. Martino, and K. K. Low. Some image modeling issues for I-line, 5 $\times$ phase shifting masks. *Proceedings of SPIE*, Vol. 2197, pp. 455–465, 1994.

87. E. Wolf. Electromagnetic diffraction in optical systems I. an integral representation of the image field. *Proceedings of the Royal Society of London A*, 253:349–357, 1959.

88. L. A. Wolsey. *Integer Programming*, Wiley-Interscience Series in Discrete Mathematics and Optimization, John Wiley & Sons, 1998.

89. A. Wong. Rigorous three-dimensional time-domain finite difference electromagnetic simulation. PhD dissertation, Electrical Engineering and Computer Sciences, University of California, Berkeley, 1994.

90. A. Wong and A. R. Neureuther. Mask topography effects in projection printing of phase shift masks. *IEEE Transactions on Electron Devices*, 41:895–902, 1994.

91. A. K. Wong. Asymmetric biasing for subgrid pattern adjustment. In C. Progler, editor, *Proceedings of SPIE*, Vol. 4346, pp. 1548–1553, 2001.

92. A. K. Wong. *Resolution Enhancement Techniques*, SPIE Press, 2001.

93. A. K. Wong. *Optical Imaging in Projection Microlithography*, SPIE Press, 2005.

94. A. K.-K. Wong. Rigorous three-dimensional time-domain finite-difference electromagnetic simulation. PhD dissertation, University of California, Berkeley, 1994.

95. P. Yao, G. J. Schneider, B. L. Miao, J. Murakowski, D. W. Prather, E. D. Wetzel, and D. J. O'Brien. Multilayer three-dimensional photolithography with traditional planar method. *Applied Physics Letters*, 85:2920–2922, 2004.

96. M. S. Yeung and E. Barouch. Limitation of the Kirchhoff boundary conditions for aerial image simulation in 157 nm optical lithography. *IEEE Electron Device Letters*, 21(9):433–435, 2000.

97. M. S. Yeung, D. Lee, R. Lee, and A. R. Neureuther. Extension of the Hopkins theory of partially coherent imaging to include thin-film interference effects. *Proceedings of SPIE*, Vol. 1927, pp. 452–463, 1993.

98. P. Yu and D. Z. Pan. TIP-OPC: a new topological invariant paradigm for pixel based optical proximity correction. *Proceedings of ACM/IEEE International Conference on Computer-Aided Design (ICCAD)*, pp. 847–853, 2005.

99. C. M. Yuan. Calculation of one-dimension lithographic aerial images using the vector theory. *IEEE Transactions on Electron Devices*, 40:1604–1613, 1993.

100. H. Yune, C. Kim, Y. Ahn, B. Nam, and D. Yim. High accurate hybrid-OPC method for sub-60 nm memory device. *Proceedings of SPIE, The International Society for Optical Engineering*, Vol. 6156, pp. 61561A.1–661561A.8, 2006.

101. J. Zhang, W. Xiong, M. Tsai, Y. Wang, and Z. Yu. Efficient mask design for inverse lithography technology based on 2D discrete cosine transformation (DCT). *Simulation of Semiconductor Processes and Devices*, 12:49–52, 2007.

Index

Computational Lithography By Xu Ma and Gonzalo R. Arce

Printed and bound by CPI Group (UK) Ltd, Croydon, CR0 4YY
16/04/2025
14658604-0001